Trends in Synthetic Carbohydrate Chemistry

547.78 T722r

ACS SYMPOSIUM SERIES **386**

Trends in Synthetic Carbohydrate Chemistry

Derek Horton, EDITOR
The Ohio State University

Lynn D. Hawkins, EDITOR
Eisai Research Institute of Boston, Inc.

Glenn J. McGarvey, EDITOR
University of Virginia

Developed from symposia sponsored
by the Divisions of Organic Chemistry
and of Carbohydrate Chemistry
at the 191st National Meeting
of the American Chemical Society,
New York, New York,
April 13–18, 1986,
and the 194th National Meeting
of the American Chemical Society,
New Orleans, Louisiana,
August 30–September 4, 1987

American Chemical Society, Washington, DC 1989

Library of Congress Cataloging-in-Publication Data

Trends in synthetic carbohydrate chemistry.
Derek Horton, Lynn D. Hawkins, Glenn J. McGarvey, editors.

p.   cm.—(ACS Symposium Series, 0097–6156; 386).

"Developed from a symposium sponsored by the Divisions of Organic Chemistry and of Carbohydrate Chemistry at the 191st National Meeting of the American Chemical Society, New York, New York, April 13–18, 1986, and the 194th National Meeting of the American Chemical Society, New Orleans, Louisiana, August 30–September 4, 1987."

Includes bibliographies and index.

ISBN 0–8412–1563–4
1. Carbohydrates—Congresses. I. Horton, Derek, 1932– . II. Hawkins, Lynn D., 1954– . III. McGarvey, Glenn J., 1951– . IV. American Chemical Society. Division of Organic Chemistry. V. American Chemical Society. Division of Carbohydrate Chemistry. VI. American Chemical Society. Meeting (191st: 1986: New York, N.Y.). VII. American Chemical Society. Meeting (194th: 1987: New Orleans, La.) VIII. Series.

QD320.T74   1989
547.7'8—dc19   88–39237
                                              CIP

ACS Symposium Series

M. Joan Comstock, *Series Editor*

1988 ACS Books Advisory Board

Foreword

The ACS SYMPOSIUM SERIES was founded in 1974 to provide a medium for publishing symposia quickly in book form. The format of the Series parallels that of the continuing ADVANCES IN CHEMISTRY SERIES except that, in order to save time, the papers are not typeset but are reproduced as they are submitted by the authors in camera-ready form. Papers are reviewed under the supervision of the Editors with the assistance of the Series Advisory Board and are selected to maintain the integrity of the symposia; however, verbatim reproductions of previously published papers are not accepted. Both reviews and reports of research are acceptable, because symposia may embrace both types of presentation.

Contents

viii

INDEXES

Preface

CARBOHYDRATES, CHIRAL ORGANIC MOLECULES found naturally or obtained by synthetic transformations, currently comprise a quarter of a million or so known compounds. The rich structural diversity of this group and the multifaceted importance of carbohydrates in biochemistry, medicinal chemistry, microbiology, technology, and many other areas have long challenged synthetic chemists toward a multitude of objectives. The potential of sugars as starting points for highly efficient, stereochemically designed syntheses of noncarbohydrate targets is now increasingly recognized by the wider chemical community. Carbohydrates also serve as excellent systems for the study of fine aspects of stereochemical influence and control of chemical transformations in multifunctional, three-dimensional matrices.

Trends in Synthetic Carbohydrate Chemistry is divided into two sections, each of which may be further subdivided into three themes. The first section, "Synthetic Transformations in Carbohydrate Chemistry", surveys a variety of synthetic methodologies useful for transformation of natural saccharides into desired target molecules. The initial three chapters focus on functional-group transformations and protective-group strategy, with special emphasis on aminodeoxy, deoxyfluoro, and deoxynitro sugars and cyclitols, as well as the formation and cleavage of cyclic acetals. Examples of applications to specific synthetic aims are given in Chapters 4–6, including the synthesis of bicyclic nucleosides, the Wittig approach to long-chain sugars, and the transformation of sugar precursors into chiral pyrrolidine alkaloids. The last three chapters of the first section address the ever-significant problem of high-yielding, stereoselective glycosidic coupling procedures, first from the standpoint of basic methodology, next in glycosylations directed toward antibiotics containing deoxy sugars, and finally in the notable applications that have provided practical syntheses of cyclodextrins and complex oligosaccharides.

The second section of the book, entitled "Total Synthesis of Carbohydrates", focuses on strategies for the generation of monomeric carbohydrates, with major emphasis on the use of nonchiral, acyclic precursors. The contributors do not "reinvent the wheel" by providing tedious synthetic access to abundant natural sugars. Rather, they show the potential of synthetic design for controlled access to molecules containing multiple chiral centers that are not readily accessible from natural precursors. Chapters 10–13 illustrate from a variety of viewpoints the great utility of Diels–Alder reactions for the direct and indirect formation of chirally functionalized tetrahydropyrans and tetrahydrofurans. Chapters 14–16 address the aldol reaction and its control, the use of boron and tin enolates, and the use of chiral auxiliaries in the stereocontrolled formation of carbon–carbon bonds generating new chiral centers. Finally, the last two chapters describe the harnessing of enzymes as synthetic tools for chiral precursors and target products, both by use of isolated enzymes and by reactions brought about by living cultures of microorganisms.

Trends in Synthetic Carbohydrate Chemistry offers the reader a wealth of contemporary ideas for the construction of complex natural molecules and their analogs, from conceptualization to practical realization. The rich legacy of the carbohydrate literature in conjunction with newer concepts in general organic synthesis has created a unified field that provides some of the most exciting challenges for today's chemist. As the molecular basis of biological concepts opens up new vistas of understanding, the synthetic chemist is presented with unique opportunities for exercising creative talent toward significant objectives of ever-increasing complexity. Emulation of the virtuosity of nature in synthesis provides an ever-present challenge for the chemist. We hope that the efforts of those who have created this book will bring broader awareness of the role of carbohydrates in modern synthetic work and stimulate others in the pursuit of great intellectual satisfaction and worthy objectives for their creative efforts in the laboratory.

This book is an international collaborative effort, with authors from Canada, France, the Federal Republic of Germany, Great Britain, Italy, Japan, Sweden, Switzerland, and the United States. It is not possible to cover all aspects of this subject in a single volume, but the contributions here are broadly representative of innovative work in the field. The order of the chapters is developed from the relationship of the topics and is not necessarily related to the sequence of contributions at the two symposia from which much of the initial material was derived.

Acknowledgments

The editors deeply appreciate the excellent work of the contributing authors that has made this book possible. We also thank the many other colleagues in the field who have given of their time to review the chapters and offer constructive criticism. Excellent support by the American Chemical Society Divisions of Carbohydrate Chemistry and

Organic Chemistry helped to make possible the original ACS symposia that provided the impetus for this book. Additional support for the symposia came from Burroughs Wellcome Company, Ciba–Geigy, ICI Americas, The Upjohn Company, Merck & Company, Sandoz Research Institute, SmithKline Beckman, Syntex Research, and Warner Lambert Company. The support and patience of Joan Comstock and Robin Giroux of the ACS Books Department is recognized. We particularly appreciate the fine support and consultation of David C. Baker at all stages of the development of this book.

DEREK HORTON
Department of Chemistry
The Ohio State University
Columbus, OH 43210

LYNN D. HAWKINS
Eisai Research Institute of Boston, Inc.
Lexington, MA 02173

September 3, 1988

SYNTHETIC   TRANSFORMATIONS
IN   CARBOHYDRATE   CHEMISTRY

Chapter 1

New Synthetic Methods Emphasizing Deoxyfluoro Sugars and Protective-Group Strategy

Walter A. Szarek

Department of Chemistry, Queen's University, Kingston, Ontario K7L 3N6, Canada

The selective introduction of fluorine is of continuing interest not only because of the synthetic challenge but also because of the possibility of a dramatic change in biological activity. The fluoride-ion displacement of carbohydrate trifluoromethanesulfonates using tris(dimethylamino)sulfonium difluorotrimethylsilicate (TASF) provides a convenient route to deoxyfluoro sugars. Partially protected monosaccharides, having the anomeric hydroxyl underivatized, react with pyridinium poly(hydrogen fluoride) to yield the corresponding glycosyl fluorides. Two new developments in protective-group strategy are also described. These are (i) a method for the selective silylation of primary hydroxyl groups in carbohydrates involving the use of N-trimethylsilyl- or N-tert-butyldimethylsilyl-phthalimide and (ii) a method for acetal cleavage in carbohydrate derivatives using the simple reagent system, iodine in methanol.

The search for new methods of synthesis of halogenated carbohydrates continues to be an active area of investigation. The compounds are of utility as synthetic intermediates, and many of them are of intrinsic value in biochemistry and pharmacology. In the present Chapter methods for the synthesis of deoxyfluoro sugars and glycosyl fluorides are discussed.

Because of the polyfunctional nature of carbohydrates, protective-group strategy plays an important role in synthetic methodology involving this class of compounds. In the present Chapter, results are described from a study of the utility of N-trimethylsilyl- and N-tert-butyldimethylsilyl-phthalimide for the selective silylation of primary hydroxyl groups in carbohydrates. Also described, is a new, facile method for cleavage of acetals and dithioacetals in carbohydrate derivatives; the method involves treatment of the derivatives with a dilute solution of iodine in methanol.

0097–6156/89/0386–0002$06.00/0
© 1989 American Chemical Society

Synthesis of Deoxyfluoro Sugars

The expanding application of deoxyfluoro sugars for the study of carbohydrate metabolism and transport in both normal and pathological states has stimulated interest in their chemical (1,2) and biological (3—5) properties. Furthermore, it has actuated intensive efforts to develop improved methods of synthesis, and especially procedures suitable for the preparation of ^{18}F-labeled carbohydrates for use in medical imaging (6,7). Included in the approaches taken to this end are addition reactions of such reagents as molecular fluorine (8—11), xenon difluoride (12—14), and acetyl hypofluorite (15—20), reaction of free hydroxyl groups with (diethylamino)sulfur trifluoride (21—23), nucleophilic ring-openings with potassium hydrogenfluoride (24—27), and nucleophilic displacement of good leaving groups by a range of fluoride salts (28—35).

We recently described (36,37) the rapid, fluoride-ion displacement of carbohydrate triflates using tris(dimethylamino)sulfonium difluorotrimethylsilicate (TASF) (38), a reagent which previously had been utilized (39) for the synthesis of 1-deoxy-1-fluoro-D-fructose. TASF is a hygroscopic solid which is freely soluble in a variety of organic solvents in which it acts as an effective fluoride-ion donor when employed under rigorously anhydrous conditions; the relatively brief reaction times (36,37) are such that it may be of interest for the potential synthesis of ^{18}F-labeled radiopharmaceuticals for positron emission tomography. Examples of the utility of the reagent are described in the present Chapter; TASF has been used to effect the displacement, with inversion of configuration, of triflate groups at each of C-2, C-3, C-4, and C-6 of suitably protected aldohexopyranosides, at C-6 of 1,2:3,4-di-O-isopropylidene-α-D-galactose, and at C-3 of 1,2:5,6-di-O-isopropylidene-α-D-allofuranose.

The synthetic results have been summarized in Table I. In most cases the displacement of triflate anion occurred rapidly (≤30 min) at or below reflux temperature; however, in two examples (see Table I, compounds 4 and 13) an elimination reaction was found to predominate.

The reaction of methyl 4,6-O-benzylidene-3-O-methyl-2-O-trifluoromethanesulfonyl-β-D-mannopyranoside (1) with TASF (an approximately 3-fold molar excess of reagent was employed per mole of triflyl group in each case) in dichloromethane occurred rapidly (<10 min) in cold solution to afford methyl 4,6-O-benzylidene-2-deoxy-2-fluoro-3-O-methyl-β-D-glucopyranoside (14) in 64% yield. The 3-O-benzyl derivative 2 underwent a rapid reaction with TASF at reflux temperature to give methyl 3-O-benzyl-4,6-O-benzylidene-2-deoxy-2-fluoro-β-D-glucopyranoside (15) in 45% yield, and a minor product (19% yield) tentatively assigned the structure of methyl 3-O-benzyl-4,6-O-benzylidene-2-deoxy-β-D-erythro-hex-2-enopyranoside. Base-catalyzed elimination reactions with triflyl derivatives are uncommon (40,41), but have been observed in certain furanoid (40,42) and, recently, in pyranoid (33,35) ring systems (see also Table I). Eliminations in glycopyranosides occurred (33,35) under conditions which decreased the ease of nucleophilic substitution (33,43,44).

Table I.   Reactions of TASF with Derivatives (R= SO_2CF_3) of
Aldohexo-pyranoses and -furanoses

Substrate	Reaction Time (min)	Reaction Temperature ($^{\circ}$C)	Product	Yield (%)
1	<10	0-20	**14**	64
2	10	∿40	**15**	45
3	∿5	0-20	**16**	11
4	50	23	**17**	∿77

Table I.   Continued

Substrate	Reaction Time (min)	Reaction Temperature (°C)	Product	Yield (%)
5	60	∿40	**18**	65
6	300	∿40	**19**	23
7	<10	0-20	**20**	65
8	20	∿40	**21**	77
9	10	0-20	**22**	67

Continued on next page.

Table I. Continued

Substrate	Reaction Time (min)	Reaction Temperature (°C)	Product	Yield (%)
10	<10	0-20	**23**	71
11	30	∿40	**24**	39
12	<10	0-20	**25**	66
13	<10	0-20	**26**	83

The reaction of benzyl 3,4,6-tri-O-benzyl-2-O-trifluoromethane-
sulfonyl-β-D-mannopyranoside (3) with TASF was complete in less than
5 min below room temperature, to give a low (11%) yield of benzyl
3,4,6-tri-O-benzyl-2-deoxy-2-fluoro-β-D-glucopyranoside (16),
whereas the isomeric α-D-mannopyranoside 4 reacted relatively slowly
with TASF at room temperature to give a product tentatively assigned
the structure of 17.
    The reactions of the β-D-glucopyranosides 5, 6, and 7 with
TASF yielded results which were very different in each case, but
which were fully consistent with the current understanding of
nucleophilic displacements in carbohydrates (41,43—46). The less
rapid and efficient displacement of triflate anion from 6, as
compared with that from 5, may be attributable to the increased
steric hindrance occasioned by the introduction of benzyl and
2,2,2-trichloroethyl groups in place of methyl groups.
Furthermore, the facile displacement of triflate from C-3 of 7
reflects the diminution of the steric and electronic effects which
render the displacement of C-2 sulfonates in β-D-glucopyranosides
much more difficult (44). It is noteworthy that the product of
the relatively difficult reaction of TASF with 6 was found to
contain, in addition to 19, two unidentified components, a result
which indicates the relative increase in competing reactions as
nucleophilic substitution is impeded. The reaction of methyl
2,3,6-tri-O-benzyl-4-O-trifluoromethanesulfonyl-β-D-glucopyranoside
(8) with TASF in dichloromethane at reflux temperature gave within
20 min methyl 2,3,6-tri-O-benzyl-4-deoxy-4-fluoro-β-D-
galactopyranoside (21) in 77% yield. Methyl 2,3,6-tri-O-benzyl-4-O-
trifluoromethanesulfonyl-α-D-galactopyranoside (9) reacted with TASF
much more rapidly than did 8; the conversion was complete within 10
min after the addition of TASF to a cold solution of 9 to afford
methyl 2,3,6-tri-O-benzyl-4-deoxy-4-fluoro-α-D-glucopyranoside (22)
in 67% yield. The enhanced ease of displacement of the axially
oriented trifloxy group of 9, as compared to that of the
equatorially oriented one of 8, is consistent with the suggestion
(45) that a galacto isomer has a higher ground-state energy than the
corresponding gluco compound.
    The nucleophilic displacement by TASF of a trifloxy group
located on a primary carbon atom occurred with great rapidity
under mild conditions. Thus, the reaction of 1,2:3,4-di-O-
isopropylidene-6-O-trifluoromethanesulfonyl-α-D-galactopyranose
(10) with TASF was complete in less than 10 min at 0—20°C and
afforded a 71% yield of 6-deoxy-6-fluoro-1,2:3,4-di-O-
isopropylidene-α-D-galactopyranose (23). In a similar manner,
methyl 2,3-di-O-benzyl-4,6-bis-O-(trifluoromethanesulfonyl)-β-D-
glucopyranoside (11) reacted rapidly with TASF at reflux
temperature to give a 39% yield of methyl 2,3-di-O-benzyl-4,6-
dideoxy-4,6-difluoro-β-D-galactopyranoside (24); this, however, was
accompanied by the formation of a slightly smaller amount of a less-
polar, unidentified compound.
    The selective introduction of fluorine into a furanose ring
was demonstrated by the reaction of TASF with 1,2:5,6-di-O-
isopropylidene-3-O-trifluoromethanesulfonyl-α-D-allofuranose (12)
which gave a 66% yield of 3-deoxy-3-fluoro-1,2:5,6-di-O-
isopropylidene-α-D-glucofuranose (25). The isomeric triflate 13,

having the D-gluco configuration, reacted with TASF under the same
conditions to give an 83% yield of the elimination product,
3-deoxy-1,2:5,6-di-O-isopropylidene-α-D-erythro-hex-3-enofuranose
(26), but no fluorine-containing product was detected.
    Very recently, there have been reports of the utility of
TASF for the synthesis of 2'-deoxy-2'-fluoroinosine (47) and of
the β-6'-fluoro analog of (±)-aristeromycin (48).

Synthesis of Glycosyl Fluorides

The utility of glycosyl fluorides in enzymology (49—51) and as
glycosylating agents (52—64) has stimulated interest in their
preparation and chemistry (2,52,65). The original synthesis (66)
of glycosyl fluorides employed the reaction of peracetylated
aldoses with hydrogen fluoride. In addition, glycosyl fluorides
have been obtained by treatment of an acylated glycosyl bromide or
chloride with silver fluoride (67,68), or with silver
tetrafluoroborate (69,70), or with 2,4,6-trimethylpyridinium
fluoride (71). Two recently reported methods involve treatment of
1-O-acetylated sugar derivatives with pyridinium poly(hydrogen
fluoride) (72) and treatment of phenyl 1-thioglycosides with
diethylaminosulfur trifluoride (DAST) and N-bromosuccinimide (59).
Also, glycosyl fluorides have been obtained by substitution of a
free, anomeric hydroxyl group by fluorine using 2-fluoro-1-
methylpyridinium tosylate (54), or diethyl 1,1,2,3,3,3-
hexafluoropropylamine (56,73), or diethylaminosulfur trifluoride
(74,75).
    We recently described (76) a method for the synthesis of
glycosyl fluorides involving treatment of partially protected
monosaccharides, having the anomeric hydroxyl group underivatized,
with pyridinium poly(hydrogen fluoride), a reagent introduced by
Olah et al. (77). Examples which illustrate the scope of the
reaction are given in Table II. The utilization of Olah's reagent
for the fluorination of carbohydrates at sites other than the
anomeric carbon failed in a series of experiments with methyl
hexopyranosides or their partially protected derivatives (see Ref.
78).
    In the case of 2,3,5-tri-O-benzoyl-D-ribofuranose (27)
pyridinium poly(hydrogen fluoride) was added to a solution of 27
in anhydrous dichloromethane and the solution was shaken at room
temperature for 10 h in an atmosphere of dry argon. Anhydrous
acetone was found to be equally effective in most reactions. The
reaction of compound 28 with Olah's reagent required the use of
anhydrous acetone or anhydrous dichloromethane—collidine [1:1
(v/v)]; in the case of compound 30 the addition of collidine was
disadvantageous, whereas in the case of compound 31 best results
were obtained using anhydrous acetone—collidine [1:1 (v/v)] as
the solvent. Compounds 32 and 33 were treated using pyridinium
poly(hydrogen fluoride) as the only solvent. Reaction times
varied from 2 h for compound 31 to more than 12 h for 32 and 33.
    The action of pyridinium poly(hydrogen fluoride) on
compounds 29—33 resembles that of anhydrous hydrogen fluoride on
peracetylated D-glucopyranose (2,66), and of silver
tetrafluoroborate in diethyl ether (when prolonged) on

Table II.   Synthesis of Glycosyl Fluorides

Substrate	Product	Yield (%)

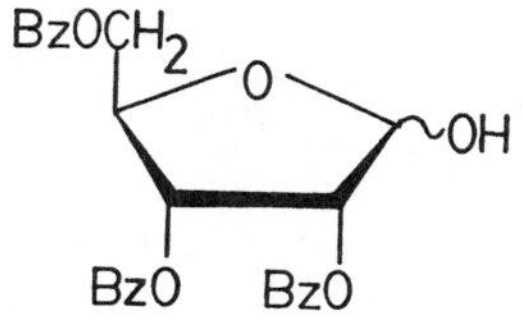

BzOCH$_2$ ... OH
BzO   BzO
27

BzOCH$_2$ ... F
BzO   BzO
34

78.8[a]

BnOCH$_2$ ... OH
BnO
OBn
28

BnOCH$_2$ ... F
BnO
OBn
35

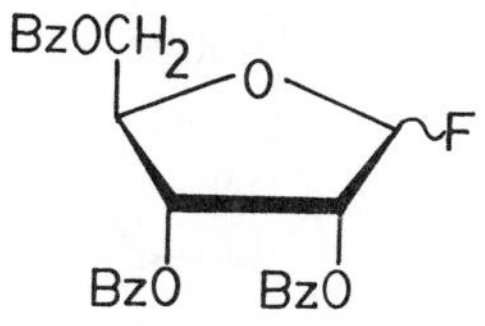

74 (α)

58.0 (total)

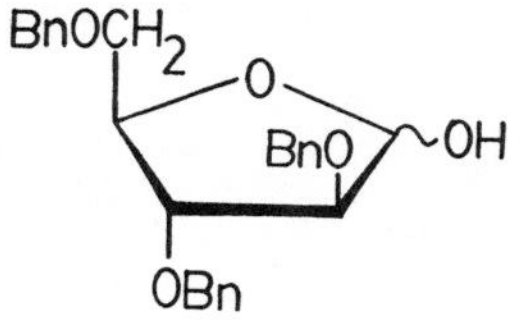

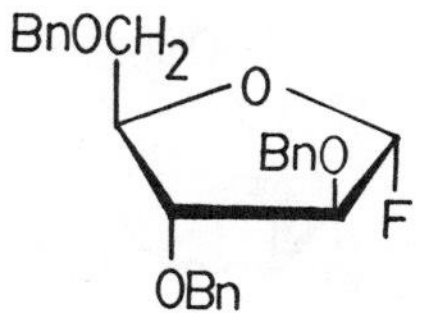

BnOCH$_2$ ... F
BnO
OBn
36

26 (β)

Me$_2$C—OCH$_2$ / OCH ... OH
Me$_2$
O—C—O
29

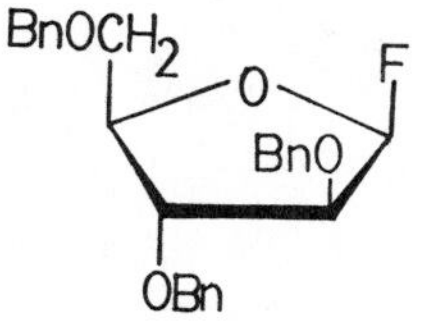

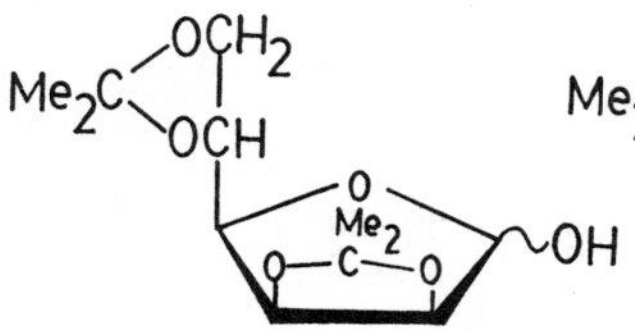

Me$_2$C—OCH$_2$ / OCH ... F
Me$_2$
O—C—O
37

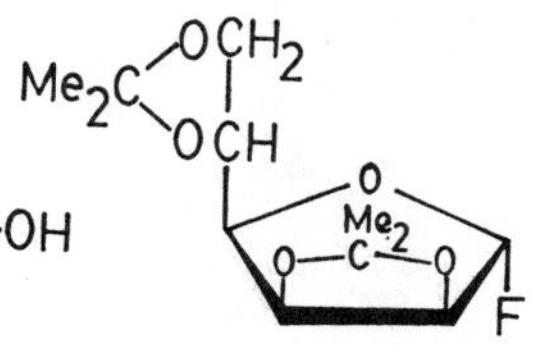

31.4

Continued on next page.

Table II.   Continued

Substrate	Product	Yield (%)

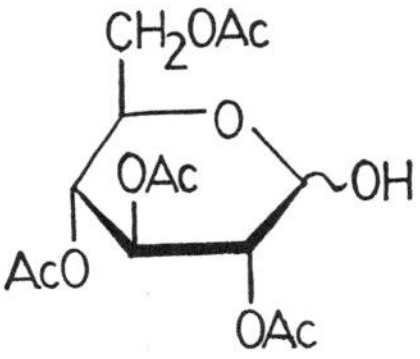

30   **38**   53

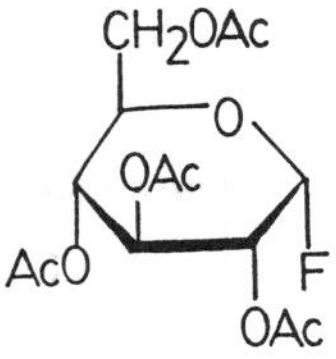

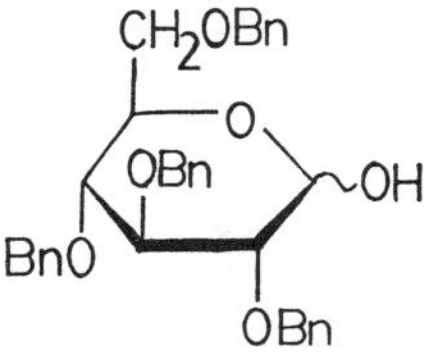

31   **39**   82[b]

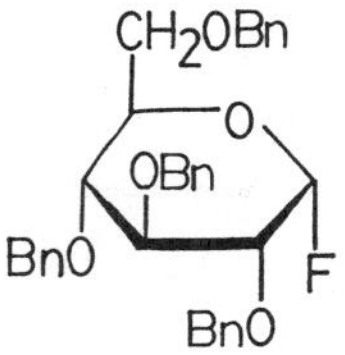

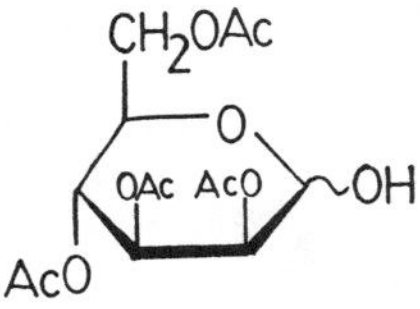

32   **40**   69

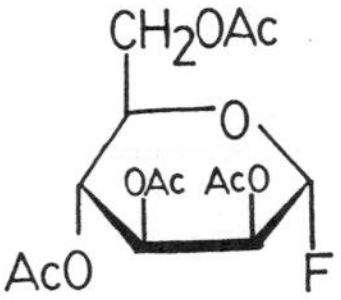

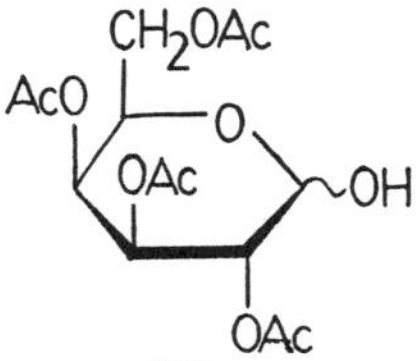

33   **41**   62

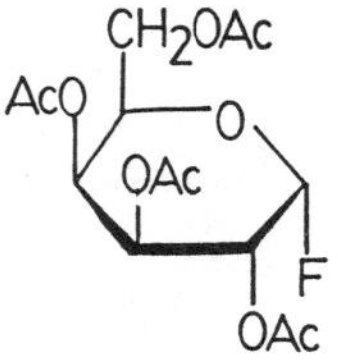

[a] ^{1}H–NMR data indicated that the ratio of α– and β–isomers was ∿1:1.

[b] A trace of the β–isomer was indicated by the ^{1}H–NMR spectrum.

peracetylated α-D-glucopyranosyl chloride (69) in that it gives
rise to the thermodynamically stable isomer. Moreover, with
compounds 30, 32, and 33, Olah's reagent contrasts with both 2,4,6-
trimethylpyridinium fluoride (71) and silver fluoride (68) in
affording the α-glycosyl fluoride from a partially acetylated
aldose regardless of whether the participating group at C-2 is
cis- or trans-related to the fluorine atom.

Reactions of N-Trimethylsilyl- or N-tert-Butyldimethylsilyl-
phthalimide with Carbohydrate Derivatives

N-Trimethylsilylphthalimide (42) (79) is a poor donor of the
trimethylsilyl group, and, hence, its application in organic
chemistry has been limited to only special cases (80,81), not
involving hydroxyl groups. However, we have found that 42 in the
presence of weak bases as catalysts provides a reagent system
capable of performing selective trimethylsilylation of primary
hydroxyl groups. As catalysts, the tertiary phosphines,
triphenylphosphine, tri-n-butylphosphine, and
methyldiphenylphosphine, and 4-N,N-dimethylaminopyridine are
suitable; the use of triethylamine affords mono- and higher-
trimethylsilylated products.

The silylation reactions were performed by treatment of a
solution of the substrate (1 mol. equiv.) in oxolane [or a 4:1
(v/v) mixture of oxolane—dimethyl sulfoxide for substrates
insoluble in oxolane] with 42 (1.4—1.5 mol. equiv.) and
triphenylphosphine (0.5 mol. equiv.). The structures of the
substrates employed and of the products obtained, and yields, are
shown in Figure 1. Under the particular reaction conditions
employed secondary hydroxyl groups are either not silylated
or are silylated distinctly slower.

Although a trimethylsilyl group blocking a primary hydroxyl
group is not very stable (82,83) and can be readily removed, for
example, under acetylation conditions (84), nevertheless, we have
been able to perform a significant reaction at a secondary
hydroxyl group leaving the (trimethylsilyloxy)methyl group
intact. Thus, ethyl 2,3-dideoxy-α-D-erythro-hex-2-enopyranoside
(48) reacted readily with 42 and triphenylphosphine to afford 49
and phthalimide; this mixture, on treatment in situ with diethyl
azodicarboxylate, gave ethyl 2,3,4-trideoxy-4-phthalimido-6-O-
trimethylsilyl-α-D-threo-hex-2-enopyranoside (50) in 71.4% overall
yield (see Figure 2). A salient feature of this synthetic process
is that two of the reagents required in the Mitsunobu amination
(85) step, namely triphenylphosphine and phthalimide, are present
already from the first step.

The tert-butyldimethylsilyl group is known to be a
particularly useful blocking group, and Ogilvie and Hakimelahi
(86) have described a method for the introduction of this group
selectively at primary hydroxyl groups. We have examined the
utility of N-tert-butyldimethylsilylphthalimide for this purpose.
Using conditions similar to those employed in the case of N-
trimethylsilylphthalimide did not lead to the transfer of the
tert-butyldimethylsilyl group; even heating at reflux temperature
for several hours was not successful. If the solvent system was

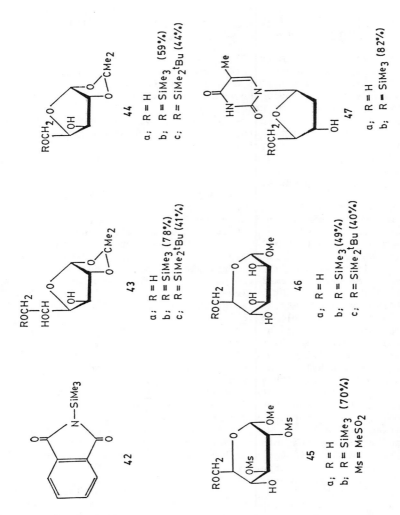

Figure 1. Substrates and products for the reactions of N-trimethylsilyl- or N-tert-butyldimethylsilyl-phthalimide with carbohydrate derivatives.

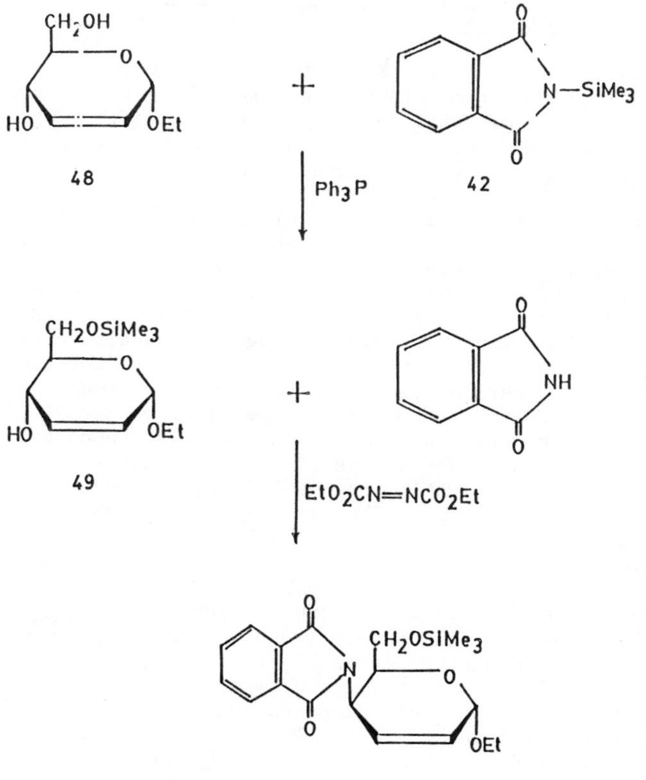

50 (71.4%)

Figure 2. Synthesis of ethyl 2,3,4-trideoxy-4-phthalimido-6-O-trimethylsilyl-α-D-*threo*-hex-2-enopyranoside.

changed to 4:1 (v/v) oxolane—hexamethylphosphoric triamide, then
43a and 44a at room temperature gave products having the primary
hydroxyl group silylated in 41 and 44% yield, respectively, and
46a at reflux temperature gave the corresponding product in 40%
yield. The non-carcinogenic solvent, 1,3-dimethyl-3,4,5,6-
tetrahydro-2(1H)-pyrimidinone (87), could not be employed as a
substitute for hexamethylphosphoric triamide.

Cleavage of Acetals and Dithioacetals in Carbohydrate Derivatives Using Iodine in Methanol

The reagent system, iodine and methanol, has been reported
(88,89) to open oxirane rings to afford β-methoxy alcohols. We
have found (90) that this reagent system is a highly efficient one
for the cleavage of acetal and dithioacetal groupings in
carbohydrate derivatives, groupings which find wide application in
synthetic carbohydrate chemistry (91,92). Benzylidine, ethylidene
and isopropylidene acetals can be cleaved at room temperature or
by heating at reflux temperature for a short period. If two
acetal groupings are present in the molecule, one of them can be
removed selectively. Simple glycosides and disaccharides do not
undergo cleavage of their glycosidic linkages under the conditions
employed. Also, acetyl groups survive the reaction conditions.
However, if the reaction mixture is heated at reflux temperature
for a prolonged period, carbohydrates having a free hydroxyl
group at the anomeric center are converted into methyl glycosides.
It is noteworthy that methyl glycofuranosides preponderate in the
mixtures of glycosides that are formed, a result that resembles
that generally observed in the case of the acid-catalyzed, Fischer
glycoside synthesis (93). The results obtained using a variety of
carbohydrate acetals are shown in Table III. The overall yields
are usually high. Recently, we were able to cleave the
isopropylidene acetal in 6-chloro-9-(3-deoxy-5,6-O-isopropylidene-
α-D-threo-hexofuranosyl-2-ulose)purine and the β-D-erythro isomer
by treatment with a dilute solution of iodine in methanol to
afford the corresponding, parent 3'-deoxy-2'-ketonucleosides, in
each case in 65% yield; these results are particularly
significant, since both of the protected ketonucleoside
derivatives were found to be labile under acidic conditions
normally required for the removal of an O-isopropylidene group.
      The acetal-cleavage reactions presumably involve initially a
complexation of an iodine species with one of the oxygen atoms; a
subsequent reaction with methanol would lead to the free alcohols.
On this basis it would be expected that dithioacetals should
undergo facile cleavage, since the soft acid, iodine, would be
expected to complex readily with the soft sulfur site. Indeed,
treatment of D-arabinose diethyl dithioacetal with a 1% solution
of iodine in methanol afforded, after ∿2 days at room temperature,
methyl α-D-arabinofuranoside in 70% yield. Other examples of the
removal of dithioacetal groupings under mild conditions are given
in Table III. It was found that cleavage of the dithioacetal
grouping in D-glucose ethylene dithioacetal required heating at
reflux temperature; it is known that mercury(II) chloride-
catalyzed hydrolysis of ethylene dithioacetals occurs slowly (94).

Table III.   Cleavage of Carbohydrate Acetals and Dithioacetals
Using Iodine and Methanol

Substrate	Reaction Conditions[a]	Compounds Obtained	Overall Yield (%)
PhCH (benzylidene acetal of methyl glycoside)	A, reflux, 30 min	(free triol methyl glycoside)	>90
MeCH (ethylidene acetal)	A, room temp, 16 h or reflux, 6.5 h	(methyl glycoside)	85
Me₂C di-O-isopropylidene furanose	B, room temp, 24 h or reflux, 1—1.5 h		80
Me₂C di-O-isopropylidene furanose	B, reflux, 4 h	OMe + OMe 4.2(α):4.6(β):1.2	80
Me₂C isopropylidene furanose	A, room temp, 24 h or reflux, 4.5 h		65—70
Me₂C isopropylidene furanose	A, room temp, 36 h or reflux, 10 h	OMe[b]	85—90

Continued on next page.

Table III.   Continued

Substrate	Reaction Conditions[a]	Compounds Obtained	Overall Yield (%)
	A, room temp, 14 h or reflux, 7 h	3(α):7(β)	90—95
	A, room temp, 6 h		35
	A, room temp, 14 h or reflux, 6 h		90
	A, room temp, 6 h		75
	A, room temp, 24 h or reflux, 6 h		90
	A, room temp, 24 h or reflux, 2.5 h		85—90

Table III. Continued

Substrate	Reaction Conditions[a]	Compounds Obtained	Overall Yield (%)
[structure: HOCH₂, Me₂C, O–CMe₂ bicyclic sugar]	A, room temp, 100 h	starting material	—
[structure: HOCH₂, Me₂C, O–CMe₂ bicyclic sugar]	A, reflux, 4 h	[structures] HOCH₂, HO, OH, OMe, OH **+** O, OMe, OH, HCOH, CH₂OH [b]	51
CH(SEt)₂ HOCH HCOH HCOH CH₂OH	A, room temp, 48 h	[structure] HOCH₂, O, HO, OMe, OH [b]	70
CH(SEt)₂ HCOH HOCH HCOH HCOH CH₂OH	A, room temp, 32 h	[structure] HOCH₂, HOCH, O, OH, OMe, OH [b] 1(α):2(β)	74
[dithiane structure] S–S HCOH HOCH HCOH HCOH CH₂OH	A, reflux, 18 h	[structure] HOCH₂, O, OH, OMe, HO, OH 4(α):1(β)	90
CH(SCH₂Ph)₂ HOCH HOCH HCOH HCOH CH₂OH	A, room temp, 24 h	[structure] HOCH₂, HOCH, O, OH, HO, OMe [b]	76

[a] Solution A: 1% iodine in methanol (w/v); solution B: 0.5% iodine in methanol (w/v).

[b] Isolated as the per-O-acetylated derivative.

Developments in protective-group strategy, including the formation and cleavage of acetals, continue to be of interest in synthetic carbohydrate chemistry (95,96). The attractiveness of the reagent system described here stems from its simplicity, convenience, versatility, and the high yields of the cleavage products.

Acknowledgments

It is a pleasure to acknowledge the financial support of the Natural Sciences and Engineering Research Council of Canada and the Medical Research Council of Canada, and the dedicated efforts of the following participants in the research: Dr. Bogdan Doboszewski, Dr. Grzegorz Grynkiewicz, Professor George W. Hay, Dr. Edward R. Ison, Dr. Ramesh K. Sood, Dr. Kamal N. Tiwari, and Professor Aleksander Zamojski.

Literature Cited

1. Kent, P.W. In Carbon—Fluorine Compounds. Chemistry, Biochemistry and Biological Activities; Ciba Found. Symp.; Elsevier: Amsterdam, 1972; pp 169—213.

2. Penglis, A.A.E. Adv. Carbohydr. Chem. Biochem. 1981, 38, 195—285.

3. Taylor, N.F. In Carbon—Fluorine Compounds. Chemistry, Biochemistry and Biological Activities; Ciba Found. Symp.; Elsevier: Amsterdam, 1972; pp 215—238.

4. Schwartz, R.T.; Datema, R. Trends Biochem. Sci. 1980, 5, 65—67.

5. Schwartz, R.T.; Datema, R. Adv. Carbohydr. Chem. Biochem. 1982, 40, 287—379.

6. De Kleijn, J.P. J. Fluorine Chem. 1977, 10, 341—350.

7. Fowler, J.S.; Wolf, A.P. Brookhaven National Laboratory, Report 1981 BNL 31222 (U.S. Department of Energy Order. No. DE82012799).

8. Ido, T.; Wan, C.-N.; Fowler, J.S.; Wolf, A.P. J. Org. Chem. 1977, 42, 2341—2342.

9. Ido, T.; Wan, C.-N.; Casella, V.; Fowler, J.S.; Wolf, A.P.; Reivich, M.; Kuhl, D.E. J. Labelled Compd. Radiopharm. 1978, 14, 175—183.

10. Takahashi, T.; Ido, T.; Shinohara, M.; Iwata, R.; Fukuda, H.; Matsuzawa, T.; Tada, M.; Orui, H. J. Labelled Compd. Radiopharm. 1984, 21, 1215—1217, and references cited therein.

11. Bida, G.T.; Satyamurthy, N.; Barrio, J.R. J. Nucl. Med. 1984, 25, 1327—1334.

12. Korytnyk, W.; Valentekovic-Horvat, S. Tetrahedron Lett. 1980, 21, 1493—1496.

13. Shiue, C.-Y.; To, K.C.; Wolf, A.P. J. Labelled Compd. Radiopharm. 1983, 20, 157—162.

14. Sood, S.; Firnau, G.; Garnett, E.S. Int. J. Appl. Radiat. Isot. 1983, 34, 743—745.

15. Shiue, C.-Y.; Salvadori, P.A.; Wolf, A.P.; Fowler, J.S.; MacGregor, R.R. J. Nucl. Med. 1982, 23, 899—903.

16. Adam, M.J. J. Chem. Soc., Chem. Commun. 1982, 730—731.
17. Adam, M.J.; Ruth, T.J.; Jivan, S.; Pate, B.D. Int. J.
    Appl. Radiat. Isot. 1984, 35, 985—986.
18. Diksic, M.; Jolly, D. Int. J. Appl. Radiat. Isot. 1983, 34,
    893—896.
19. Ehrenkaufer, R.E.; Potocki, J.F.; Jewett, D.M. J. Nucl.
    Med. 1984, 25, 333—337.
20. Van Rijn, C.J.S.; Herscheid, J.D.M.; Visser, G.W.M.;
    Hoekstra, A. Int. J. Appl. Radiat. Isot. 1985, 36,
    111—115.
21. Straatmann, M.G.; Welch, M.J. J. Labelled Compd.
    Radiopharm. 1977, 13, 210.
22. Card, P.J.; Reddy, G.S. J. Org. Chem. 1983, 48, 4734—4743.
23. Somawardhana, C.W.; Brunngraber, E.G. Carbohydr. Res. 1983,
    121, 51—60, and references cited therein.
24. Pacák, J.; Točík, Z.; Cerný, M. Chem. Commun. 1969, 77.
25. Barford, A.D.; Foster, A.B.; Westwood, J.H.; Hall, L.D.;
    Johnson, R.N. Carbohydr. Res. 1971, 19, 49—61.
26. Beeley, P.A.; Szarek, W.A.; Hay, G.W.; Perlmutter, M.M. Can.
    J. Chem. 1984, 62, 2709—2711.
27. Szarek, W.A.; Hay, G.W.; Doboszewski, B.; Perlmutter, M.M.
    Carbohydr. Res. 1986, 155, 107—118.
28. Christman, D.R.; Orhanovic, Z.; Wolf, A.P. J. Labelled
    Compd. Radiopharm. 1977, 13, 283.
29. Levy, S.; Elmaleh, D.R.; Livni, E. J. Nucl. Med. 1982, 23,
    918—922.
30. Olesker, A.; Dessinges, A.; Thang, T.T.; Lukacs, G. C.R.
    Acad. Sci. Ser. 2 1982, 295, 575—577.
31. Tewson, T.J. J. Org. Chem. 1983, 48, 3507—3510.
32. Dessinges, A.; Olesker, A.; Lukacs, G.; Thang, T.T.
    Carbohydr. Res. 1984, 126, C6—C8.
33. Haradihara, T.; Maeda, M.; Yano, Y.; Kojima, M. Chem. Pharm.
    Bull. 1984, 32, 3317—3319.
34. Haradihara, T.; Maeda, M.; Omae, M.; Yano, Y.; Kojima, M.
    Chem. Pharm. Bull. 1984, 32, 4758—4766.
35. Haradihara, T.; Maeda, M.; Kai, Y.; Omae, H.; Kojima, M.
    Chem. Pharm. Bull. 1985, 33, 165—172.
36. Szarek, W.A.; Hay, G.W.; Doboszewski, B. J. Chem. Soc.,
    Chem. Commun. 1985, 663—664.
37. Doboszewski, B.; Hay, G.W.; Szarek, W.A. Can. J. Chem. 1987,
    65, 412—419.
38. Middleton, W.J. U.S. Patent 3 940 402, 1976.
39. Card, P.J.; Hitz, W.D. J. Am. Chem. Soc. 1984, 106,
    5348—5350.
40. Binkley,, R.W.; Ambrose, M.G.; Hehemann, D.G. J. Org. Chem.
    1980, 45, 4387—4391.
41. Binkley,, R.W.; Ambrose, M.G. J. Carbohydr. Chem. 1984, 3,
    1—49.
42. Su, T.L.; Klein, R.S.; Fox, J.J. J. Org. Chem. 1981, 46,
    1790—1792.
43. Richardson, A.C. Carbohydr. Res. 1969, 10, 395—402.
44. Miljković, M.; Gligorijević, M.; Glišin, D. J. Org. Chem.
    1974, 39, 3223—3226.

45. Stevens, C.L.; Taylor, K.G.; Valicenti, J.A. J. Am. Chem. Soc. 1965, 87, 4579—4584.
46. Szarek, W.A. Adv. Carbohydr. Chem. Biochem. 1973, 28, 225—306.
47. Pankiewicz, K.W.; Nawrot, B.; Watanabe, K.A. Abstr. Pap. 193rd ACS Natl. Meeting, Denver, April 5—10, 1987, Carb-14.
48. Madhavan, G.V.B.; Prisbe, E.J.; Verheyden, J.P.H.; Martin, J.C. Abstr. Pap. 193rd ACS Natl. Meeting, Denver, April 5—10, 1987, Carb-11.
49. Barnett, J.E.G.; Jarvis, W.T.S.; Munday, K.A. Biochem. J. 1967, 105, 669—704.
50. Ariki, M.; Fukui, T. J. Biochem. (Tokyo) 1975, 788, 1191—1199.
51. Palm, D.; Blumenauer, G.; Klein, H.W.; Blanc-Muesser, M. Biochem. Biophys. Res. Commun. 1983, 111, 530—536.
52. Micheel, F.; Klemer, A. Adv. Carbohydr. Chem. 1961, 16, 85—103.
53. Mukaiyama, T.; Murai, Y.; Shoda, S. Chem. Lett. 1981, 431—432.
54. Mukaiyama, T.; Hashimoto, Y.; Shoda, S. Chem. Lett. 1983, 935—938.
55. Hashimoto, S.; Hayashi, M.; Noyori, R. Tetrahedron Lett. 1984, 25, 1379—1382.
56. Araki, Y.; Watanabe, K.; Kuan, F.-H; Itoh, K.; Kobayashi, N.; Ishido, Y. Carbohydr. Res. 1984, 127, C5—C9.
57. Nicolaou, K.C.; Dolle, R.E.; Chucholowski, A.; Randall, J.L. J.. Chem. Soc., Chem. Commun. 1984, 1153—1154.
58. Nicolaou, K.C.; Chucholowski, A.; Dolle, R.E.; Randall, J.L. J. Chem. Soc., Chem. Commun. 1984, 1155—1156.
59. Nicolaou, K.C.; Dolle, R.E.; Papahatjis, D.P.; Randall, J.L. J. Am. Chem. Soc. 1984, 106, 4189—4192.
60. Voznyi, Ya. V.; Kalicheva, I.S.; Galoyan, A.A. Bioorg. Khim. 1984, 10, 1256—1259; Chem. Abstr. 1985, 102, 95930q.
61. Dolle, R.E.; Nicolaou, K.C. J. Am. Chem. Soc. 1985, 107, 1695—1698.
62. Nicolaou, K.C.; Randall, J.L.; Furst, G.T. J. Am. Chem. Soc. 1985, 107, 5556—5558.
63. Kunz, H.; Sager, W. Helv. Chim. Acta 1985, 68, 283—287.
64. Voznyi, Ya. V.; Galoyan, A.A.; Chizhov, O.S. Bioorg. Khim. 1985, 11, 276—278; Chem. Abstr. 1985, 102, 167047q.
65. Paulsen, H. Adv. Carbohydr. Chem. Biochem. 1971, 26, 127—195.
66. Brauns, D.H. J. Am. Chem. Soc. 1923, 45, 833—835.
67. Helferich, B.; Gootz, R. Ber. 1929, 62, 2505—2507.
68. Hall, L.D.; Manville, J.F.; Bhacca, N.S. Can. J. Chem. 1969, 47, 1—17.
69. Igarashi, K.; Honma, T.; Irisawa, J. Carbohydr. Res. 1969, 11, 577—578.
70. Igarashi, K.; Honma, T.; Irisawa, J. Carbohydr. Res. 1970, 13, 49—55.
71. Voznyi, Ya. V.; Kalicheva, I.S.; Galoyan, A.A. Bioorg. Khim. 1981, 7, 406—409; Chem. Abstr. 1981, 95, 43491q.
72. Hayashi, M.; Hashimoto, S.; Noyori, R. Chem. Lett. 1984, 1747—1750.

73. Takaoka, A.; Iwakiri, H.; Ishikawa, N. Bull. Chem. Soc. Jpn. 1979, 52, 3377—3380.
74. Rosenbrook, W., Jr.; Riley, D.A.; Lartey, P.A. Tetrahedron Lett. 1985, 26, 3—4.
75. Posner, G.H.; Haines, S.R. Tetrahedron Lett. 1985, 26, 5—8.
76. Szarek, W.A.; Grynkiewicz, G.; Doboszewski, B.; Hay, G.W. Chem. Lett. 1984, 1751—1754.
77. Olah, G.A.; Welch, J.T.; Vankar, Y.D.; Nojima, M.; Kerekes, I.; Olah, J.A. J. Org. Chem. 1979, 44, 3872—3881.
78. Olah, G.A.; Welch, J. Synthesis 1974, 653—654.
79. Janzen, A.F.; Kramer, E.A. Can. J. Chem. 1971, 49, 1011—1018.
80. Dickopp, H. Ph.D. Thesis, University of Cologne, 1966; cited in: Pierce, A.E. Silylation of Organic Compounds; Pierce Chem. Co.: Rockford, 1968; p 23.
81. Kozyukov, V.P.; Kozyukov, V.P.; Mironov, V.F. Zh. Obshch. Khim. 1983, 53, 2091—2097; Chem. Abstr. 1984, 100, 22696s.
82. Hurst, D.T.; McInnes, A.G. Can. J. Chem. 1965, 43, 2004—2011.
83. Hengstenberg, W.; Morse, M.L. Carbohydr. Res. 1968, 7, 180—183.
84. Fuchs, E.-F.; Lehmann, J. Chem. Ber. 1974, 107, 721—724.
85. Mitsunobu,, O.; Wada, M.; Sano, T. J. Am. Chem. Soc. 1972, 94, 679—680.
86. Ogilvie, K.K.; Hakimelahi, G.H. Carbohydr. Res. 1983, 115, 234—239.
87. Mukhopadhyay, T.; Seebach, D. Helv. Chim. Acta 1982, 65, 385—391.
88. Jewell, J.S.; Szarek, W.A. Carbohydr. Res. 1971, 16, 248—250.
89. Kocór, M.; Kurek, A.; Tomaszewska, L. Abstr. Pap. 11th IUPAC Int. Symp. Chem. Nat. Prod., Golden Sands, Bulgaria, 1978, 2, 136—138; Chem. Abstr. 1979, 91, 211651e.
90. Szarek, W.A.; Zamojski, A.; Tiwari, K.N.; Ison, E.R. Tetrahedron Lett. 1986, 27, 3827—3830.
91. De Belder, A.N. Adv. Carbohydr. Chem. Biochem. 1977, 34, 179—242.
92. Wander, J.D.; Horton, D. Adv. Carbohydr. Chem. Biochem. 1976, 32, 15—124.
93. Capon, B. Chem. Rev. 1969, 69, 407—498.
94. Zimmer, H.; Brandner, H.; Rembarz, G. Chem. Ber. 1956, 89, 800—813.
95. Guindon, Y.; Yoakim, C.; Morton, H.E. J. Org. Chem. 1984, 49, 3912—3919, and references cited therein.
96. Albert, R.; Dax, K.; Pleschko, R.; Stütz, A.E. Carbohydr. Res. 1985, 137, 282—290.

RECEIVED May 31, 1988

Chapter 2

New Approaches to the Synthesis of Nitrogenous and Deoxy Sugars and Cyclitols

Hans H. Baer

Department of Chemistry, University of Ottawa, Ottawa, Ontario K1N 9B4, Canada

Chiral syntheses based on nitroalkane cyclization, of 3-amino-2,3-dideoxy-D-myo- and D-epi-inositol and of some asymmetrically substituted derivatives and preparative precursors of 2-deoxystreptamine are reported. A new mode of formation of carbohydrate nitrocyclopropane derivatives by an internal, cyclizing displacement is disclosed, with a discussion of the potential of such compounds in chiral synthesis. Methods for the preparation of amino and deoxy analogs of α,α-trehalose are outlined, including nitromethane cyclization, oxyamination, selective triflate displacement, reductive amination and desulfonyloxylation, palladium-catalyzed allylic substitution, and iron carbonyl-mediated chain elongation.

Ongoing concerns in this laboratory include the study of methods for functional and stereochemical modification of carbohydrates, the adaptation of synthetic tools largely developed outside the field to the special requirements in carbohydrate chemistry, and the synthesis of sugar derivatives of potential value in biochemical research. For the advancement of biological and medicinal research there is a need not only for efficient syntheses of natural products of recognized significance, but also for a steady supply of synthetic derivatives, stereoisomers and other analogs which may, potentially, display bio-activities of their own or may serve as probes for unraveling mechanisms of biological action. Aminocyclitols, because of their central standing in the field of antibiotics, constitute a particularly important case in point. The present article describes (a) chiral syntheses of molecules related to 2-deoxystreptamine, a pivotal component of aminocyclitol antibiotics; (b) discloses a new, preparative route to carbohydrate nitrocyclopropane derivatives having potential for conversion into various chiral, branched-chain synthons; and (c) outlines several current methods being developed for the synthesis of amino and deoxy analogs of the biologically important disaccharide, α,α-trehalose.

0097–6156/89/0386–0022$06.50/0
© 1989 American Chemical Society

Chemical Synthesis of an Intermediate in the Biosynthesis of 2-Deoxy-
streptamine

2-Deoxystreptamine ($\underset{\sim}{1}$) is a central building block in the structures
of many important aminocyclitol antibiotics, including the neomycins,
kanamycins, and gentamicins ($\underline{1}$). The pathway for its biosynthesis from
$\underline{D}$-glucose has been proposed by Rinehart ($\underline{2},\underline{3}$) to involve an amino-
cyclohexanetetrol, most probably 1$\underline{L}$-(1,3,$\overline{5/2}$,4)-5-amino-1,2,3,4-cyclo-
hexanetetrol, also designated as 3-amino-2,3-dideoxy-$\underline{D}$-$\underline{myo}$-inositol
($\underset{\sim}{2}$), which was then unknown. This assumption was subsequently proved
correct when $\underset{\sim}{2}$ was isolated from culture media of certain microorganisms ($\underline{4}$), its structure confirmed through comparison with a semisyn-
thetic sample obtained from degradation of chemically modified kana-
mycin-A ($\underline{5}$), and its bioconversion into $\underset{\sim}{1}$ demonstrated ($\underline{6}$).

$\underset{\sim}{\underline{1}}$ R = NH$_2$
$\underset{\sim}{\underline{2}}$ R = OH

Convenient preparative access to such chiral aminocyclitols as
$\underset{\sim}{2}$ should facilitate chemical analog synthesis and biochemical muta-
synthesis in the field of aminocyclitol antibiotics. A short and
economical synthesis of $\underset{\sim}{2}$ and its hitherto unknown 1$\underline{L}$-(1,3,4,5/2), or
$\underline{D}$-$\underline{epi}$, stereoisomer $\underset{\sim}{3}$ was therefore devised ($\underline{7}$). It is based on the
nitroalkane cyclization method ($\underline{8}$), and is delineated in Figure 1.
The nitro sugar $\underset{\sim}{4}$, available ($\underline{9},\overline{10}$) from $\underline{D}$-glucose in four **steps** with
high yields, gives in a single operation (Ac$_2$O—NaOAc; 84%) the nitro-
alkenic acetate $\underset{\sim}{7}$, via the intermediary diacetate $\underset{\sim}{5}$ ($\underline{10}$). Reduction
of $\underset{\sim}{7}$ by sodium borohydride had been reported ($\underline{11}$) to furnish the
nitroalkanol $\underset{\sim}{9}$ directly, but only in 49% yield, on a small scale.
Scaled-up operation under carefully controlled conditions, with isol-
ation of the intermediate $\underset{\sim}{8}$, now gave $\underset{\sim}{9}$ in 92% yield from $\underset{\sim}{7}$. Various
attempts to prepare $\underset{\sim}{9}$ directly from the 3,5-diacetate $\underset{\sim}{5}$, by reductive
dehydroacetoxylation and O-deacetylation with sodium borohydride,
proved unsatisfactory. The same was true when the 3,5-bis(trifluoro-
acetate) $\underset{\sim}{6}$ was used instead; although some $\underset{\sim}{9}$ could be isolated, a
major (and undesired) by-product was in this case identified as the
trifluoroethylidene acetal $\underline{11}$.

Hydrolytic removal of the isopropylidene group in $\underset{\sim}{9}$ then gave
5,6-dideoxy-6-nitro-$\underline{D}$-glucose ($\underset{\sim}{10}$, not isolated), which was caused to
cyclize immediately by rendering its aqueous solution slightly basic.
The mixture of 4-epimeric, 2,3-dideoxy-3-nitroinositols ($\underset{\sim}{12}$) could
not be separated but their tetraacetates $\underset{\sim}{13}$ and $\underset{\sim}{14}$ could be isolated
pure by fractional crystallization (29 and 23%). Standard, catalytic
hydrogenation followed by deacetylation procedures finally furnished
the target compounds $\underset{\sim}{2}$ and $\underset{\sim}{3}$ via their acetylated derivatives $\underset{\sim}{15-18}$.
Hydrogenation of the mixture $\underset{\sim}{12}$, followed by sequential N- and O-
acetylation of the amines produced, proved to be more economical for
larger-scale preparations. This permitted clean separation of the
epimers by virtue of a high tendency for $\underset{\sim}{18}$ and $\underset{\sim}{15}$ to crystallize.

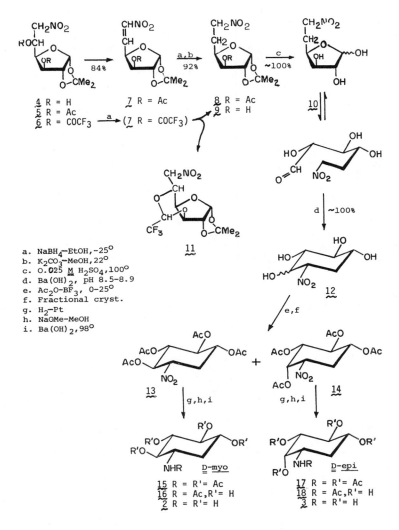

a. NaBH$_4$-EtOH,-25°
b. K$_2$CO$_3$-MeOH,22°
c. 0.025 M H$_2$SO$_4$,100°
d. Ba(OH)$_2$, pH 8.5-8.9
e. Ac$_2$O-BF$_3$, 0-25°
f. Fractional cryst.
g. H$_2$-Pt
h. NaOMe-MeOH
i. Ba(OH)$_2$,98°

Figure 1. Synthesis of 3-amino-2,3-dideoxy-D-*myo*- and D-*epi*-inositols (**2** and **3**).

A Chiral Synthesis of 2-Deoxystreptamine

Although 2-deoxystreptamine (1) is itself achiral (meso), it is ren-
dered chiral by monosubstitution at either of the enantiotopic OH-4
or OH-6 groups, or at either amino group, or by unequal disubstitu-
tion at these sites. Such patterns are typical for aminocyclitol
antibiotics. For instance, the neomycins and ribostamycins possess a
glycosylated OH-4 (and OH-5) group, whereas OH-6 is unsubstituted; in
the seldomycins, kanamycins, and gentamicins, both enantiotopic positions are occupied but unequally so. The stereoisomers destomycin A
and hygromycin B differ solely by being N-monomethylated in positions
1 and 3, respectively. Total syntheses that employ, as building blocks,
unsymmetrically modified but racemic derivatives of 1 must rely on
diastereoselectivities which may be modest, and on diastereomer separation at an advanced stage, which may be cumbersome and inefficient.
It would clearly be advantageous to have available, as synthons,
some chiral derivatives of 1 (or suitable, preparative precursors)
in pure enantiomeric forms rather than as racemates. A new approach
to 1 was therefore designed, starting from D-mannose and proceeding
entirely through chiral stages, thereby providing a number of optically active intermediates of the type desired (Figures 2-4) (12).

The known 1-deoxy-1-nitro-D-glycero-D-galacto-heptitol hexaacetate (19), obtained from D-mannose by the nitromethane method
(13), was reductively dehydroacetoxylated, and the product (20) O-
deacetylated. The resultant dideoxynitropentol 21 was acetonated
under thermodynamic control to give, exclusively, the 4,5;6,7-diiso-
propylidene acetal 22. The latter is the favored regioisomer as it
incorporates a trans-disubstituted dioxolane structure. The mesylate
23 and triflate 24 then prepared were the keystones for later introduction of a second nitrogenous function (Figure 2).

From this point onward, two alternative routes leading to the
same targets were pursued. They differed in the sequence whereby the
carbocyclic system was established and the second nitrogen group incorporated. In the first variant (Figure 3), compound 23 was selectively deacetonated with 90% trifluoroacetic acid in toluene at -20°
to furnish the 6,7-diol 25, which was cleaved by periodate to give
the substituted, 5,6-dideoxy-6-nitro-aldehydo-D-arabino-hexose 26.
Base-catalyzed cyclization of this nitro sugar led to a separable
mixture of epimeric cyclitols, 27 and 28, with the latter slightly
preponderating. When the mixture was fractionally crystallized from
ethanol in the presence of a trace of alkali, part of the less stable
(and more soluble) 27 present was converted into the desired, less
soluble epimer 28, which crystallized in 70% yield. Compounds 27 and
28 were readily hydrolyzed to the respective triols 30 and 32, characterized as highly crystalline triacetates (31 and 33). Next, both
28 and its progeny 32 were subjected to displacement reactions with
azide ion, which gave high yields of azidonitrocyclitols. However,
both products had lost their stereochemical integrity, as partial
epimerization at the carbinol position vicinal to the nitro group
occurred during the process, vitiating the preceding epimer separation. Nevertheless, pure triol 34 was isolated in 72% yield, and
purification of the isopropylidene derivative 29 was also possible
(see later).

In the second approach, the sequence of ring closure and azide

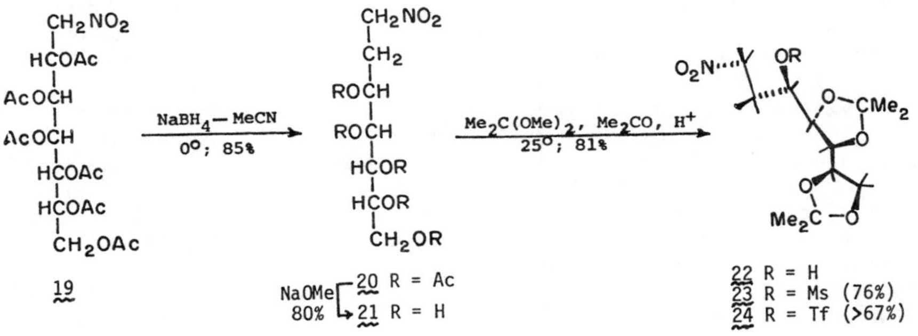

Figure 2. Synthesis of 1,2-dideoxy-1-nitroheptitol 3-sulfonates.

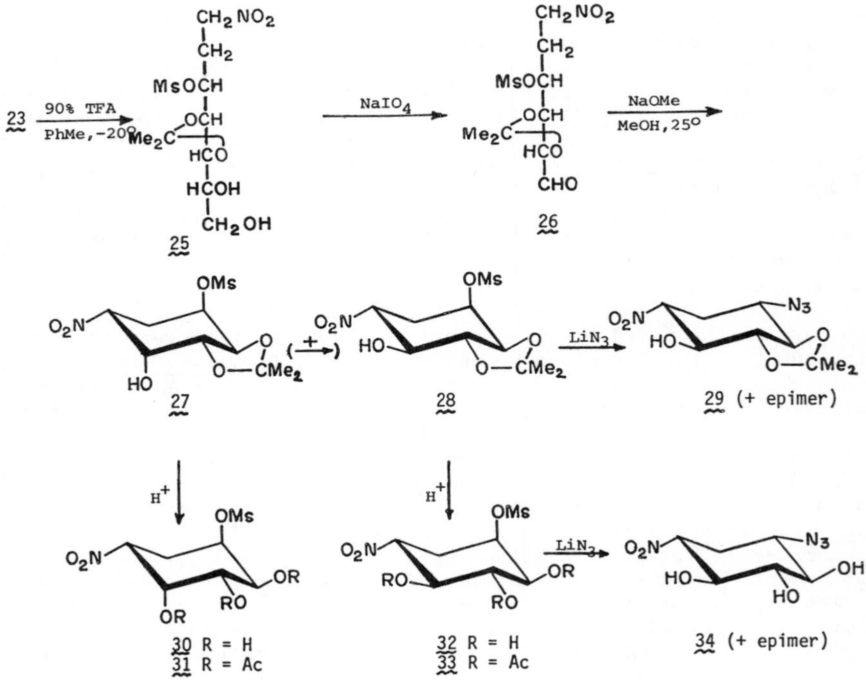

Figure 3. Synthesis of azidonitrocyclitols **29** and **34**.

displacement were reversed (Figure 4). First experiments to displace
the mesyloxy group in 23 under conventional conditions gave the protected azidonitroheptanetetrol 35 in low yields only, as an unexpected side-reaction intervened (to be discussed in the next section).
Although pase-transfer conditions were eventually found that produced
35 from 23 in 76% yield, the reaction was inordinately slow (6 days
at 56°). The 3-triflate 24, on the other hand, proved far superior in
that regard, affording 35 (> 85%) within 6 h at 25°. Selective 6,7-
deacetonation, followed by periodate oxidation of the diol 36, gave
the aldehydo-azidonitrohexose 37, which was cyclized by base catalysis to provide a mixture of 29 and epimer 38. Again, fractional
crystallization under epimerizing conditions (as for 27 + 28) allowed
the less-soluble 29 to be isolated in ~70% yield. Finally, platinum-
catalyzed hydrogenation converted the triol 34 into 2-deoxystreptamine
(1), and the acetal 29 into the optically active 4,5-O-isopropylidene
derivative (39) of 1, isolated as the crystalline diacetamide 40.
     Compounds 29, 34, 39, and 40 constitute chiral synthons suitable
for use in stereospecific aminocyclitol synthesis. Thus, 40 or other,
appropriately N-protected derivatives of 39 may be employed for
stereospecific substitution at OH-6; alternatively, after temporary
protection of OH-6 followed by removal of the acetal, the molecule
should be amenable to manipulation at OH-4. In 29 and 34, the two
unequal nitrogenous functions may be reduced stepwise to amino groups,
thus offering possibilities for stereospecific introduction of an N-
substituent at either position. In order to demonstrate that such a
strategy is feasible, reaction sequences leading to the enantiomers
of mono-N-methyl-2-deoxystreptamine were performed, as illustrated
in Figure 5.

Formation and Potential Utility of Carbohydrate Nitrocyclopropanes

As mentioned in the foregoing section, an unexpected side-reaction
was observed when various conditions for azide displacement in the
mesylate 23 were studied. Homogeneous-phase reaction with tetrabutylammonium azide in boiling toluene consumed 23 completely within 1 h,
but gave only 46% of 35. Three byproducts, isolated in yields of 25,
4.5 and 3%, were elucidated (14) as the 1-epimeric nitrocyclopropanes
41 and 42, and the branched-chain azidonitro compound 43 (Figure 6).
Evidently, 41 and 42 arose from internal displacement initiated by
proton abstraction from the nitromethylene group, caused by the basi-
city of azide ion (Equation 1), and 43 seems to stem from a slow,
subsequent nucleophilic substitution on the ring, with nitronate
anion functioning as the leaving group. The results of using phase-
transfer conditions are also shown in Figure 6.

$$23 \xrightarrow{-H^+} \quad \text{(structures)} \quad \longrightarrow \quad \text{(structures)} \quad \equiv \quad \text{(structures)} \qquad (1)$$

With solid sodium hydrogencarbonate instead of the azide, compound 23 gave 41 as the main product (85%), together with a trace of
42 (14) This recalls a related precedent, namely, γ-elimination of

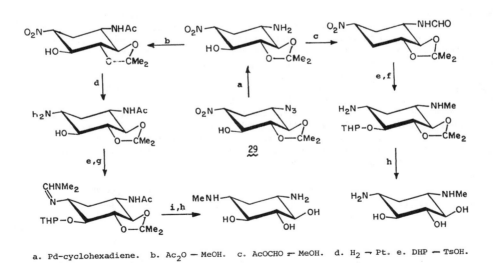

Figure 4. Synthesis of chiral acetals **39** and **40** derived from 2-deoxystreptamine.

a. Pd-cyclohexadiene.   b. Ac$_2$O — MeOH.   c. AcOCHO $\rightleftharpoons$ MeOH.   d. H$_2$ — Pt.   e. DHP — TsOH.
f. LiAlH$_4$.   g. Me$_2$NCH(OMe)$_2$.   h. HCl.   i. MeOTf.

Figure 5. Synthesis of the enantiomeric mono-*N*-methyl-2-deoxystreptamines from **29**.

hydrogen bromide by the action of potassium acetate from (α-alkyl- or
aryl-β-nitroethyl)bromomalonates (15,16); see Equation 2. Our reaction
took place under conditions similar to those of the well-known, nitro-
alkene-forming dehydroacetoxylation of β-nitro esters (the Schmidt –
Rutz reaction) and might therefore be dubbed a "homo" variant of the
latter.

$$O_2N-CH_2-CHR-CBr(CO_2Me)_2 \rightarrow O_2N-CH\overset{CHR}{-\!\!-\!\!-}C(CO_2Me)_2 \qquad (2)$$

Spurred by these observations, we examined a relevant application
of the method for nitrocyclopropane synthesis from nitroalkenes and
dimethylsulfoxonium methylide (17), Equation 3. It had previously been
employed for the synthesis of a 2,3-dideoxy-2,3-C-methylene-3-nitro-
hexopyranoside, the first one of the small number of carbohydrates
containing the nitrocyclopropane structure thus far known (18).

$$R-CH=CH-NO_2 + Me_2\overset{+}{\underset{\underset{O}{\|}}{S}}-\overset{-}{C}H_2 \rightarrow R-CH\overset{CH_2}{-\!\!-\!\!-}CH-NO_2 + Me_2S \qquad (3)$$

Treatment of the known nitroalkene 44 with the ylide indeed gave
41, but it was accompanied by a small proportion of the stereoisomer
45 (14). Although the preparative yield was low (~30%), the high
diastereofacial selectivity of the methylene addition was remarkable.
It becomes plausible on inspection of a molecular model, which points
to hindered approach from one face, and unhindered approach from the
other (Figure 7).

Another mode of formation of carbohydrate nitrocyclopropanes,
studied by us earlier (19), consists of nitrogen extrusion from
pyranosidic, 2,3-dideoxy-3-nitro sugars bearing a methyleneazo bridge
in the 2,3-positions. Such fused-ring 1-pyrazolines had been obtained
(19) by 1,3-dipolar cycloaddition of diazomethane to 3-nitro-2-eno-
pyranosides (Figure 8). In order to explore a possible application to
carbohydrates possessing a terminal nitroalkene grouping, the compounds 7, 44, and 46 were treated with ethereal diazomethane. They
reacted rapidly at low temperatures, but not in altogether clear ways.
(More than 1 molar equivalent of CH_2N_2 was consumed.) Crystalline,
yellow products isolated in moderate yields were determined to be
4-substituted, 3-nitro-2-pyrazolines (48), probably formed by tauto-
merization of the 1-pyrazolines expected as the primary adducts (20);
see Figure 9.

In the readily available cyclopropane 41, C-2 comprises a center
of chain branching whose configuration is rigorously defined (as R)
thanks to the manner in which it was engendered. It should be possible
to take advantage of this circumstance for generating, by scission of
the 3-membered ring, some stereospecifically functionalized molecules
which, after appropriate manipulations in the sugar moiety, could
serve as useful synthons for general purposes. Compound 43, apparently
originating from ring opening after azide-promoted formation of 41
(or 42), offers itself as a first candidate for such transformations,
provided it can be prepared in an acceptable yield. Thus, sequential
reductions of the two nitrogenous functions, each followed by individual N-substitution as desired — in analogy to the transformations of
29 illustrated in Figure 5 — would provide avenues to compounds of
type 49, in which the residue Z, originally representing the polyol

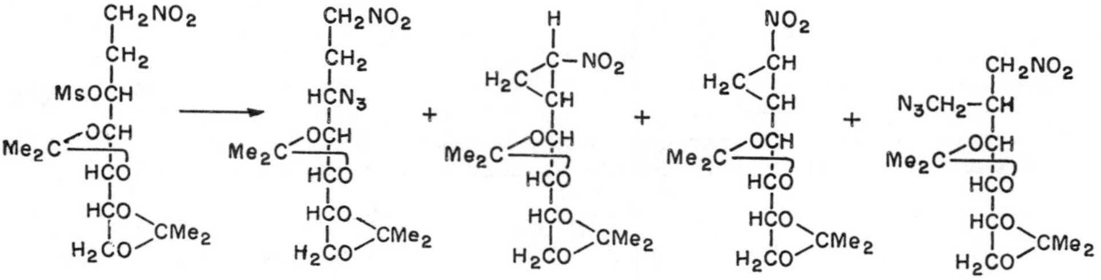

24 35 41 42 43

Reaction conditions	Product ratio 35:(41+42+43)
a. $Bu_4N^+N_3^-$ — PhMe; 100°, 1 h	1.4
b. NaN_3 — Bu_4NHSO_4, H_2O — PhMe; 100°, 2 d	~0.5
c. as in b, but at 80°	~0.2
d. as in b, but at 56°	10

Figure 6. Formation of nitrocyclopropanes **41** and **42** from mesylate **24**.

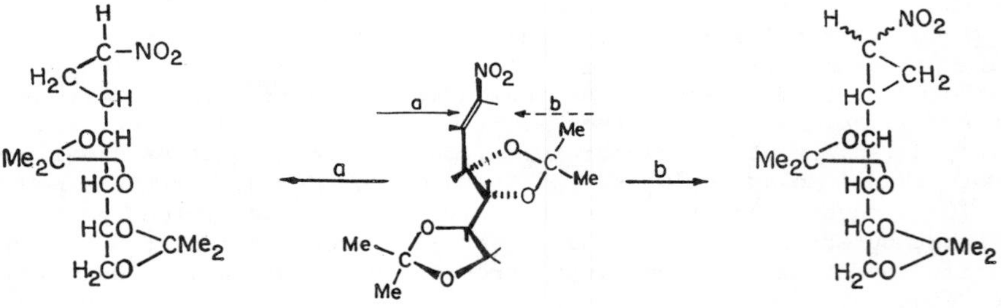

41, D-manno 44, D-arabino 45, D-gluco

a: Re,Re face attack
b: Si,Si face attack

Ratio 41:45 = 20:1

Figure 7. Formation of nitrocyclopropanes from nitroalkene **44**.

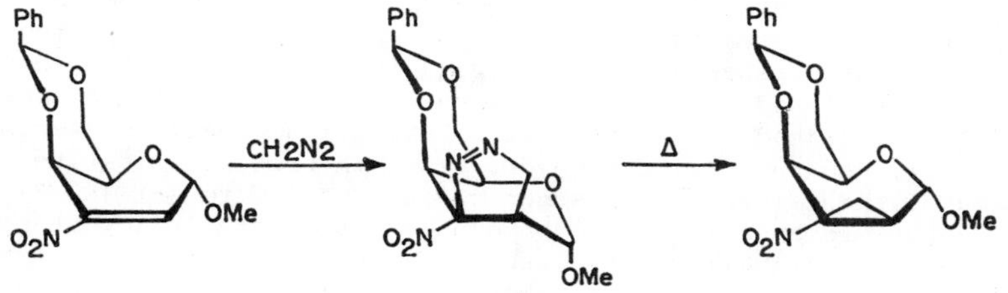

Figure 8. Formation of a carbohydrate nitrocyclopropane from a nitroalkenic sugar via a pyrazoline.

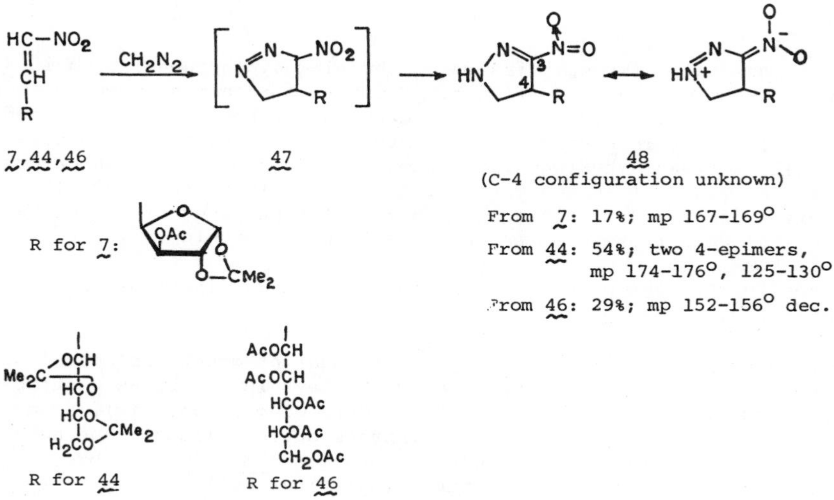

From $\underline{7}$: 17%; mp 167–169°

From $\underline{44}$: 54%; two 4-epimers, mp 174–176°, 125–130°

From $\underline{46}$: 29%; mp 152–156° dec.

Figure 9. Preparation of some carbohydrate pyrazolines by cycloaddition of diazomethane to nitroalkenic sugars.

chain, may be modified by degradation to comprise functionalities
(e.g., —CHO, —CH$_2$Br) suitable for carbon — carbon bond formation.
Furthermore, conversion of the nitromethyl group into a formyl, hydr-
oxymethyl, or carboxyl group by existing procedures (21), deamination
of a selectively-formed amino group, and other functional modifications may be contemplated, which should make 43 an exceedingly versatile stepping stone in syntheses of chiral, ω,ω'-disubstituted isoalkyl
structures. We have not yet achieved a preparative conversion of 41
into 43 by azide, but sodium thiophenoxide in boiling oxolane reacted
readily with 41 to give the thioether 50a and, interestingly, the thio-
hydroximic phenyl ester 50b (R as for 44 in Figure 9). Both products
were desulfurized and reduced with Raney nickel, yielding the same
amine (50c), isolated as a crystalline N-acetyl derivative (Baer,H.H.,
Williams, U., Radatus, B., Carbohydr. Res., in press). Oxidative degradation of the (deprotected) sugar chain then led to (−)-(R)-3-
amino-2-methylpropanoic acid, a compound of considerable importance
(22) in thymine metabolism and elsewhere in biochemistry. Its carbo-
hydrate-based, stereospecific synthesis exemplifies the concept just
propounded.

$$\underset{\underset{49}{\sim\sim}}{R^3R^4NCH_2}-\underset{Z}{\overset{CH_2NR^1R^2}{\overset{|}{\underset{|}{C}}}}-H \qquad \underset{\underset{50a}{\sim\sim}}{PhSCH_2}-\underset{R}{\overset{CH_2NO_2}{\overset{|}{\underset{|}{C}}}}-H \qquad \underset{\underset{50b}{\sim\sim}}{PhSCH_2}-\underset{R}{\overset{PhS\diagdown_{C}\diagup^{NOH}}{\overset{|}{\underset{|}{C}}}}-H \qquad \underset{\underset{50c}{\sim\sim}}{CH_3}-\underset{R}{\overset{H_2CNH_2}{\overset{|}{\underset{|}{C}}}}-H$$

In general, however, fission of the nitrocyclopropane ring in 41
appears to be unusually difficult. A number of reagents known to open
cyclopropane rings were found ineffective under the conditions tried;
they included hydrochloric and hydrobromic acids, bromine, and cata-
lytically activated hydrogen. Hydrogenation over palladium-on-carbon
readily reduced the amino group but failed to cleave the ring. The
exploitation of chiral nitrocyclopropanes for purposes of stereospecific synthesis therefore remains a challenging problem.

Methods for the Synthesis of Aminodeoxy and Deoxy Disaccharides
Related to α,α-Trehalose

The disaccharide α,α-trehalose (51) is a sugar of great biological significance because of the varied roles it plays in Nature as a structural constituent (e.g., in the cord factor of mycobacteria), as a
reserve source of energy (e.g., in insects during flight), and as a
possible intermediate in D-glucose resorption (23). Furthermore, its
2-, 3-, and 4-aminodeoxy derivatives (52—54) as well as an α-D-manno
stereoisomer of 52 (55, with OH-2' inverted) occur as actinomycetal
metabolites, reported to show (modest) antibiotic activity (24—27).
The chemical synthesis of derivatives and analogs has long commanded
much interest, as such compounds are required for the study of struc- .
ture—activity relationships in the action of trehalases (23,28), may
serve as substitutes for 51 in the synthesis of cord-factor analogs
to be used as probes in the field of mycobacterial biochemistry (29,
30), and could possibly prove to possess interesting properties as
enzyme inhibitors or antibiotic agents.
     Several groups of investigators have reported syntheses of numer-
ous, aminated α-D-hexopyranosyl α-D-hexopyranosides. These include

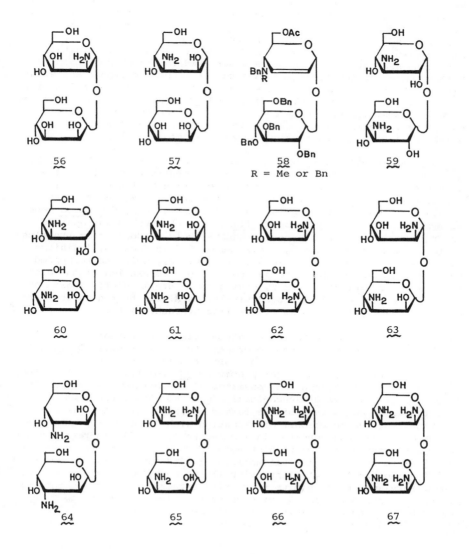

51 $R^1 - R^4$ = OH

52 R^1 = NH_2, $R^2 - R^4$ = OH (also with inverted OH-2')

53 R^1 = R^3 = R^4 = OH, R^2 = NH_2

54 R^1 = R^2 = R^4 = OH, R^3 = NH_2

55 $R^1 - R^3$ = OH, R^4 = NH_2

56

57

58
R = Me or Bn

59

60

61

62

63

64

65

66

67

the natural products 52 (31,32) and its aforementioned D-manno isomer
(32), the 6-amino isomer 55 (33,34), the α-D-altro,α-D-gluco stereo-
isomer(35)of 53, and 6,6'-diamino-6,6'-dideoxy-α,α-trehalose (33,36)
as well as its 2,2'-diamino-2,2'-dideoxy-α-D-altro,α-D-altro isomer
(35,37), the 3,3'-diacetamido-3,3'-dideoxy-α-D-allo,α-D-allo analog
(38), the 3,6,6'-triacetamido-3,6,6'-trideoxy-α-D-allo,α-D-gluco
analog (39), and the 4,4',6,6'-tetraamino-4,4',6,6'-tetradeoxy-α-D-
galacto,α-D-galacto analog (40). Many further 2,2'-, 3,3'-, 4,4'-,
and 6,6'-diamines and 4,4',6,6'-tetraamines that possess additional
(unsubstituted) deoxy groups were described by Hough, Richardson, and
their associates in an extensive series of articles published during
1970—1973 (see 35, 41 for lead references). Although some of the
syntheses here referred to were based on Koenigs—Knorr condensations
of monosaccharidic components, the majority comprised chemical transformations starting from commercial 51. Our own efforts in this domain
were focused partly on the latter approach, and partly on a third
alternative; the methods employed for preparing the new, aminated
trehalose analogs 53 and 56— 67 are surveyed in the subsections that
follow. Only the key steps relevant to introducing nitrogen and establishing the regio- and stereo-chemical patterns in the molecules will
be highlighted, whereas well-precedented procedures for generating the
requisite starting compounds and for elaborating the final targets
cannot be reviewed here. The last two subsections deal with the preparation of some non-nitrogenous deoxy analogs.

Nitromethane Cyclization. In the realm of disaccharides, the first
application of the nitromethane method for cyclization of "sugar di-
aldehydes" had afforded, from sucrose selectively cleaved in the
fructofuranosyl moiety with lead tetraacetate, a mixture of nitro
sugars from which reduction furnished 4-amino-4-deoxy-β-D-gluco-hept-
ulopyranosyl α-D-glucopyranoside (42). When the method was applied
later (43) to the tetraaldehyde quantitatively produced from trehalose
(51) by periodate oxidation, a mixture of 3,3'-dideoxy-3,3'-dinitro
disaccharides was obtained (in 90% crude yield), which corresponded,
configurationally, chiefly to 59, 60, and 61. Separation of the nitro
isomers was laborious, however, giving them pure in yields (based on
51) of only 14, 16 and 6%, respectively. Catalytic hydrogenation then
afforded 59— 61, isolated as crystalline hydrochlorides.

Osmylation. Osmium tetraoxide-catalyzed cis-oxyamination of the disaccharidic diene 70 was used to prepare the regioisomeric D-manno,D-
manno diamines 61— 63 (44) and the corresponding monoamines 56 and 57
(45). A practical procedure for preparing 70 from tri-O-acetyl-D-glucal
(68) was first elaborated. It consisted of converting one part of 68
into "4,6-di-O-acetyl-D-pseudoglucal" (69) and then condensing the
latter with a second part of the former by catalysis with boron trifluoride (Ferrier reaction). Oxyamination with chloramin-T then gave
mixtures of bis- and mono-tosylamido sugars (71 + 72), in ratios dependent on the proportion of reagent used (Figure 10). Separation of the
regioisomers gave the bis(sulfonamides) (2:4:7 ratio) and the mono-
(sulfonamides) (~1:1 ratio) individually in pure form. The latter were
cis-hydroxylated under osmium tetraoxide catalysis in the presence of
either N-methylmorpholine N-oxide or triethylamine N-oxide, to give
the isomers 73 in high yields. Each of the five tosylamides (71 and
73) was finally converted into the corresponding amine by treatment

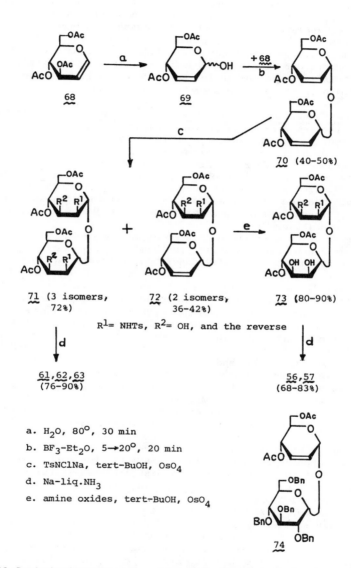

a. H$_2$O, 80°, 30 min
b. BF$_3$-Et$_2$O, 5→20°, 20 min
c. TsNClNa, tert-BuOH, OsO$_4$
d. Na-liq.NH$_3$
e. amine oxides, tert-BuOH, OsO$_4$

Figure 10. Synthesis of trehalose-type amino and diamino sugars having the D-manno,D-manno configuration.

with sodium in liquid ammonia. The osmylations appeared to occur with complete diastereofacial selectivity, with the reagent attacking the face of the alkene opposite from the allylic substituents present (at C-1 and C-4); no evidence for the formation of allo products was found.

The mono-unsaturated disaccharide 74 was also prepared (46). It was obtained in 93% yield by condensation of 68 with 2,3,4,6-tetra-O-benzyl-α-D-glucose under catalysis (47) with stannic chloride. Preliminary experiments indicated that it can likewise be oxyaminated (48). If this is borne out in further studies, a route would be open to the D-manno,D-gluco stereoisomers of 56 and 57.

Trifluoromethanesulfonate Displacements by Azide. We were first persuaded of the superior leaving-group characteristics of the trifluoromethylsulfonyloxy group, as compared to the mesyloxy group, during our synthesis (49) of 3-amino-3-deoxy-α,α-trehalose (53) which, incidentally, anteceded by a year the discovery of this compound as an antibiotically active metabolite of Norcardiopsis species (25). Displacement with lithium azide in the mesylate 75 was extremely sluggish, and was accompanied by a great deal of 3,4-elimination, whereas the triflate 76 reacted smoothly under mild conditions, with elimination being avoided (Figure 11).

However, even triflate displacement may be problematical, as was realized in a new, stereospecific synthesis of the diamine 61 (50), undertaken because the two aforedescribed syntheses had furnished 61 as a minor product only. The partially blocked α-D-altro,α-D-altro disaccharide 77, obtained (51) from 51 in 4 steps (53% overall), gave the bis(triflate) 78, which was subjected to the action of azide ion under a large variety of conditions. Invariably, the desired diazide 79 was accompanied by elimination products (80 and 81), which often preponderated (Figure 12). This was perhaps to be expected as 78, unlike its counterpart 76, is axially substituted at C-2; displacement at C-3 is thereby impeded and competing elimination becomes more favored. Nevertheless, optimal conditions involving phase-transfer reaction were worked out that permitted isolation of 79 in 51% yield. Hydrolytic removal of the benzylidene acetals, followed by catalytic transfer hydrogenolysis for simultaneous reduction of the azide groups and O-debenzylation, provided 61 in 30% overall yield from 77 (16% from 51).

The unsymmetrical ditriflate 82, identical with 78 in one half of the molecule but regio- and stereo-isomeric in the other (Figure 13), showed interesting behavior in azide displacement performed under the same phase-transfer conditions (52). Mainly the axial 3-triflate group in the altro moiety reacted, with a ratio for displacement (to give 55% of monoazide 83) and elimination (to give 22% of enol ether 84) qualitatively similar to that observed for 78; only 8% of diazide 85 was formed, indicating a much lower reactivity of the equatorial 2-triflate group in the gluco moiety. However, 83 was convertible into 85 (yield, 78%) by azide displacement in homogeneous phase. The discovery that displacements in a ditriflate may be performed regioselectively in a stepwise fashion was exploited for an alternative route to the monoamine 57. Thus, the 3-azido-2'-triflate 83 was subjected to a second displacement, but with sodium benzoate in homogeneous phase, and the product (86) was then by standard procedures converted into 57, obtained in 14% yield over 6 steps from 82 (52).

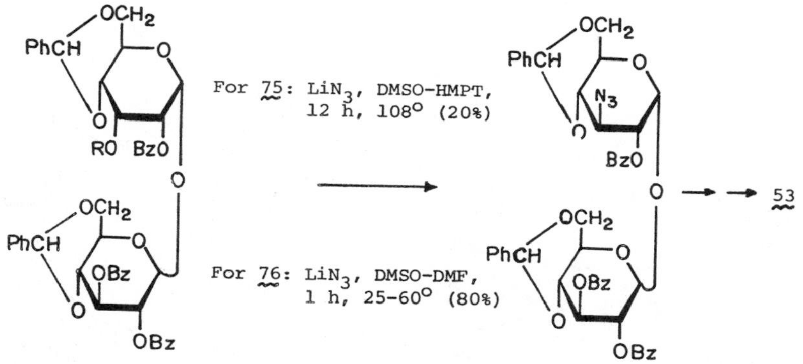

For 75: LiN₃, DMSO-HMPT, 12 h, 108° (20%)

For 76: LiN₃, DMSO-DMF, 1 h, 25-60° (80%)

75 R = Ms
76 R = Tf

Figure 11. The key step in the synthesis of 3-amino-3-deoxy-α,α-trehalose (53).

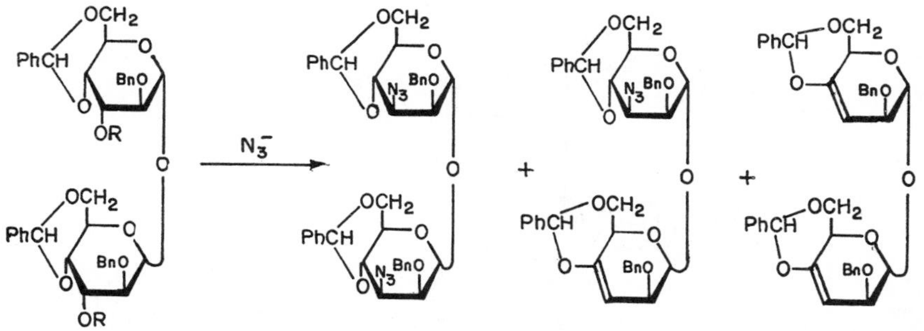

77 R = H
78 R = Tf

79
(51% max.)

80

81

Figure 12. Azide displacement in a disaccharidic 3,3'-ditriflate (78) having the D-*altro*,D-*altro* configuration.

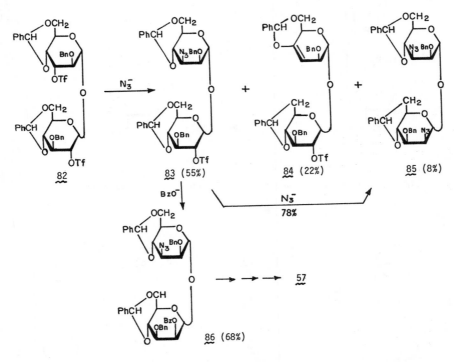

Figure 13. Azide displacements in the unsymmetrical, 3,2'-ditriflate (**82**) having the D-*altro*,D-*gluco* configuration.

Differential triflate reactivities as just described similarly gave access to the regioisomeric triamines 65 and 66 from a single precursor, the unsymmetrical diazido sugar 89 (Figure 14). Action of sodium azide upon the diepoxide 87 (readily prepared from 51) had been known known to give mainly the symmetrical counterpart 88 (35), but was now found (53) to yield additionally the (partial) anti-Fürst — Plattner product, 89, on a practical scale. Sequential displacements in its ditriflate, with azide followed by benzoate and in reverse order, furnished 2,3,3'- and 2,3,2'-triazido D-manno,D-manno derivatives, respectively, which were used to prepare 65 and 66 by standard manipulations. The tetraamine 67 was synthesized via double displacement, with azide, from the ditriflate of 88 (53).

Reductive Amination. Reductive amination of ketones with sodium cyanoborohydride in the presence of ammonium acetate constitutes a useful, general method for amine synthesis (54). In order to explore its potential merits in disaccharide functionalization, the diketone 90 was prepared from the diol 77 by Pfitzner—Moffatt oxidation, and subjected to that reaction (55). The resultant rather complex mixture was treated with acetic anhydride in methanol, in order to N-acetylate any amines present, and then partially fractionated by chromatography, furnishing in < 40% yield the α-D-altro,α-D-altro diacetamide 91 as the main product. Functional-group adjustment led to the diamino sugar 64, which was identical with 64 concurrently prepared for comparison through debenzylidenation and hydrogenation of the known (35) diazide 92 (Figure 15).

About one-half of the product-weight could not be recovered in chromatography of the reductive-amination mixture, and more work is clearly needed to unravel the complexity of the process. Thus far, 4 minor byproducts were isolated crystalline in yields of 1.5—8%. All possessed one 3-acetamido-3-deoxy-α-D-altro residue, in common with 91, but were not aminated in the second residue. Two of them were epimeric 3'-carbinols resulting from nonaminative reduction, and the other two were epimeric 3'-cyanohydrins.

Palladium-catalyzed, Allylic Amination. Allylic substitution of monosaccharidic hex-2-enopyranoside 4-acetates with secondary amines in the presence of tetrakis(triphenylphosphine)palladium(0) had led to a large variety of 4-aminated 2-enosides, with retention of configuration (56-58). The method was applied to the disaccharidic enoside 74 to give, with benzylmethylamine or dibenzylamine, the 4-amino sugar derivatives 58 in yields of 92 and 67% (46). Studies concerning hydroxylation of the double bond and subsequent deprotection are incomplete.

Deoxy Analogs by Reductive Desulfonyloxylation with Lithium Triethyl borohydride. Lithium triethylborohydride (LTBH) is an efficient reagent for the reductive desulfonyloxylation of certain secondary p-toluenesulfonates of monosaccharidic glycosides (59). Its action differs mechanistically and, consequently, with respect to the stereochemistry of the major products formed, from that of lithium aluminum hydride, a less efficient reagent previously used for similar purposes. The newer method of deoxygenation has been applied successfully (60) to various sulfonic esters of α,α-trehalose (Figure 16). Thus, treatment of 4,6;4',6'-di-O-benzylidene-2,3,2',3'-tetra-O-tosyl-α,α-trehalose (93) with LTBH in boiling oxolane for 3 h caused preferential O-desul-

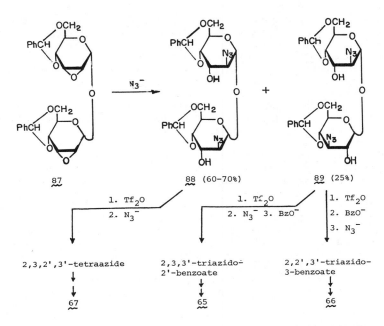

Figure 14. Synthesis of the triamino and tetraamino derivatives **65–67** of α-D-mannopyranosyl α-D-mannopyranoside.

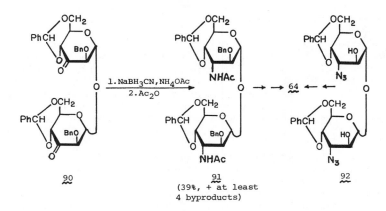

Figure 15. Synthesis of 3-amino-3-deoxy-α-D-altropyranosyl 3-amino-3-deoxy-α-D-altropyranoside (**64**).

fonylation at O-2 and O-2', followed by internal displacement of the
3- and 3'-tosyloxy groups and reduction of the resulting epoxide rings,
to give directly the 2,2'-dideoxy disaccharide 98 (yield, 65%), previously obtained (61,62) from the diepoxide 97 by its reduction (18 h)
with lithium aluminum hydride (yield, 52%).

Action of LTBH upon the 2,2'-ditosylate 94 afforded the known
3,3'-dideoxy sugar 100 less efficaceously, evidently because of competing, partial O-desulfonylation which, however, provided as a benefit
the new, unsymmetrical disaccharides 102 and 103 (11 and 8%) that were
formed via the monoepoxide 101. Compound 102 was the main product of
deoxygenation when the 2-monotosylate 95 was treated with LTBH, although anti-Fürst—Plattner opening of intermediary oxirane 101 is
believed to have taken place as a minor side-reaction in this instance.

The 2,3,2'-tritosylate 96 produced a complex mixture from which
the main products, namely, the new unsymmetrical 2,3'-dideoxy and 3-
deoxy sugars 104 and 105 were isolated crystalline, each in ~20% yield
after chromatography.

For possible future applications of the LTBH method with other
oligosaccharide derivatives it must be borne in mind that regio- and
stereo-chemical reaction courses as just described are contingent on
a sterically rigid protection of the 4,6-positions (as is present in
trans-fused, cyclic acetals), and on a 2,3-**trans** substituent **arrangement** (that permits formation of intermediary epoxides). In the **absence**
of the former constraint, contraction of the pyranoid to a branched-
chain, furanoid ring may occur (63), whereas for 2,3-cis orientations,
as for example in 4,6-O-benzylidene-α-D-mannopyranosides, the regiochemistry of deoxygenation is reversed owing to intervention of intramolecular hydride shifts (64).

Chain Elongation by Means of an Iron Carbonyl Reagent. The method of
chain elongation at the nonreducing terminal of 6-deoxy-6-halo and 6-
O-tosyl-hexopyranosides, recently developed (65) on the basis of
organoiron chemistry, was applied to 6,6'-di-O-tosyl-α,α-trehalose
hexaacetate (106). Reaction of 106 with sodium dicarbonyl-η⁵-cyclo-
pentadienyliron in oxolane at 25° led to substitution of the tosyloxy
substituents by the anionic iron complex (Figure 17). Without its
isolation, the product was treated with bromine in the presence of
methanol, effecting oxidative carbonyl insertion followed by methanolysis, to give dimethyl (2,3,4-tri-O-acetyl-6-deoxy-α-D-gluco-hepto-
pyranosyluronate) (2,3,4-tri-O-acetyl-6-deoxy-α-D-gluco-heptopyrano-
siduronate) (107), the first derivative of a trehalose homolog con-
sisting of two seven-carbon sugars. The dibenzyl ester was obtained
in similar fashion, and studies are underway to prepare the free dicarboxylic acid and, by its reduction, the neutral 6-deoxyheptopyran-
osyl 6-deoxyheptopyranoside (Baer, H. H., Breton, R.; unpublished).
The goal of these investigations is to synthesize novel "pseudo cord-
factors" by esterifying the 7,7'-positions of the neutral sugar with
lipid acids, and "mirror pseudo cord-factors", by esterifying the
dicarboxylic acid with lipid alcohols. Similar syntheses have been
performed (29,30,66,67) with other trehalose analogs including (α-D-
glucopyranosyluronic acid) (α-D-glucopyranosiduronic acid), with a
view to providing probes for correlating structure with biological
function in mycobacterial lipids.

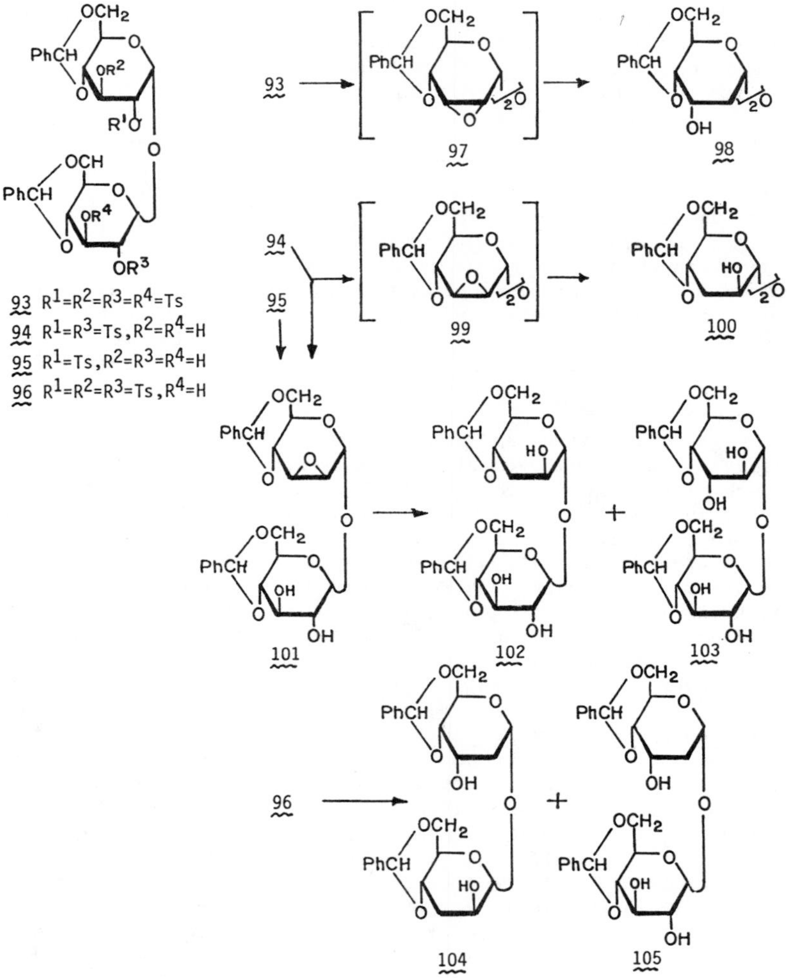

Figure 16. Synthesis of 2-deoxy, 3-deoxy, and 2,3-dideoxy derivatives of α,α-trehalose.

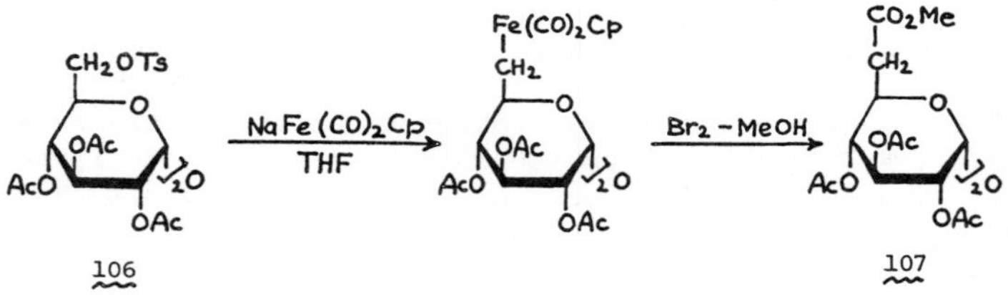

Figure 17. Synthesis of 6-deoxy-α-D-*gluco*-heptopyranosyluronate 6-deoxy-α-D-*gluco* heptopyranosiduronate dimethyl ester **107**.

Acknowledgments

Support of these studies from the Natural Sciences and Engineering Research Councils of Canada, and from the United States Public Health Service is gratefully acknowledged.

Literature Cited

1. Rinehart, Jr., K. L.; Suami, T. (Eds) Aminocyclitol Antibiotics; ACS Symposium Series No. 125; American Chemical Society: Washington, DC, 1980.
2. Rinehart, Jr., K. L.; Stroshane, R. M. J. Antibiot. 1976, 29, 319.
3. Rinehart, Jr, K. L. In ref. 1, pp 335-370.
4. Fujiwara, T.; Takahashi, Y.; Matsumoto, K.; Kondo, E. J. Antibiot. 1980, 33, 824.
5. Igarashi, K.; Honma, T.; Fujiwara, T.; Kondo, E. J. Antibiot. 1980, 33, 830.
6. Fujiwara, T; Kondo, E. J. Antibiot. 1981, 34, 13.
7. Baer, H. H.; Siemsen, L.; Astles, D. J. Carbohydr. Res. 1986, 156, 247.
8. Baer, H. H. Adv. Carbohydr. Chem. Biochem. 1969, 24, 67.
9. Grosheintz, J. M.; Fischer, H. O. L. J. Am. Chem. Soc. 1948, 70, 1476.
10. Whistler, R. L.; Pyler, R. E. Carbohydr. Res. 1970, 12, 201.
11. Szarek, W. A.; Lance, D. G.; Beach, R. L. Carbohydr. Res. 1970, 13, 75.
12. Baer, H. H.; Arai, I.; Radatus, B.; Rodwell, J.; Nguyen, C. Can. J. Chem. 1987, 65, 1443.
13. Perry, M. B.; Williams, D. T. Methods Carbohydr. Chem. 1976, 7, 44.
14. Radatus, B.; Williams, U.; Baer, H. H. Carbohydr. Res. 1986, 157, 242.
15. Sopova, A. S.; Yurchenko, O. I.; Perekalin, V. V. J. Org Chem. USSR 1965, 1, 1732.
16. Kohler, E. P.; Darling, S. F. J. Am. Chem. Soc. 1930, 52, 1174.
17. Asunskis, J.; Shechter, H. J. Org. Chem. 1968, 33, 1164.
18. Sakakibara, T.; Sudoh, R. J. Chem. Soc. Chem. Commun. 1977, 7.
19. Baer, H. H.; Linhart, F.; Hanna, H. R. Can. J. Chem. 1978, 56, 3087.
20. Baer, H. H.; Gilron, I. Carbohydr. Res. 1987, 146, 486.
21. Nielsen, A. T. In The Chemistry of the Nitro and Nitroso Groups; Feuer, H., Ed.; Interscience: New York, 1969; Part 1, pp 349-486.
22. Gani, D.; Hitchcock, P. B.; Young, D. W. J. Chem. Soc. Perkin Trans. 1 1985, 1363.
23. Elbein, A. D. Adv. Carbohydr. Chem. Biochem. 1974, 30, 227.
24. Arcamone, F.; Canevazzi, G.; Ghione, M. Giorn. Microbiol. 1956, 2, 205.
25. Dolak, L. A.; Castle, T. M.; Laborde, A. L. J. Antibiot. 1980, 33, 690.
26. Naganawa, H.; Usui, N.; Takita, T. Hamada, M.; Maeda, K.; Umezawa, H. J. Antibiot. Ser. A 1974, 27, 145.
27. Uramoto, M.; Otake, N.; Yonehara, H. J. Antibiot. Ser. A 1967, 20, 236.
28. Defaye, J.; Driguez, H.; Henrissat, B.; Bar-Guilloux, E. In Mechanisms of Saccharide Polymerization and Depolymerization; Marshall, J. J., Ed.; Academic Press: New York, 1980, pp 331-353.

29. Goren, M. B. Am. Rev. Respir. Dis. 1982, 125, 50.
30. Liav, A.; Goren, M. B. Carbohydr. Res. 1983, 123, c22.
31. Umezawa, S.; Tatsuta, K.; Muto, R. J. Antibiot. Ser. A 1967, 20, 388.
32. Paulsen, H.; Sumfleth, B. Chem. Ber. 1979, 112, 3203.
33. Jezo, I. Chem. Zvesti 1971, 25, 364.
34. Hanessian, S.; Lavallée, P. J. Antibiot. Ser. A 1972, 25, 683.
35. Hough, L.; Munroe, P. A.; Richardson, A. C.; Ali, Y.; Bukhare, S. T. K. J. Chem. Soc. Perkin Trans. 1 1973, 287.
36. Umezawa, S.; Tsuchiya, T.; Nakada, S. Tatsuta, K. Bull. Chem. Soc. Jpn 1967, 40, 395.
37. Jezo, I. Chem. Zvesti 1973, 27, 381.
38. Jezo, I. Chem. Zvesti 1971, 25, 369.
39. Jezo, I. Chem. Zvesti 1973, 27, 634.
40. Ali, Y.; Hough, L.; Richardson, A. C. Carbohydr. Res. 1970, 14, 181.
41. Richardson, A. C.; Tarelli, E. J. Chem. Soc. Perkin Trans. 1 1973, 1520.
42. Baer, H. H.; Ahammad, A. Can. J. Chem. 1966, 44, 2893.
43. Baer, H. H.; Bell, A. J. Can. J. Chem. 1978, 56, 2872.
44. Baer, H. H.; Siemsen, S.; Defaye, J.; Burak, K. Carbohydr. Res. 1984, 134, 49.
45. Baer, H. H.; Siemsen, S. Carbohydr. Res. 1986, 146, 63.
46. Hanna, Z. S. Ph.D. Thesis, University of Ottawa, Ottawa, 1981.
47. Grynkiewicz, G.; Priebe, W.; Zamojski, A. Carbohydr. Res. 1979, 68, 33.
48. Astles, D. J. Ph.D. Thesis, University of Ottawa, Ottawa, 1985.
49. Baer, H. H.; Bell, A. J. Carbohydr. Res. 1979, 75, 175.
50. Baer, H. H.; Radatus, B. Can. J. Chem. 1985, 63, 440.
51. Baer, H. H.; Radatus, B. Carbohydr. Res. 1984, 128, 165.
52. Baer, H. H.; Radatus, B. Carbohydr. Res. 1985, 144, 77.
53. Baer, H. H.; Radatus, B. Carbohydr. Res. 1986, 146, 43.
54. Borch, R. F.; Bernstein, M. D.; Durst, H. D. J. Am. Chem. Soc. 1971, 93, 2897.
55. Baer, H. H.; Radatus, B. Carbohydr. Res. 1986, 157, 65.
56. Baer, H. H.; Hanna, Z. S. Can. J. Chem. 1981, 59, 889.
57. Baer, H. H.; Hanna, Z. S. Carbohydr. Res. 1981, 94, 43.
58. Baer, H. H.; Hanna, Z. S. Carbohydr. Res. 1980, 78, c11.
59. Baer, H. H.; Hanna, H. R. Carbohydr. Res. 1982, 110, 19.
60. Baer, H. H.; Mekarska, M.; Bouchard, F. Carbohydr. Res. 1985, 136, 335.
61. Hough, L.; Munroe, P. A.; Richardson, A. C. J. Chem. Soc., C 1971, 1090.
62. Hough, L.; Richardson, A. C.; Tarelli, E. J. Chem. Soc., C 1971, 1732.
63. Baer, H. H.; Astles, J. D.; Chin, H. C.; Siemsen, L. Can. J. Chem. 1985, 63, 432.
64. Baer, H. H.; Mekarska-Falicki, M. Can. J. Chem. 1985, 63, 3043.
65. Baer, H. H.; Hanna, H. R. Carbohydr. Res. 1982, 102, 169.
66. Goren, M. B.; Jiang, K.-S. Carbohydr. Res. 1980, 79, 225.
67. Liav, A.; Das, B.C.; Goren, M.B. Carbohydr. Res. 1981, 94, 230.

RECEIVED May 31, 1988

Chapter 3

New Method of Orthoesterification Under Kinetic Control

Formation and Selective Hydrolysis of Methoxyethylidene Derivatives of Carbohydrates

Mohamed Bouchra, Pierre Calinaud, and Jacques Gelas

École Nationale Supérieure de Chimie, Université de Clermont-Ferrand, Boîte Postale 45, 63170 Aubière, France

The reaction of ketene acetals upon mono- and oligo-saccharides is a new route to unusual cyclic orthoesters essentially under kinetic control . The reaction proceeds by preferential attack of the reagent on the primary hydroxyl group if any, and the prefered tautomeric form in solution for free sugars. Obtention of strained- and medium-sized rings is possible.

The methoxyethylidene derivatives thus obtained are very sensitive to hydrolysis and very mild conditions can be used to obtain selectively α- or β-hydroxyacetates useful as intermediates in structural modifications of sugars or in glycosidic syntheses.

The use of enol ethers is now a well-established method for acetonation of carbohydrates under kinetic control (1,2). Specially, 2-methoxypropene reacts with diols (Figure 1) to give isopropylidene derivatives which may be significantly different from those obtained under classical thermodynamic conditions. The reaction has been successfully applied to free sugars and deoxysugars (3-7), glycosides (8), oligosaccharides (9-12), diethyldithio-acetals of free sugars (13-15) and polyols (16). Its extension to the use of other enol ethers has also been exploited: ethylidene (from ethyl vinyl ether) (13, 14) and cyclohexylidene (from 1-ethoxycyclohexene) (17) derivatives were obtained under the same conditions. Specifically protected sugars thus available, usually in high yields, have been prepared in various other laboratories and employed in structural modification of sugars or in glycosydic synthesis (see, for instance, refs. 18- 22). Scaling-up the reaction presents no major difficulties [see, for instance, ref. 23 for the synthesis of 1,2:5,6-di-O-isopropylidene-D-mannitol on more than a 100-gram basis following ref.16, notwithstanding some difficulties apparently experienced by some authors (24,25)].

Having at hand this strategy for specific acetalation of sugars, we considered the use of ketene acetals instead of enol ethers might afford a good mean of access to orthoesters under kinetically controlled conditions (Figure 2). At the time we initiated this work, at least two reactions of ketene acetals with an alcohol were known (Figure 2). The first one described the intramolecular addition of the primary hydroxyl group at position 4' of 4-hydroxymethyl-2-methylene-1,3-dioxolane (26); the second one was the production (27) of a bicyclic orthoester of the bicyclo[2.2.2]octane series from dehydrohalogenation of a bifunctionnal 1,3-dioxane [probably through the intermediacy of the (non-isolated) methylene dioxane generated by β-elimination of HCl and subsequent intramolecular addition of the hydroxyl group].

By analogy with the reaction using enol ethers (which generally does not cause any attack on the hydroxyl group at the anomeric center), we expected that the reaction using ketene acetals would lead to orthoesters at the non-anomeric positions. Although 1,2-orthoesters of sugar are well-known (28-30) and widely used, especially for glycoside synthesis (31-34), few orthoesters are know in which the anomeric center is not involved. Four different examples are given in Figure 3, respectively in the nucleoside series (35-37) (ethanolysis of the 2,3-orthoester gave a mixture of regioisomers of formates), for methyl α-D-arabinopyranoside (38) (reduction of the 3,4-orthoester provided an unusual approach to ethylidene derivatives), for a D-galacto derivative (39) (hydrolysis of the 3,4-orthoester was regiospecific, giving the axial ester, affording OH-3 free for glycosylation in the synthesis of blood-group substances) and, finally, for the 3-O-methyl-1,2-O-isopropylidene-α-D-glucofuranose(40) (thermal degradation of the 5,6-orthoester led to an ethylenic sugar). Very few other examples can be found in the literature: they concern (i) the hydrolysis of the 3,4-orthoacetate of methyl 2,6-dideoxy-α-D-lyxo-hexopyranoside (41); (ii) the formation of the 2,3:5,6-diorthoformate derivative of methyl α-D-mannofuranoside (42); and (iii) the isolation of the 4,6-orthoacetate of D-idopyranose in the reaction of antimony pentachloride with β-D-glucopyranose pentaacetate (43)

ORTHOESTERIFICATION OF PYRANOSES AND PYRANOSIDES

The reagent chosen for this study was 1,1-dimethoxyethene prepared by dehydrochlorination of chloroacetaldehyde dimethyl acetal according to McElvain (44). The dry reagent can be stored for several weeks or months in small vials, with molecular sieves, in a refrigerator.

A magnetically stirred solution of methyl α-D-glucopyranoside 1 (Figure 4) in dry N,N-dimethylformamide

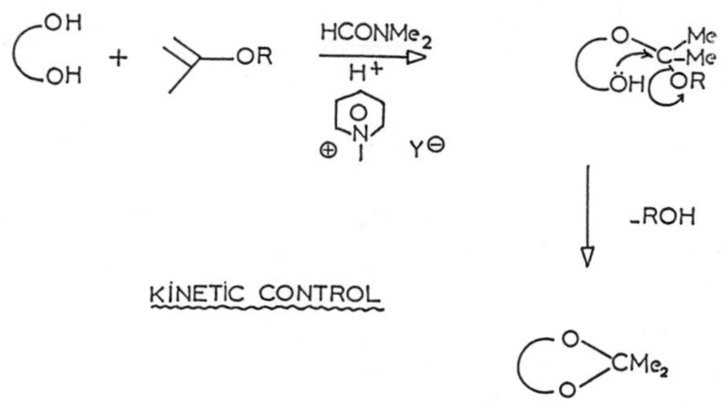

Figure 1. Synthesis of acetals from enol ethers and diols.

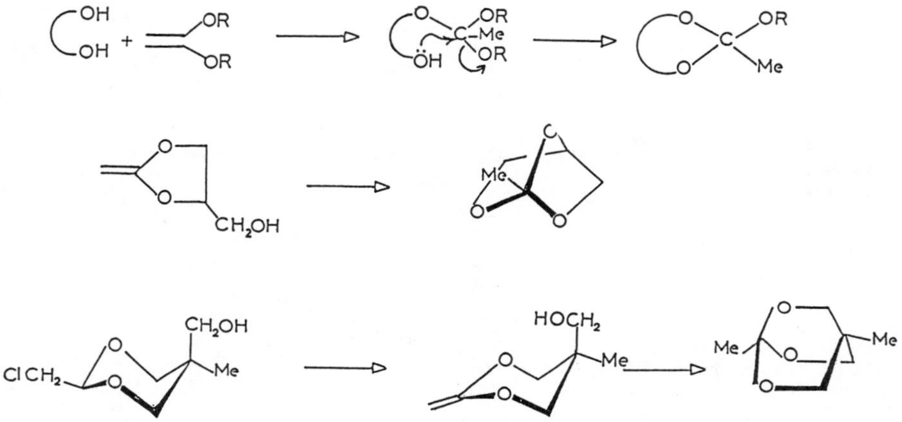

Figure 2. Synthesis of orthoesters from ketene acetals and diols.

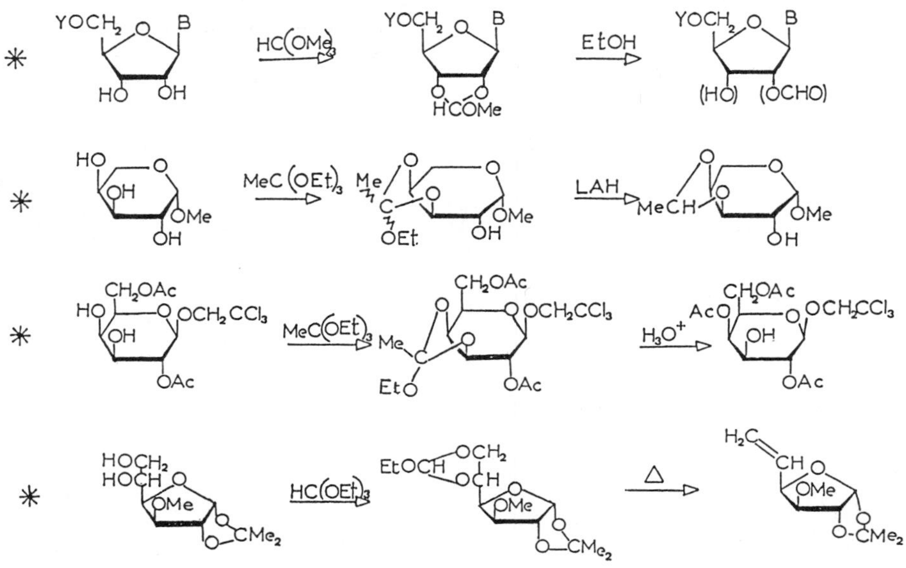

Figure 3. Orthoesters at non-anomeric position.

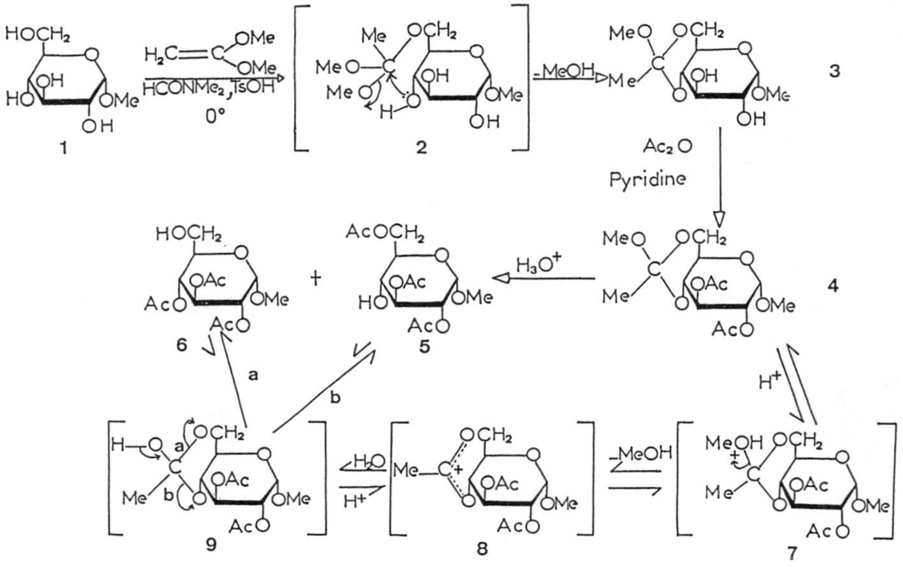

Figure 4. Orthoesterification of Methyl α-D-glucopyranoside.

containing a small quantity of a desiccant (Sikkon, Drie-
rite, or molecular sieves) was maintained at a temperature
below 5°C (ice-bath). Twice the stoichiometric amount of
1,1-dimethoxyethene and a few crystals (5 to 20 mg) of p-
toluenesulfonic acid (a pyridinium salt, such as pyridinium chloride or pyridinium p-toluenesulfonate, is also an
efficient catalyst) were added. After a few hours (3--
4h) all of the starting material had disappeared (t.l.c.
monitoring). The solution was stirred for one hour with
sodium carbonate, filtered and the filtrate evaporated
under diminished pressure (1 mm Hg, bath below 45°C). The
residue showed essentially a single spot on t.l.c. (crude
yield 92%) and could be used directly in a multistep
synthesis. Purification of this compound by column chromatography could not be performed without decomposition if
the silica gel and eluent were not carefully dried (a
small quantity of triethylamine could also be added to the
eluent). The product was identified (n.m.r. and mass spectra) as methyl 4,6-0-methoxyethylidene-α-D-glucopyranoside
3 [m.p. 97--98°C, [α]$_D^{20}$ +112° (acetone)]. Depending on the
method of purification (crystallization or column chromatography), the ratio between the two possible diastereoisomers [identified by the ^{1}H-n.m.r. chemical shift of
substituents according to criteria published (45) for 2-
alkoxy-1,3-dioxanes] at the orthoester carbon atom was in
the range 70:30 to 99:1 in favor of the axial position of
the methoxyl group and provides an illustration of the
influence of the anomeric effect. The diol 3 was readily
transformed into the diacetate 4 [m.p. 57--58°C, [α]$_D^{20}$
+85° (chloroform)] which was submitted to partial hydrolysis performed by one of the three following methods: (i)
action of 1:3 acetic acid--water at room temperature;
(ii) addition of 1-5 mg of p-toluenesulfonic acid to a
stirred solution of 4 in a 19:1 chloroform water; (iii)
adsorption of a solution of 4 (in a mixture of chloroform
and water) onto a small column of silica gel for a few
hours, followed by classical elution. In each experiment,
a mixture of acetates 5 and 6 (ratio 65:35) was obtained
quantitatively. These derivatives having respectively OH-4
or OH-6 free were separated in high yield.

　　　　The formation of orthoester 3 may be explained
by the preferential addition of the ketene acetal to the
most reactive alcohol function (primary hydroxyl group)
giving the (non-isolated) acyclic orthoester 2, which is
attacked by the neighbouring OH-4 with subsequent elimination of methanol. The partial hydrolysis of the diacetate
4 is assumed to proceed through protonation of the methoxyl group (7), via the dioxocarbenium ion 8 and the orthoacid 9, collapse of 9 by either path b or path a according to the mechanism generally proposed (see, for
instance, ref. 29 and refs. cited therein) affords the
compounds 5 or 6 respectively.

　　　　Having demonstrated the feasibility of the reaction of ketene acetals for the synthesis of unusual orthoesters from glycosides, it was essential to test the

behavior of free sugars in the same reaction to answer the question of the reactivity of the anomeric hydroxyl group. Using conditions already described for its methyl glycoside, D-glucose gave the 4,6-O-methoxyethylidene derivative 10 (Figure 5), which was easily transformed into the β-triacetate 11 [yield 77% from D-glucose after purification, syrup, $[\alpha]_D^{20}$ -37° (chloroform)]. Partial hydrolysis of the orthoester function gave quantitatively the mixture of regioisomers 12 and 13 (ratio 65:35). This experiment confirmed the results obtained in the glycoside series and showed that the anomeric group did not compete with OH-6 for addition of ketene acetal. This conclusion was also confirmed by the reaction performed on D-mannose (Figure 6) which led to the orthoester 14 [crude yield quantitative, 60% after purification, m.p. 67-68°C, $[\alpha]_D^{20}$ +17°, final (chloroform)]. Acetylation afforded quantitatively a mixture of anomers of the triacetate 15, which could be separated into the pure α [yield 55%, m.p. 42-43°C, $[\alpha]_D^{20}$ +48° (chloroform)] and pure β anomer [yield 20% m.p. 49-51°C, $[\alpha]_D^{20}$ -17.5° (chloroform)]. Starting from the α anomer, partial hydrolysis of acetate 15 gave a mixture of regioisomers 16 and 17 (ratio 65:35).

The question that then arose was the possible use of ketene acetals in the synthesis of 2,3-orthoesters if position 6 is not available. Starting from 4,6-O-isopropylidene-D-mannopyranose 18 (readily prepared by the acetonation of D-mannopyranose with 2-methoxypropene according to ref. 5), a diastereoisomeric mixture of the orthoesters 19 [yield 80%, m.p. 115--116°C, $[\alpha]_D^{20}$ -14° final (chloroform)] was obtained without any significant participation of the anomeric hydroxyl group. Acetylation of this mixture gave essentially the α anomer of diastereoisomeric mixture of acetates 20 [yield 73%, m.p. 108--109°C, $[\alpha]_D^{20}$ +26° (chloroform)]. Partial hydrolysis was highly regioselective and essentially only the derivative 21 [yield 60% m.p. 125--126°C, $[\alpha]_D^{20}$ +43.5°(chloroform)] having OH-3 free and AcO-2 axial was obtained (only 2-3% of the regioisomer having OH-2 free was detected in the ^{1}H-n.m.r. spectrum of the crude product). This high regioselectivity (or regiospecificity) was not unexpected as other examples have been previously described in the cyclohexane (46) and carbohydrate (39, 47, 48) series, and have been discussed (49).

Although the strategy of using ketene acetals for the synthesis of 4,6- and 2,3-orthoesters, especially from free sugars, is new, the preparation of orthoesters from vicinal cis diols of glycosides having all other positions subsituted, using orthoacetates or orthoformates (transorthoesterification), was already available (see, for instance, the preparation of a 3,4-methoxyethylidene derivative in the D-galacto series in ref. 39). It was thus of interest to try to extend our results to the situation where transorthoesterification does not readily

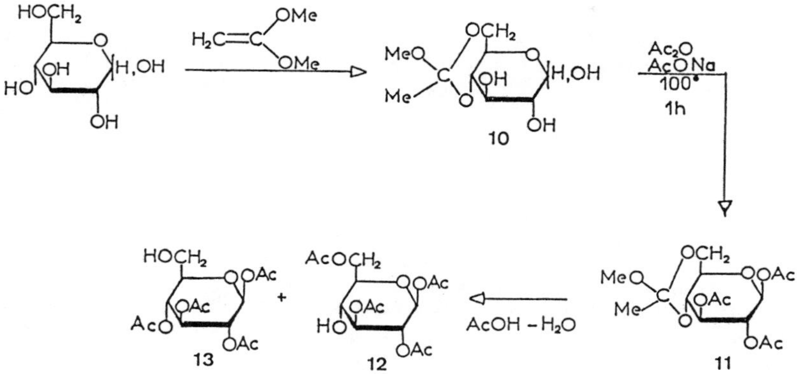

Figure 5. Orthoesterification of free sugars: example of D-glucose.

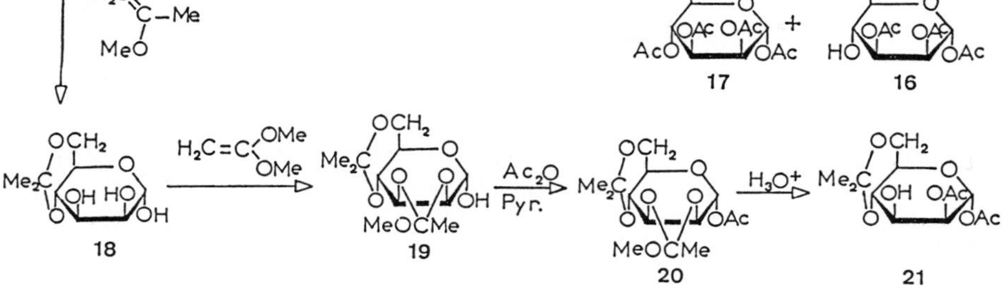

Figure 6. Orthoesterification of D-mannose.

give an orthoester, namely, with a vicinal trans-diol.
We compared (Figure 7) the behavior of the 2,3-diol for
the D-manno and the D-gluco series .In the first case (D-
manno), from methyl 4,6-O-isopropylidene-α-D-mannopyrano-
side 22 (prepared according to ref. 5) results were similar to those described in Figure 6 for the free sugar 18
[23: syrup, [α]$_D^{20}$ +20° (chloroform); 24: yield 85%, m.p.
97-98°C, [α]$_D^{20}$ +32.6 (chloroform)]. Starting from the
protected glucoside 25 (3, 8), an excellent yield of the
trans 2,3-orthoester 27 [80%, syrup, [α]$_D^{20}$ +78° (chloroform)] was obtained. The use of such reagents as ketene
acetals for the preparation of five-membered strained
cyclic orthoesters, where two equatorial bonds are en-
gaged, is comparable to the synthesis of similar isopropylidene derivatives from enol ethers (8). Assuming that OH-
2 is more reactive than OH-3, the intermediacy of the
acyclic orthoester 26 may be hypothesized.
          As expected, compounds 23 and 27 (and also com-
pounds 20 and 21) were mixtures of diastereoisomers
corresponding to both possible configurations of methoxyl
and methyl groups. It is noteworthy that although the
ratio between stereoisomers (n.m.r. determination) depended on the origin of the sample (crude material, sample
from column chromatography, or from crystallization), in
no instance we did observe a large displacement in favor
of one of the two isomers (often the ratio varied between
50:50 and 60:40). This may be an indication that the
anomeric effect is not as influential in five-membered
rings as it is in six-membered rings.
          Finally, partial hydrolysis of compound 27 [ad-
sorption of a solution in chloroform—water onto a column
of silica-gel and standard elution) gave regiospecifically the acetate 28 [yield 60%, plus recovery of 10% of
starting material, m.p. 83—84°C, [α]$_D^{20}$ +56° (chloroform)]
with OH-2 free. This orientation was not affected if we
started from samples of compound 27 containing different
proportions of diastereoisomers.

          At this stage, it was not easy to determine
whether the results of partial hydrolysis of such compounds as 4, 15 and 27 corresponded to kinetic or thermodynamic control. It is well known that selective acyla-
tions may favor either OH-2 or OH-3, according to the
structure of the starting material and the conditions of
the reaction (see ref. 50 and references cited therein).
For instance, acyl migration from position 3 to position 2
is known to occur in the D-galacto series under alkaline
conditions (51).

ORTHOESTERIFICATION OF FURANOSES

          The readily accessible (52) 1,2-O-isopropylidene-
D-glucofuranose 29 was subjected (Figure 8) to addition of

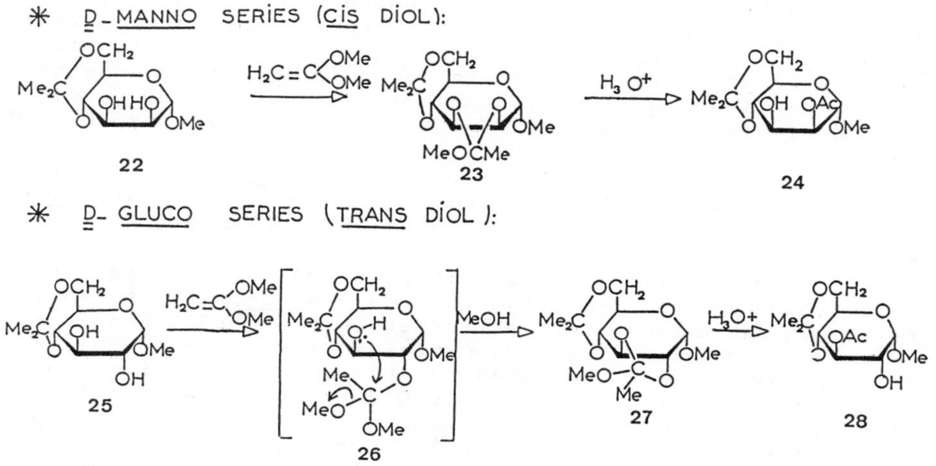

Figure 7. Orthoesterification of pyranosides at position 2,3.

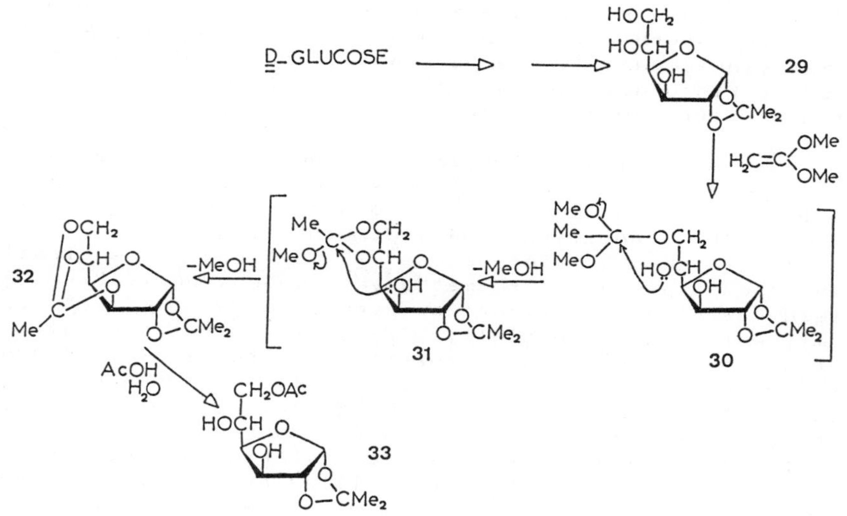

Figure 8. Orthoesterification of 1,2-0-isopropylidene-α-D-glucofuranose.

the ketene acetal under standard conditions . Only one
compound was obtained rapidly and quantitatively. It was
identified as the quadricyclic orthoester 32 [m.p. 131—
134°C, [α]$_D^{20}$ -40° (chloroform)], which had been previously
isolated (53) from the reaction of ethyl orthoacetate with
triol 29. In this case it appeared that the expected
orthoester 31 (obtained through compound 30, through preferential involvment of the primary hydroxyl group) could
not be isolated because of the spatial proximity of OH-3
and the orthoester function.

          Partial hydrolysis of compound 32 was found to
be regiospecific, leading only to the 6-monoacetate 33
[yield 95%, m.p. 143-144.°C, [α]$_D^{20}$ -5° (ethanol); compare
ref. 53].

          In order to obtain information concerning the
hydrolysis of a 5,6-O-orthoester such as 31, it was thus
necessary to prevent the participation of OH-3. Starting
(Figure 9) from compound 34 (obtained from 1,2:5,6-di-O-
isopropylidene-α-D-glucofuranose through acetylation and
selective hydrolysis of the 5,6-O-isopropylidene protective group), the expected orthoester 35 was isolated
[yield 82%, m.p. 79--82°C, [α]$_D^{20}$ -38° (chloroform)] as a
diastereoisomeric mixture. Hydrolysis gave a quantitative
yield of a mixture, analysis (^{1}H-n.m.r. spectroscopy) of
which revealed the presence of the two regioisomers 36 and
37 (ratio 87:13) from which the known (57) 6-OAc derivative was isolated [yield 80%, syrup, [α]$_D^{20}$ +11.2° (chloroform)]. This result compared with a previous study (55,
56) of the hydrolysis of 5,6-O-dimethylaminoethylidene and
-dimethylaminobenzylidene derivatives of 1,2-O-isopropyli-
dene-α-D-glucofuranose which gave regiospecifically the
corresponding 6-O-acyl compounds (Figure 9).

          A logical extension was then to test the beha-
vior of a 3,5-orthoester. This was performed with convenient precursors in the D-xylo and the D-gluco series
(Figure 10). Addition of 1,1-dimethoxyethene to 1,2-O-
isopropylidene-α-D-xylofuranose 38 (57) gave quantitatively the expected 3,5-O-methoxyethylidene derivative 39
[yield 83% after purification, syrup, [α]$_D^{20}$ +6.1° (chloroform)] which was essentially (^{1}H-n.m.r. spectroscopy) one
of the two possible diastereoisomers. Its partial hydrolysis gave a single compound, identified as the acetate 40
[yield 92%, m.p. 100--100.5°C, [α]$_D^{20}$ +23.7° (chloroform)].
This 6-OAc derivative had already been reported (57), but
its preparation by selective deacetylation of the corresponding 3,5-diacetate (58) as well as by selective acylation of the diol (59) gave 40 in much lower yields.

          Having prepared the 3,5-diol 33 in the D-gluco
series (vide supra, Figure 8), we found that the ketene
acetal converted it quantitatively into the orthoester 41
in more than 95% purity (t.l.c., ^{1}H-n.m.r. spectroscopy).
This syrupy compound was partly degraded during purification by column chromatography, and final yields depended critically upon the experimental conditions [60-80%,

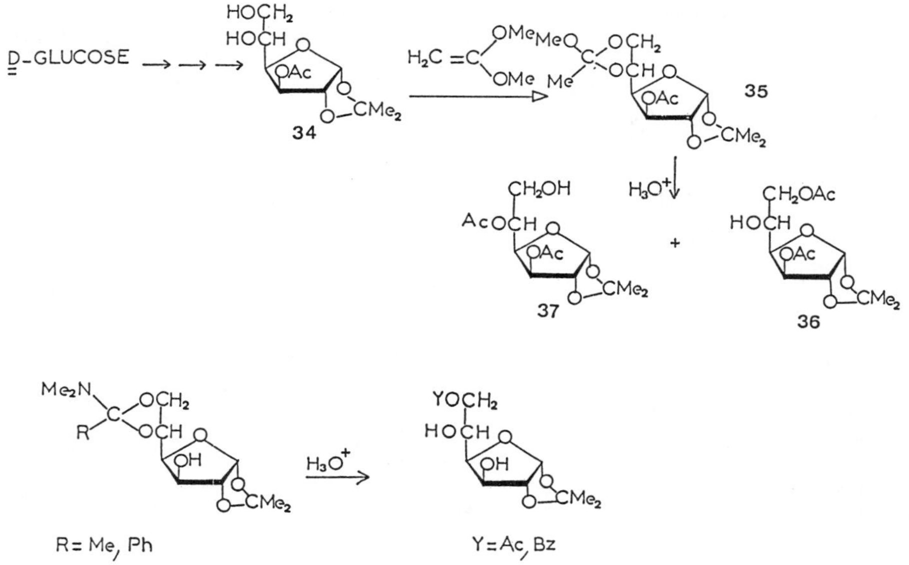

Figure 9. Orthoesterification of D-glucofuranose at position 5,6.

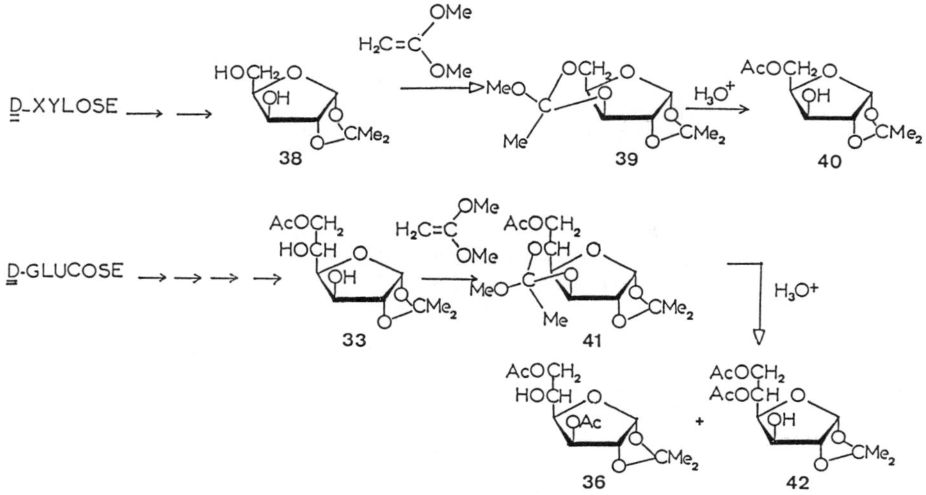

Figure 10. Orthoesterification of furanoses at position 3,5.

syrup, $[\alpha]_D^{20}$ +50° (chloroform)]. The partial hydrolysis of this 3,5-orthoester, expected to be similar to that of its analog 39, surprisingly gave a mixture of two compounds (ratio 30:70): the minor one was identical to compound 36 previously isolated from hydrolysis of the orthoester 35; the major product of the reaction was the known (60) regioisomer 42 [yield 45% after separation, syrup, $[\alpha]_D^{20}$ +29.2° (chloroform)]]

Once again the question arose concerning kinetic or thermodynamic control of these hydrolysis, and the problem of the possible migration of acyl groups initially formed. Examination of the three possible dioxocarbenium ions that could be intermediates during the hydrolysis of orthoester 32 (Figure 11) suggested that that the seven-membered-ring ion A was the intermediate in the regiospe-cific formation of acetate 33: competition between five- and six-membered-ring ions B and C would have given mix-tures (compare the hydrolysis of compound 35 or compound 41). An explanation for this preferential (or exclusive) protonation of O-5 leading to A has still to be found; hypotheses could be made of (i) attack on the more-acces-sible site of the molecule, (ii) attack on the more-basic oxygen atom, and (iii) better relief of strain in the bicyclic moiety of the molecule in giving the 1,3-dioxe-pane-like ion A (rather than the 1,3-dioxane, and 1,3-dioxolane-like ions) Relevant here is the study (61) of the behavior of monoacetates of 1,2-O-isopropylidene-D-glucofuranose: it was found that either during recrystal-lization from hot solvents or during contact with silica gel, the 3-OAc group in acetate 44 migrated to position 6 (acetate 33). The 3-O-acetyl-5-O-methyl derivative was prepared and shown to rearrange to the 6-O-acetyl-5-O-methyl isomer in very dilute basic medium. This experiment showed that an orthoacid containing a seven-membered ring is a possible intermediate in the reaction 44 ⟶ 33. Stereoelectronic control (49) could then explain the con-version of this orthoacid into the 6-acetate.

As we had previously demonstrated the ability of enol ethers to give seven-membered-ring acetals (4), we attempted to prepare an analog in the orthoester series. The action of 1,1-dimethoxyethene on 2,3-O-isopropylidene-D-ribofuranose 45 (62) (Figure 12) gave a high yield of the seven-membered-ring 1,5-orthoester 47 [72%, m.p. 85-86°C, $[\alpha]_D^{20}$ -30.6° (chloroform)], probably through the non-isolated (but detected by t.l.c.) acyclic orthoester 46. This type of participation of OH-1 is the only possi-bility for obtaining an orthoester having the anomeric center involved, as the ketene acetal does not react directly with this group. Compound 47 was extremely sen-sitive to atmospheric humidity; its purification by column

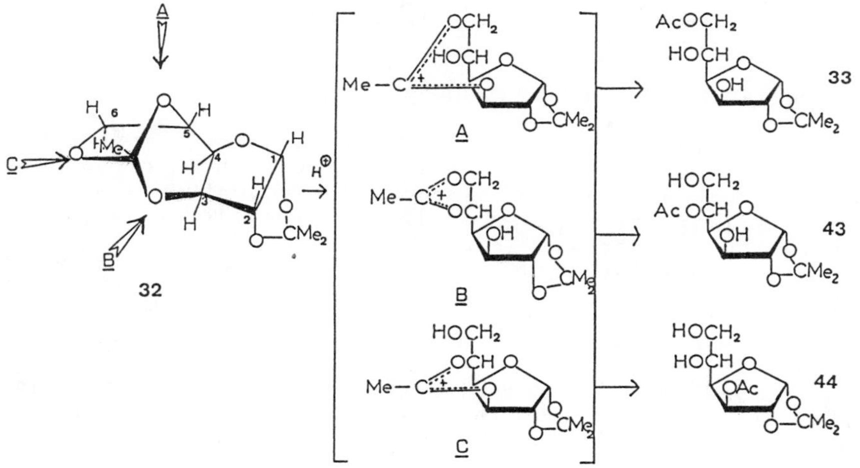

Figure 11. Dioxocarbenium ions from orthoester 32.

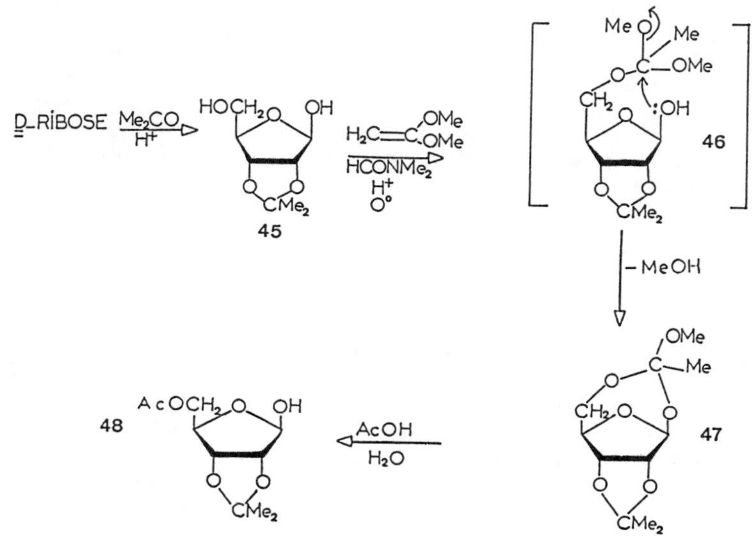

Figure 12. Obtention of a seven-membered ring orthoester in the ribofuranose series.

chromatography without special care caused partial hydrolysis to the 6-acetate 48. The latter was obtained regiospecifically by treatment of 47 with chloroform--water in
the presence of p-toluene sulfonic acid [yield 80%, syrup,
[α]$_D^{20}$ -17.1° (chloroform)].

ORTHOESTERIFICATION OF OLIGOSACCHARIDES

Finally we extended this study of monosaccharides to some oligosaccharides and report here preliminary
results concerning the disaccharide α,α-trehalose, for
which there is high interest in the specific protection or
activation of hydroxyl groups (63-66).

Addition of an excess of 1,1'-dimethoxyethene to
α,α-trehalose (49, Figure 13) led quantitatively to the
symmetrical 4,6-diorthoester 50 [yield 98%, syrup,
m.p.94--95°C, [α]$_D^{20}$ +125.6° (chloroform)]. When this
acetate was purified on a column of silica gel, only part
of the material was recovered as a pure compound (yield
47%). Another compound that migrated more slowly was isolated, and was identified as the asymmetrical OH-4 52
[yield 25%, m.p. 161--162°C, [α]$_D^{20}$ +138.7° (chloroform)].
This remarkable selective hydrolysis was unexpected, as
the regioisomer having a free hydroxyl group at C-6 was
not detected, and as methyl α-D-glucopyranoside (assumed
a priori to be a good model for trehalose) gave a mixture
of regioisomers (Figure 4). Partial hydrolysis of compound
51 (in 1:3 acetic acid--water) gave essentially one compound identified as the symmetrical OH-4,4' alcohol 53
[yield 70%, m.p. 175-176°C, [α]$_D^{20}$ +146.2° (chloroform)],
confirming the preferential isolation of derivatives acetylated at position 6. Attempts to effect asymmetrical
orthoesterification of only one glucose moiety of trehalose (following asymmetric acetalation described previously (9)) are in progress.

ASSIGNMENT OF RING SIZE IN ORTHOESTERS

The different structures prepared (five-, six-,
and seven-membered-ring orthoesters) allowed us to propose
the ^{13}C-n.m.r. chemical shift of the orthoester carbon
atom as a probe for the assignment of ring size, following
the now well-established criteria for acetals (62, and
references cited therein). Table I compares chemical-shift
ranges for ethylidene acetals (13), isopropylidene acetals
(62), and orthoesters. As expected, a downfield shift was
observed in going from acetals to orthoesters (+10 to 12
p.p.m. from isopropylidene acetals to methoxyethylidene
derivatives).

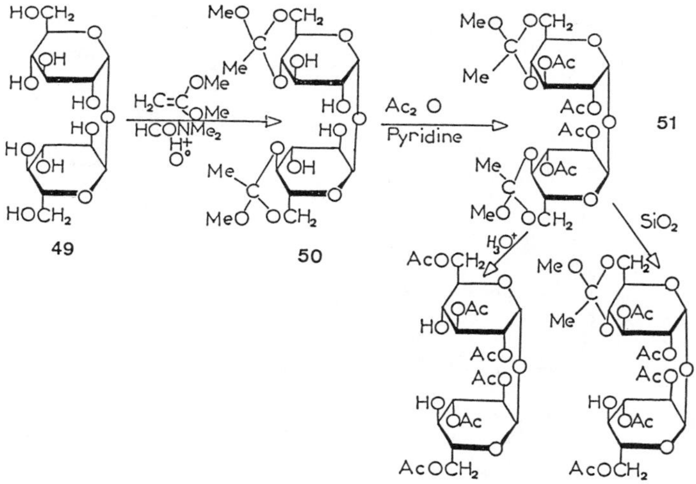

Figure 13. Orthoesterification of α,α-trehalose.

Table I. Assignment of ring size by the observation of the ^{13}C-n.m.r. chemical shift of the orthoester (or acetal) carbon atom

		Ring size		
Function al group		5	6	7
O-isopropylidene (acetal)	δ ppm	108-116	97-101	100-102
O-ethylidene (acetal)		101-102	99-100	
O-methoxymethylidene (orthoester)		121-123	111-113	115

CONCLUSIONS

        We have shown that the action of 1,1-dimethoxye-thene on sugars affords an efficient method for orthoeste-rification of diols, essentially under kinetic control. This method is characterized by the following criteria: (i) primary hydroxyl groups are the favored sites for initial attack of the ketene acetal; (ii) free monosaccha-rides, which exist in solution as pyranoses, react without change of ring size; (iii) the anomeric hydroxyl group of free sugars does not participate at least during the initial step of the reaction (it can however react with an intermediate acyclic orthoester); (iv) stoichiometric control of the reaction is possible to give either mono- or di-orthoesters; (v) extension to labile molecules is possible and the method may be valuable for oligo-saccharides; (vi) unusual rings (strained or medium-sized rings) can be obtained.
        This work has shown the extreme sensitivity of the O-methoxyethylidene group to hydrolysis under very mild conditions. This is not a handicap if the formation of these orthoesters is not intended to be a protective-group strategy but rather a way to obtain (one-flask reactions are feasible) hydroxyacetates, possibly with regioselectivity or regiospecificity. These acetates are useful synthons for glycoside synthesis.
        The reactivity of the methoxyethylidene deriva-tives with nucleophiles other than water has been exa-mined. It has been already found (13) that they may be readily opened by halogenating reagents (N-bromosuccini-mide for instance) to give α or β-haloacetates (respec-tively from five- or six-membered-ring orthoesters), and this reaction may be compare with the familiar Hanessian procedure for conversion of benzylidene acetals into bro-mo-deoxybenzoates (68).

Finally this method opens the way to the study
of other orthoesters, substitued by groups different from
methoxyl and methyl, in order to modulate their stability
and their reactivity.

Literature Cited

1. *Gelas, J.; Horton, D. Heterocycles 1981, 16, 1587 and references therein.*
2. *Gelas, J.; Horton, D. Methods Carbohydr. Chem. 9 under press.*
3. *Wolfrom, M.L.; Diwadkar, A.B.; Gelas, J.; Horton, D. Carbohydr. Res. 1974, 35, 87.*
4. *Gelas, J.; Horton, D. Carbohydr. Res. 1975, 45, 181.*
5. *Gelas, J.; Horton, D. Carbohydr. Res. 1978, 67, 371.*
6. *Gelas, J.; Horton, D. Carbohydr. Res. 1979, 71, 103.*
7. *Barbat, J.; Gelas, J; Horton, D. Carbohydr. Res. 1983, 116, 312.*
8. *Debost, J.L.; Gelas, J.; Horton, D. Carbohydr. Res. 1984, 125, 329.*
9. *Bar-Guilloux, E.; Defaye, J.; Driguez, H.; Gelas, J.; Henrissat, B. Carbohydr. Res. 1978, 63, 41.*
10. *Fanton, E.; Gelas, J.; Horton, D. J. Chem. Soc. Chem. Comm. 1980, 21.*
11. *Fanton, E.; Gelas, J.; Horton, D.; Karl, K.; Khan, R.; Kuan-Lee, C.; Patel, G. J. Org. Chem. 1981, 46, 4057.*
12. *Barbat, J. Thèse Doctorat, Clermont-Ferrand, (1985) n° 1.*
13. *Dehbi, A. Thèse Doctorat, Clermont-Ferrand, (1985) n° 8.*
14. *Dehbi, A.; Fanton, E.; Gelas, J.; Lebouc, A.; Simonet, J. Abstr. Papers 3rd European Symp. Carbohydr. 1985, 136.*
15. *Grindley, T.B.; Cote, J.C.P.; Wrickramage, C. Carbohydr. Res. 1985, 140, 215.*
16. *Debost, J.L.; Gelas, J.; Horton, D. J. Org. Chem. 1983, 48, 1381.*
17. *Garegg, P.J.; Iversen, T.; Norberg T. Carbohydr. Res. 1979, 73, 313.*
18. *Chrétien, F.; Castro, B.; Gross, B. Synthesis 1979, 937.*
19. *Chrétien, F.; Chapleur, Y.; Castro, B.; Gross, B. J. Chem. Soc. Perkin I 1980, 381.*
20. *Garegg, P.J.; Iversen T.; Johansson, R. Acta Chem. Scand. 1980 B34, 505.*
21. *Iversen, T.; Bundle, D.R. Carbohydr. Res. 1982, 103, 29.*
22. *Baer, H.H.; Abbas, S.A. Carbohydr. Res. 1980, 84, 53.*
23. *Hirth, G.; Walther, W. Helv. Chim. Acta 1985, 68, 1863.*
24. *Kuszmann, J.; Tomori, E.; Meerwald, I. Carbohydr. Res. 1984, 128, 87.*
25. *Kuszmann, J.; Tomori, E.; Dvortsak, P. Carbohydr. Res. 1984, 132, 178.*
26. *Yasnitskii, B.G.; Sarkisyants, S.A.; Ivanuk E.G. Zh. Obsch. Khim. 1964, 34, 1940.*

27. Gelas, J.; Petrequin, D. Bull. Soc. Chim. Fr. 1972, 3471.
28. Pacsu, E. Adv. Carbohydr. Chem. 1935, 57, 537.
29. DeWolfe, R.H. In Carbocyclic Orthoacid derivatives; Acad. Press. N.Y. 1970, 298.
30. Wolfrom, M.L.; Szarek, W.A. In The Carbohydrates Chemistry, Biochemistry, Pigman, W.; Horton, D., Acad. Press. 1972 IA, 217.
31. Wulf, G.; Röhle, G. Angew. Chem. Int. Ed. 1974, 13, 157.
32. Flowers, H.M. Methods Enzymol. 1978, 50, 93.
33. Bochkov, A.F. and Zaikov, G.E. In Chemistry of the O-glycosidic bond; Pergamon press, 1979.
34. Paulsen H. Angew. Chem. Int. Ed. 1982, 21, 155.
35. Zemlicka, J. Chem. Ind. 1964, 158.
36. Reese, C.B.; Sulston, J.E. Proc. Chem. Soc. 1964, 214.
37. Griffin B.E.; Jarman, M.; Reese, C.B.; Sulston, J.E. Tetrahedron 1967, 23, 2301.
38. Buchanan, J.G.; Edgar, A.R. Carbohydr. Res. 1976, 49, 289.
39. Lemieux, R.U.; Driguez, H. J. Am. Chem. Soc. 1975, 97, 4069.
40. Josan, J.S.; Eastwood, F.W. Carbohydr. Res. 1968, 7, 161.
41. Brimacombe, J.S.; Porsmouth, D. Carbohydr. Res. 1965, 1, 128.
42. Battacharjee, S.S.; Gorin, P.A.J. Carbohydr. Res. 1970, 12, 57.
43. Paulsen, H.; Trautwein, W.P.; Garrido-Espinosa, F.; Heyns, K. Chem. Ber. 1967, 100, 2822.
44. McElvain, S.M. J. Am. Chem. Soc. 1952, 74, 2665.
45. Eliel, E.L.; Giza, C.A. J. Org. Chem. 1968, 33, 3574.
46. King, J.F.; Allbutt, A.D. Canad. J. Chem. 1970, 48, 1754.
47. Lemieux, R.U. Lecture at the Institute for Natural Compounds, Moscow (URSS), March 1963.
48. Morgan, A.R. Ph.D. Thesis, University of Alberta, Edmonton (Canada), 1964.
49. Deslonchamps, P. in Stereoelectronic effects in organic chemistry; Baldwin, J.E. Ed.; Pergamon press 1983.
50. Haines, A.H. Adv. Carbohydr. Chem. Biochem. 1976, 33, 11.
51. Chittenden, G.J.F.; Buchanan, J.G. Carbohydr. Res. 1969, 11, 379.
52. Schmidt, O.T. Methods Carbohydr. Chem. 1963, 2, 318.
53. Battacharjee, S.S.; Gorin, P.A.J. Can. J. Chem. 1969, 47, 1195.
54. Freudenberg, K.; Von Oertzen, K. Justus Liebigs Ann. Chem. 1951, 574, 37.
55. Hanessian, S.; Moralioglu, E. Tetrahedron Lett. 1971, 813.
56. Hanessian, S.; Moralioglu, E. Can. J. Chem. 1972, 50 233.
57. Levene, P.A.; Raymond, A.L. J. Biol. Chem. 1933, 102, 317.

58. Ishido, Y.; Sakairi, N.; Sekiya, M.; Nakazaki, N. *Carbohydr. Res.* **1981**, *97*, 51.

59. Rana, S.S.; Barlow, J.J.; Matta, K.L. *Tetrahedron Lett.* **1981**, 5007.

60 Tronchet, J.M.J.; Bourgeois, J.M. *Helv. Chim. Acta* **1971** *54* 167.

61. Buchanan, J.G. personal communication; Oakes, E.M. *Ph. D. Thesis*, 1965.

62. Levene, P.A.; Stiller, E.T. *J. Biol. Chem.* **1933**, *102*, 187.

63. Birch, G.G. *Adv. Carbohydr. Chem. Biochem.* **1963**, *18*, 201.

64. Elbein, A.D. *Adv. Carbohydr. Chem. Biochem.* **1974**, *30*, 227.

65. Lee, C.K. In *Developments In Food Carbohydrates* **1980**, *2*, 1.

66. Defaye, J.; Driguez, H.; Henrissat, B.; Bar-Guilloux E. In *Mechanism of Polymerization and Depolymerization; Marshall, J.J., Ed.* Acad. Press **1980**, 331.

67. Buchanan, J.G.; Edgar, A.R.; Rawson, D.I.; Shahidi, P.; Wightman, R.H. *Carbohydr. Res.* **1982**, *100*, 75.

68 Gelas, J. *Adv. Carbohydr. Chem. Biochem.* **1981**, *39*, 71.

RECEIVED May 31, 1988

Chapter 4

Synthetic Approaches to Bicyclic Nucleosides

Total Synthesis of Octosyl Acid A and Octosyl Acid C Dimethyl Acetal

Stephen Hanessian, John Kloss, and Tamio Sugawara

Department of Chemistry, Université de Montréal, Montréal, Québec H3C 3J7, Canada

An overview of various synthetic approaches to bicyclic nucleosides such as the octosyl acids and the ezomycins is presented. Details of the total synthesis of octosyl acids A and C are disclosed.

The discovery of the polyoxin group (1) of antifungal nucleoside antibiotics (2) spurred the attention of synthetic chemists as well as biologists for a number of reasons. Their unusual structural features combined with unique biological activity fostered studies on many fronts (3). During their studies on the biosynthesis of the polyoxins, Isono, Crain and McCloskey (4) discovered three novel acidic nucleosides which they called octosyl acid A, B and C, 1-3 (Figure 1). Their structures, which were elucidated by spectroscopic and chemical means, revealed several unusual features, the most prominent being the presence of a trans-fused bicyclic ring system in A, B but not C. These have been considered to be anhydrooctose uronic acid nucleosides and an analogy has been derived with cyclic nucleotides. Thus, viewed in a different perspective, the octosyl acids may be considered as "carba" analogs of nucleoside 3',5'-cyclic phosphates (4) (Figure 2). This feature may have some bearing on their biological activity since the adenine nucleoside analog 5 inhibited c-AMP phosphodiesterases from various sources (5).

The trans-fused perhydrofuropyran system is also encountered in the ezomycin group of nucleoside antibiotics (6,7). The structures of ezomycin A_1 and A_2 are shown in Figure 3 for comparison with the octosyl acids.

Until recently, these unusual classes of bicyclic nucleosides had eluded the grasp of synthetic chemists. It is clear that the main challenge in the synthesis of these unusual nucleosides resides in the method of bicyclic ring formation. If one also considers the type of functionality that adorns these molecules, then the whole exercise becomes one of judicious choice of reactions and protective

0097–6156/89/0386–0064$08.25/0

groups. A fundamental issue is concerned with the "timing" of certain operations. Does one start with a pyrimidine nucleoside derivative and build from there, or should the bicyclic ring be constructed first and the heterocycle introduced at a later stage in the assembly process? The two recently completed syntheses of octosyl acid A address these issues independently (8,9).

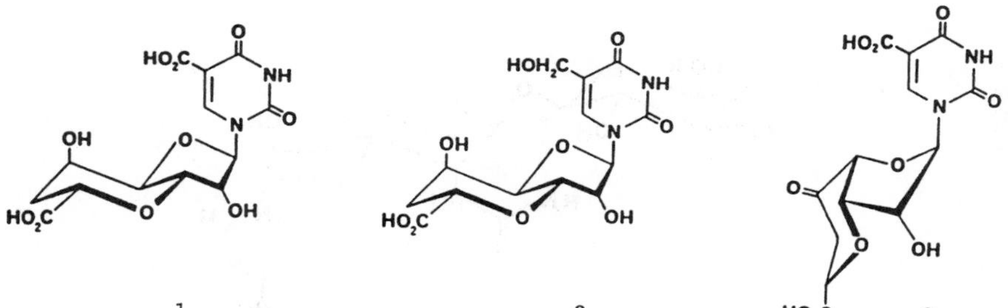

1 OCTOSYL ACID A

2 OCTOSYL ACID B

3 OCTOSYL ACID C

Figure 1. Structures of the octosyl acids.

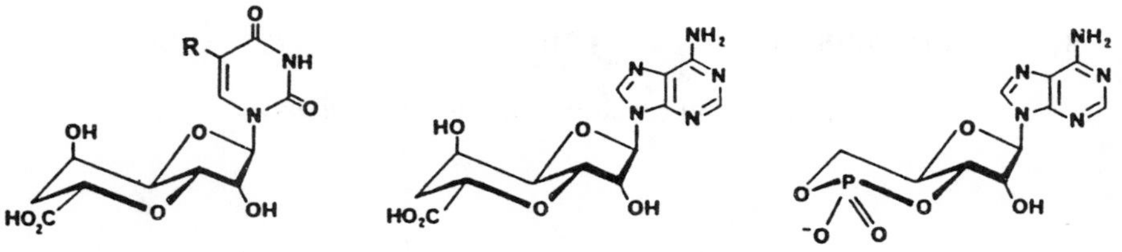

4

5

6

Figure 2. Functional analogies with cyclic-AMP.

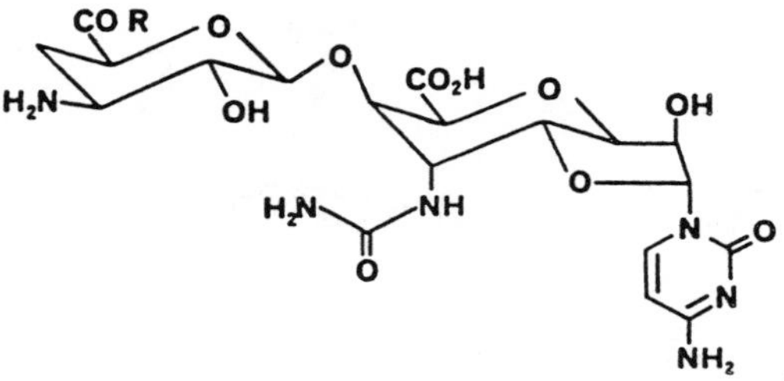

7,   EZOMYCIN A$_1$ $R = HO_2C-\underset{\underset{H}{|}}{\overset{\overset{NH_2}{|}}{C}}-CH_2SCH_2CH_2-\underset{\underset{NH}{|}}{CH}-CO_2H$ L-

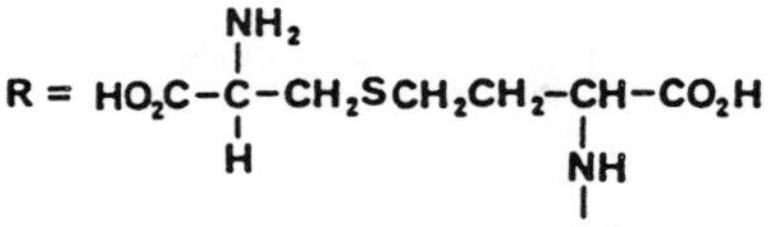

8,   EZOMYCIN A$_2$ R = OH

Figure 3.   Structures of the ezomycins.

Since there have been several approaches to the construction of the 3',7'-anhydrooctofuranose ring system, an overview of the strategies will be presented here.

The Anzai-Saita Approach

The first attempted synthesis of compounds related to the octosyl acids was reported by Anzai and Saita (10). The strategy entailed building the pyranose ring from a suitably protected D-allofuranose (Figure 4). The critical cyclization step was conceived as an intramolecular attack of a glycolate anion on a primary sulfonate ester. However the only product to form was epoxide 14. Alternatively, treatment of the 6-0-tosyl derivative 13 with NaH gave approximately a 1.5 to 1 mixture of the cyclized product 16 and the epoxide 15. Attempts to remove the 0-isopropylidene group under acidic conditions resulted in the degradation of the molecule to give 18. The sensitivity to acid was ascribed to the highly strained trans-fused anhydrooctose ring system.

Subsequently it was thought that nucleoside formation prior to ring closure would be a viable alternative (11). Towards this end, nucleoside formation from 19 gave 20 which was further transformed into 21, although no details were given.

The Szarek-Kim Approach

In 1976, Szarek and Kim (12) published their work related to the octosyl acids. Like the previous route, the formation of the 3',7'-anhydrooctose was based on a ring closure strategy which built the pyran portion of the molecule onto a pre-existing furanose (Figure 5). This approach started with aldehyde 23. Condensation with the readily available Wittig reagent gave the α,β-unsaturated ketone 24. Borohydride reduction followed by hydrogenation gave epimeric C-7' hydroxy compounds which were not separated but instead tosylated to give 25. Removal of the acetonide and intramolecular cyclization via S_N2 reaction produced the D-allo (25%) and the L-talo (21%) bicyclic derivatives, 27 and 28 respectively.

Next, a more highly functionalized Wittig reagent was employed which would lead to a 3',7'-anhydrooctose more closely resembling octosyl acid A (Figure 6). Wittig reaction with the protected α-hydroxy ketone gave the corresponding α,β-unsaturated ketone 29 in 72% yield. The reaction sequence then followed a reduction-hydrogenation methodology to give the tosyl derivative 30. Hydrolysis followed by attempted ring closure gave a complex mixture of products, which included the diastereomeric epoxides 32. It was assumed that the complex mixture arose from the instability of the terminal C-8' acetate to the conditions of ring closure. This problem was overcome by an exchange of protective groups prior to cyclization. Thus intermediate 30 was transformed into 34 by ring opening of epoxide 33. Hydrolysis, followed by intramolecular ring closure, afforded 36 as a mixture of cyclized products in 50% yield. The two diastereomers were separated after isomerization to the prop-1-enyl derivatives 37. Selective oxidation and esterification then completed the synthesis of the model anhydrooctose uronic acid, 39.

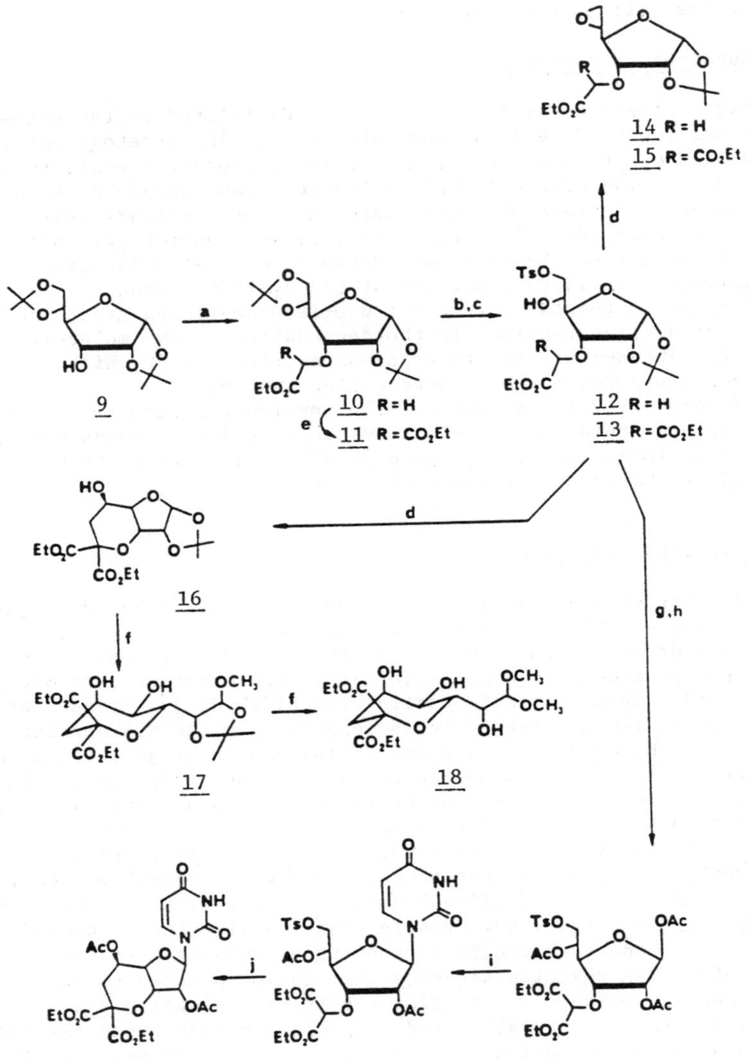

a. BrCH$_2$CO$_2$Et, NaH; b. AcOH, H$_2$O; c. TsCl, pyr.; d. NaH, DMF;
e. CO(OEt)$_2$, NaH; f. H$^+$, MeOH; g. H$^+$; h. Ac$_2$O, pyr.;
i. trimethylsilyluracil, SnCl$_4$; j. no conditions given.

Figure 4.   The Anzai–Saita approach to octosyl acid A.

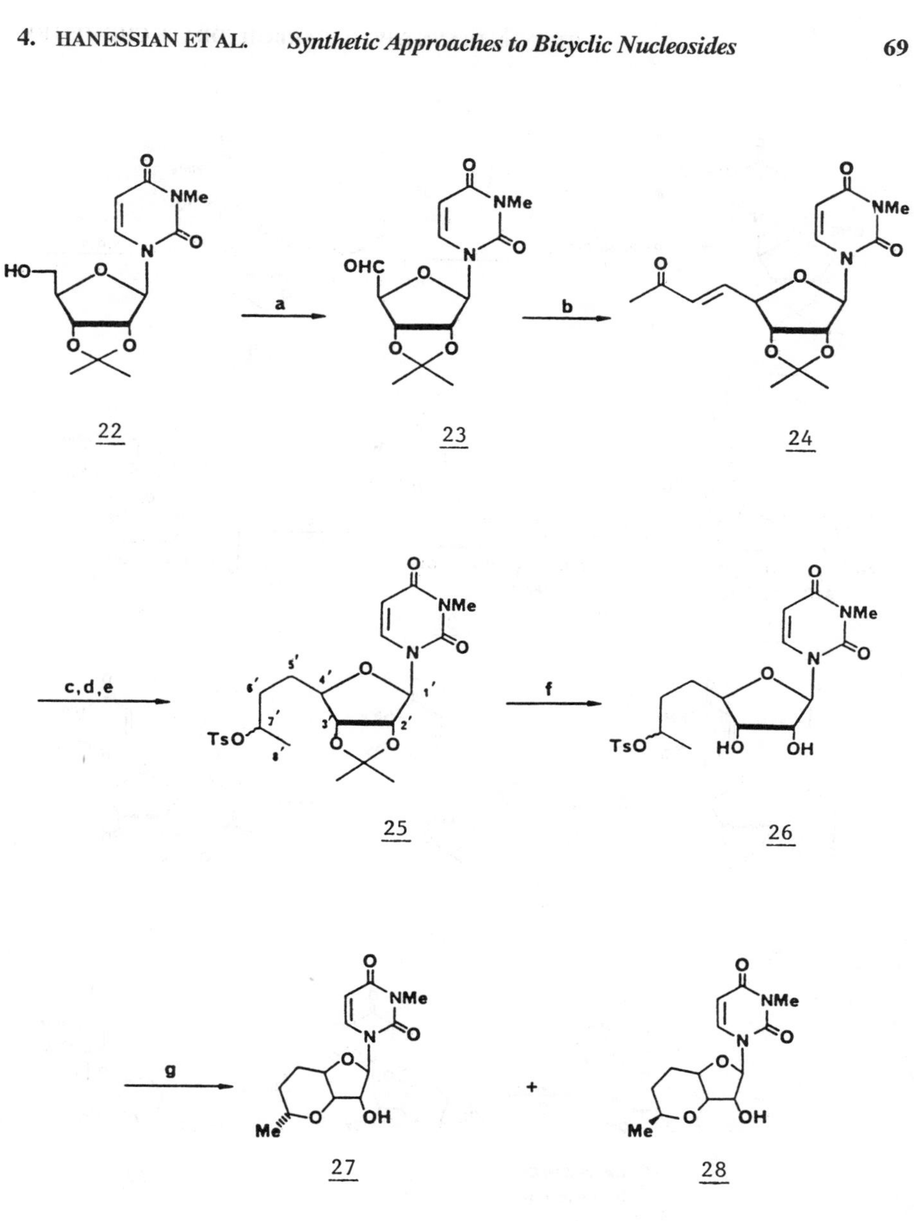

a. DCC, DMSO; b. Ph$_3$PCHCOCH$_3$; c. NaBH$_4$; d. Pd/C; e. TsCl, pyr
f. 90% HCO$_2$H, 0°C; g. NaH, DMF.

Figure 5. The Szarek-Kim approach to octosyl acid A. (Part I)

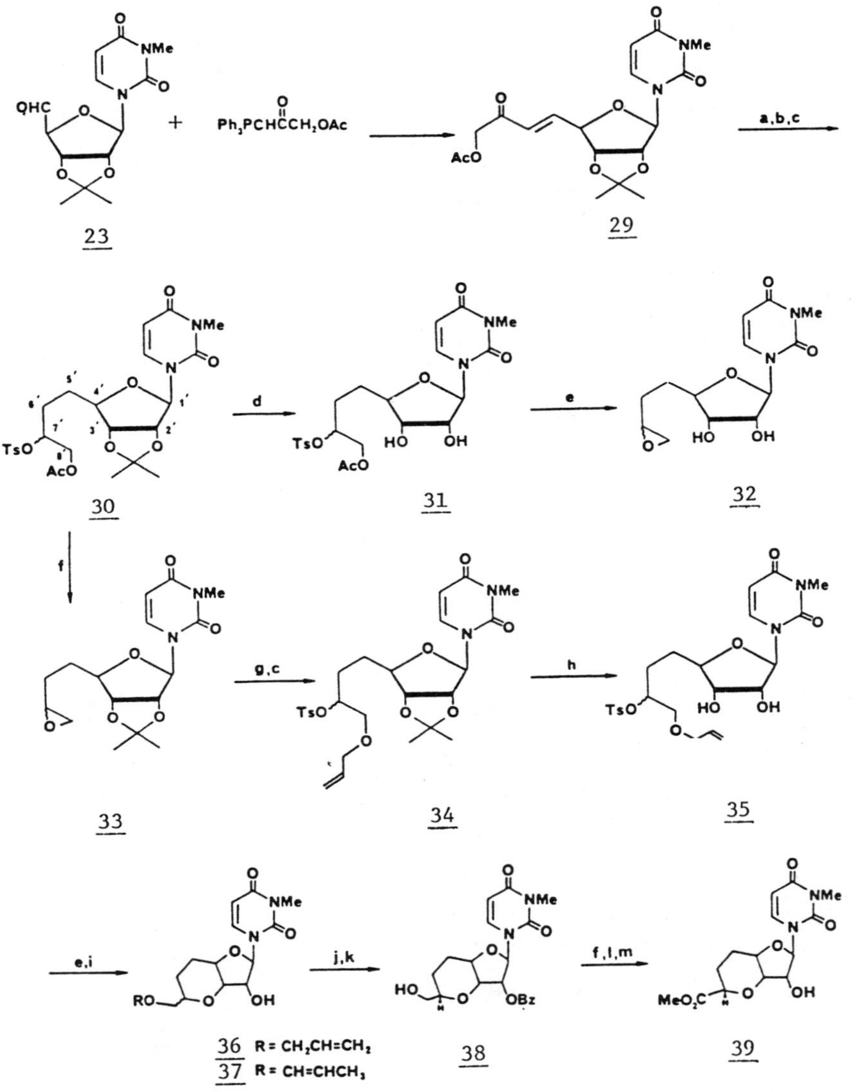

a. NaBH₄;   b. Pd/C;   c. TsCl, pyr.;   d. 90% HCO₂H;   e. NaH, DMF;
f. NaOMe, MeOH;   g. CH₂=CHCH₂ONa, CH₂=CHCH₂OH;   h. 90% HCO₂H, 0°C;
i. RhCl(PPh₃)₃, EtOH, reflux;   j. BzCl, pyr.;   k. HgCl₂, HgO;
l. Pt, O₂, NaHCO3, H2O, 90°C;   m. H⁺, MeOH.

Figure 6.   The Szarek-Kim approach to octosyl acid A.   (Part II)

Model Studies Directed at the Octosyl acids, Ezomycins and Quantamycin

In connection with the total synthesis of quantamycin (13), we adopted a different approach to the preparation of the 3',7'-anhydrooctose skeleton in these bicyclic nucleosides, namely the formation of the furanose ring onto a pre-existing pyranose ring (14). Model studies from a suitably protected D-galactose, showed that this new strategy was indeed feasible (Figure 7).

The C-allyl glycoside 40 was epoxidized to give a diastereomeric mixture of epoxides 41 which upon treatment with acid gave a diastereomeric mixture of cyclized products 42 resulting from an intramolecular epoxide opening. This strategy worked equally well on the C-vinyl glycoside (14) (Figure 8). Epoxidation of 43 afforded 44 which, when subjected to the intramolecular epoxide opening, gave 45 which contained the basic skeleton that could be adapted toward a synthesis of bicyclic nucleosides. In order for this strategy to be applicable to either the octosyl acids or the ezomycins, however, a more highly functionalized epoxide would have to be prepared in which the terminal position was substituted with a nucleic acid base.

A slightly modified strategy (14) was used to prepare a 3',7'-anhydrooctofuranosyl nucleoside. D-Galactose was transformed to the glycosyl bromide 50 and then treated with allylmagnesium bromide to give the C-allyl D-galactopyranoside 51 (Figure 9). Oxidation gave aldehyde 52 which was converted to the corresponding dithioacetal 53. Treatment of 53 with N-benzoyladenine and bromine gave a 1:1 mixture of anomeric bicyclic nucleosides 54 and 55. This key transformation entailed a sequential displacement of a sulfonium intermediate. Presumably an acyclic nucleoside derivative (15) is formed first, which, after further activation with bromine, undergoes intramolecular ether formation. Unequivocal proof for the structure of the β-adeninyl nucleoside 54 was secured by X-ray crystallographic analysis (14). Although this methodology was applicable to the formation of bicyclic nucleosides, the lack of stereocontrol at C-1' precluded its further use. In addition, a hydroxy group had to be incorporated at the 2'-position.

An extension of this methodology was used in another approach (16) to the octosyl acids and ezomycins (Figure 10). In this sequence, D-galactose was transformed into the 2-0-acetyl derivative 57. Transformation to the acyclic nucleoside derivative and selective oxidation then gave sulfoxide 58. Elimination afforded the trans olefin 59 whereupon solvolysis followed by epoxidation and acid-catalyzed cyclization produced 60 and 61 in a 1:2 ratio respectively. The ¹H-NMR spectra showed each to contain a 1',2'-trans configuration, and that the minor isomer 60 was the required β-D-nucleoside, while the major product 61 was the α-D-nucleoside.

Since octosyl acid A has the C-5'(R) configuration, the β-D-nucleoside 60 was subjected to configurational inversion. After protective group manipulations to give 62, oxidation gave ketone 63. It was hoped that reduction would give a net inversion of the

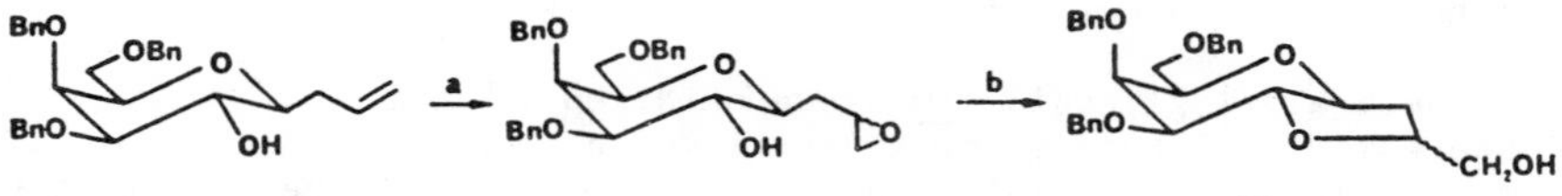

40 41 42

a. MCPBA; b. CSA, $ClCH_2CH_2Cl$, reflux.

Figure 7. Model studies for perhydrofuropyran systems.

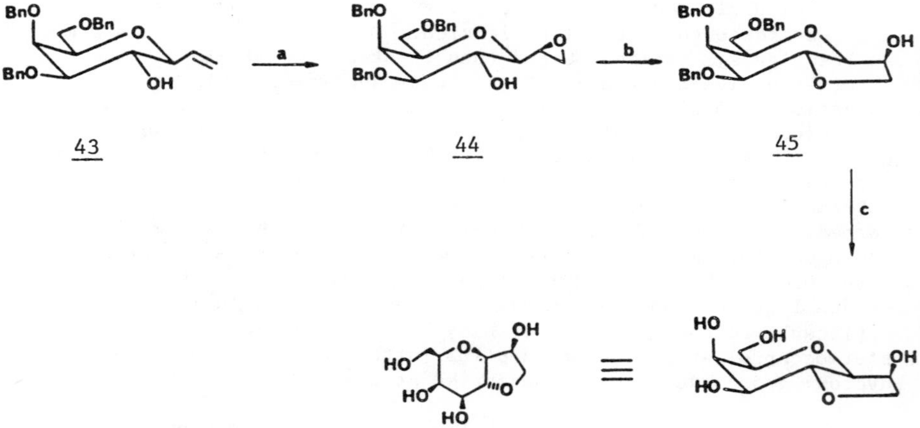

43 44 45

46

a. MCPBA; b. CSA, $ClCH_2CH_2Cl$, reflux; c. Pd/C.

Figure 8. Model studies for perhydrofuropyran systems.

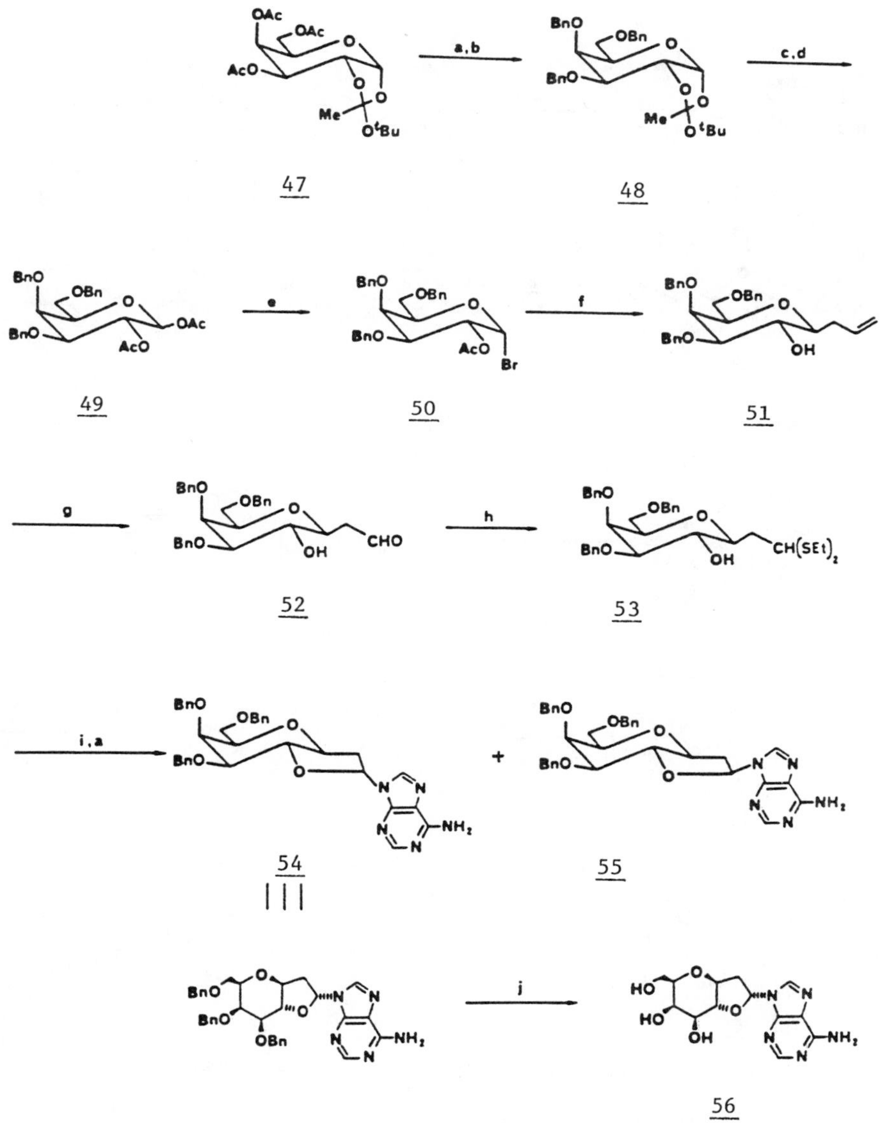

a. NaOMe, MeOH;   b. BnBr, NaH, DMF;   c. AcOH, H_2O;   d. Ac_2O, pyr.;
e. HBr, CH_2Cl_2;   f. AllylMgBr, THF;   g. OsO_4, $NaIO_4$, tBuOH, H_2O;
h. EtSH, HCl, 0°C;   i. N-Bz adenine, Br_2, DMF, CH_2Cl_2;   j. Pd/C,
cyclohexene, EtOH, reflux.

Figure 9.   Synthesis of a bicyclic adenine nucleoside related to
the octosyl acids.

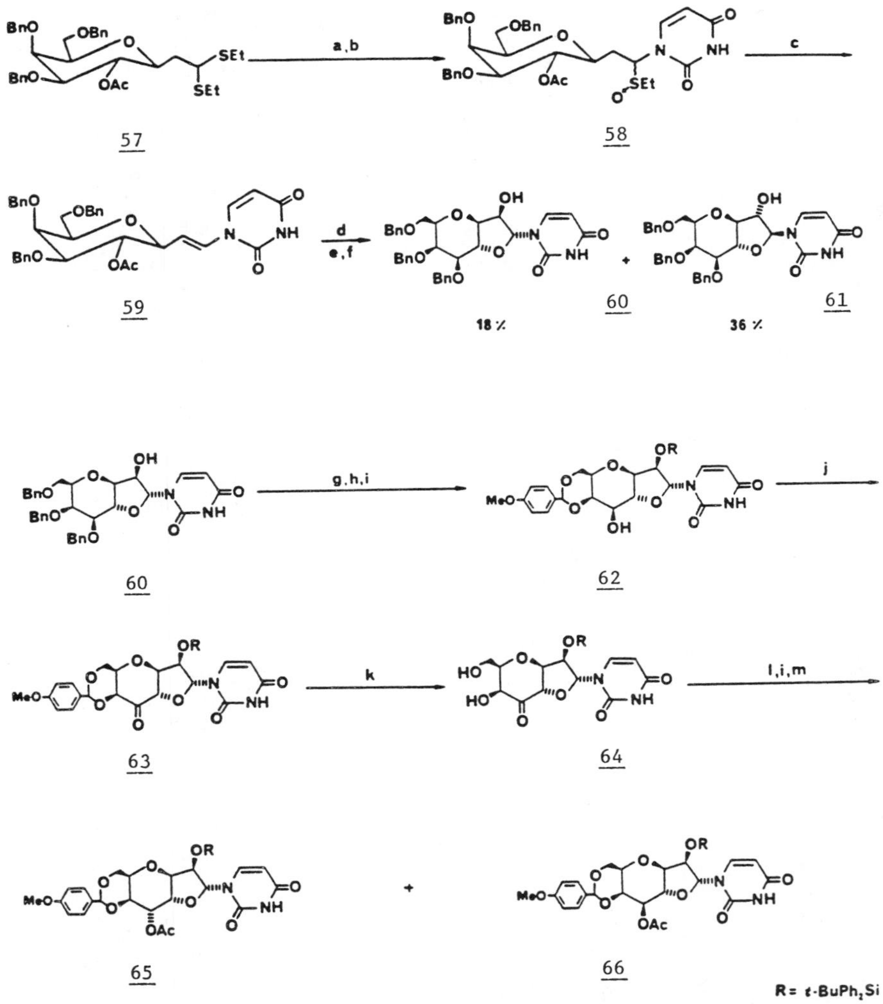

a. 2,4-bistrimethylsilyloxypyrimidine, I_2, THF; b. MCPBA, CH_2Cl_2,
-78°C;   c. $PhCH_3$, pyr., reflux; d. NaOMe, MeOH; e. MCPBA,
$ClCH_2CH_2Cl$; f. CSA, $ClCH_2CH_2Cl$; g. t-$BuPh_2SiCl$, imidazole, DMF
h. 20% $Pd(OH)_2/C$, H_2, MeOH; i. p-methoxybenzaldehyde, $ZnCl_2$;
j. PCC, molecular sieves; k. 80% AcOH, 60°C; l. $NaBH_4$, MeOH;
m. Ac_2O, pyr.

Figure 10.   Synthesis of a bicyclic uracil nucleoside related to
the octosyl acids.

stereochemistry at C-5'. However, reduction of 64 followed by
reintroduction of the 6',7'-p-methoxybenzylidene group and
acetylation at C-5' gave a 1:1 mixture of C-5' epimeric acetates 65
and 66. With these disappointing results, this approach to octosyl
acid A was discontinued.

Synthesis of Quantamycin

Quantamycin, 77 is a computer-generated structure (13) which was
found to show ribosomal binding activity (Figure 11). The main
challenge in the synthesis of quantamycin was the construction of
the strained, highly functionalized trans-fused 3',7'-anhydro-
octofuranosyl system. Towards this end, the high degree of stereochemical and functional overlap with lincosamine 67, a readily
available degradation product of lincomycin (17), made the former an
attractive starting material. Standard methodology produced the
bromide 68 which was transformed into the C-vinyl glycoside 70.
Oxidation gave aldehyde 71 which was condensed with the lithium
anion of bis(thiophenyl)methane to give a 7:1 ratio of C-2' epimers
72. Treatment of the major isomer with bromomethylsulfonium bromide
followed by benzylation gave a 3:1 mixture of α- and β-phenylthio
glycosides 73. Treatment with N-benzoyladenine in the presence of
bromine gave a 62% yield of the two β- and α-adeninyl nucleosides 74
and 75 in a ratio of 2:1 respectively. Although this glycosylation
gave only a 41% yield of the desired isomer, the unwanted
α-derivative 75 could be "recycled" by mercaptolysis to give 72
which in turn could be used to give more of 74. The desired
β-isomer 74 was then used to complete the synthesis of quantamycin.

The Danishefsky-Hungate Synthesis of Octosyl Acid A

Contemporary with our studies, Danishefsky and Hungate (8) reported
a total synthesis of octosyl acid A, the details of which are
included in a separate lecture (18). Their approach started with a
hetero Diels-Alder cycloaddition product 78 which was transformed
into the eight-carbon sugar derivative 79 (Figure 12). Nucleoside
formation led to the β-D-nucleoside 81. Conventional ring closure
strategies using 81 and metal alkoxides resulted in either no
reaction or decompositon. The successful ring closure strategy was
achieved through the use of a 2',3'-0-stannylene protective group
(19). Treatment of the 2',3'-0-stannylene derivative 82 with cesium
fluoride followed by deprotection, gave octosyl acid A, 1, [α]$_D$
+9.1° (1N NaOH), whose 250 MHz ^{1}H-NMR spectrum was reported to be
identical with that obtained from a sample of natural octosyl acid
A. The optical rotation of authentic octosyl acid A was reported as
+13.3° (c 0.5, 1N NaOH) (4).

Synthetic Approaches to the Octosyl Acids from Uridine

Our approach to the synthesis of octosyl acid A is illustrated using
the retrosynthetic analysis shown in Figure 13. We envisaged a
cyclization process that not only would give the 3',7'-anhydrooctose
bicyclic system, but would also establish the desired

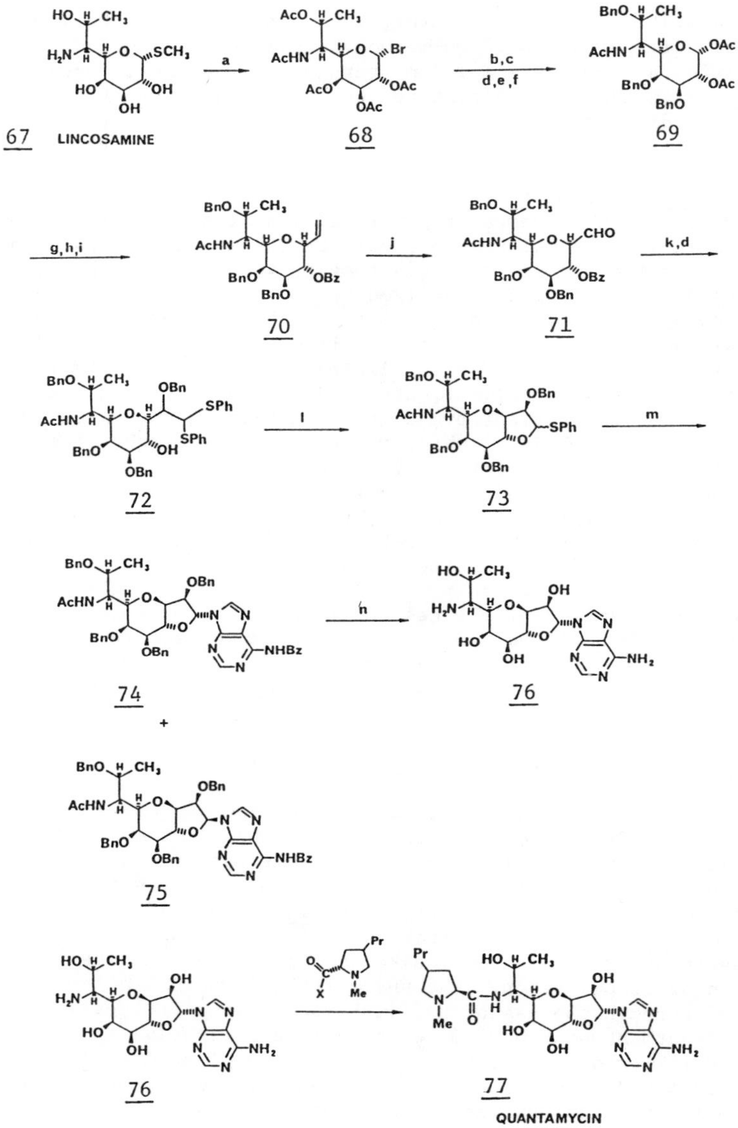

Figure 11.   Synthesis of quantamycin.

a. several steps;   b. Bu$_3$SnOMe, Bu$_4$NBr, CHCl$_3$;   c. NaOMe, MeOH;
d. BnBr, NaH, DMF;   e. HCl;   f. Ac$_2$O, pyr.;   g. HBr, AcOH;
h. vinylMgBr, THF;   i. Bz$_2$O, DMAP, pyr.;   j. OsO$_4$, NaIO$_4$, aq.
acetone;   k. (PhS)$_2$CH$_2$, nBuLi, -78°C;   l. Me$_2$SBr$^+$Br$^-$, CH$_2$Cl$_2$, 0°C;
m. N-Bz adenine, Br$_2$;   n. deprotect.

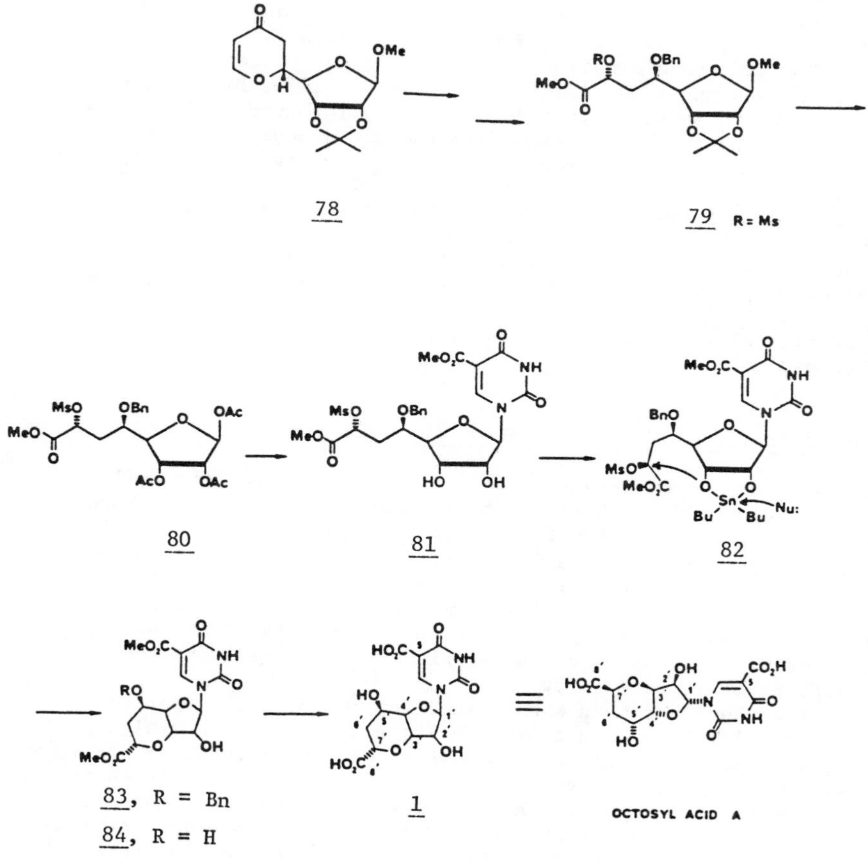

Figure 12. The Danishefsky-Hungate synthesis of octosyl acid A.

stereochemistry at C-7'. The high degree of stereochemical and
functional overlap between the precursor and uridine made the latter
an attractive starting material.

Studies in Ring Closure - The Epoxide Opening Route

Through known methodology, uridine was transformed into the aldehyde
88 (20) (Figure 14). Grignard reaction with allyl magnesium
chloride gave a 7:1 mixture of C-5' isomers 89 and 90, with the
crystalline C-5'(R) isomer 89 being preponderant.
    The key transformation necessary for the success of this
approach was the ring closure to give the 3',7'-anhydrooctofuranosyl
ring system. Initial cyclization studies involved an intramolecular
opening of an epoxide derived from the N,O-dibenzyl derivative 91.
Since the C-5' hydroxy group was homoallylic, it was rationalized
that some stereoselection in the epoxidation might occur.
Furthermore, intramolecular epoxide opening at C-7' would not only
accomplish the tetrahydropyran ring formation, but also
functionalize C-8'. Basic and Lewis acid conditions failed to
cyclize the epoxide, while acidic conditions gave the cyclized
intermediate 92 as a minor product along with a number of
unidentifiable by-products. An alternate ring closure strategy was
then explored.

The Intramolecular Oxymercuration Approach

We next focused on an electrophilic addition across the C-7',8'
double bond. It was believed that treatment with an electrophile
might induce cyclization to give the trans-fused
3',7'-anhydrooctofuranose necessary for octosyl acid A. A number of
different electrophiles were used in order to attempt this
transformation [NBS, I_2 (21-22); RSe$^+$, (21-23) and others].
Multiple products were produced with halogen electrophiles, while
with metal electrophiles, no reaction was observed with the
exception of mercury (II).
    Precedent had been established for mercury (II) mediated
cycloetherification reactions (24,25) but not in such a highly
functionalized molecule and to give a strained system. Since it was
also known that oxidative removal of the mercury transformed the
alkylmercurial into an alcohol (26), this method would not only
allow access to the tetrahydropyran portion of the molecule, but the
criterion of a functionalized terminus (C-8') would also be met.
    This ring closure-functionalization sequence entailed three
distinct steps. Oxymercuration of 93 gave an unstable
alkylmercurial acetate which was transformed to the alkylmercurial
bromide 94. Oxidative removal of the mercury then gave 95 (Figure
15). Further oxidation at C-8' followed by esterification and
hydrogenolysis gave the ester 96. Unfortunately, attempts to
de-N-benzylate compounds 95 or 96 either gave no reaction or
resulted in degradation. In an identical sequence, we synthesized
the N-MEM derivative corresponding to 95. Again, we encountered
problems in selective deprotection. Figure 16 shows the behavior of

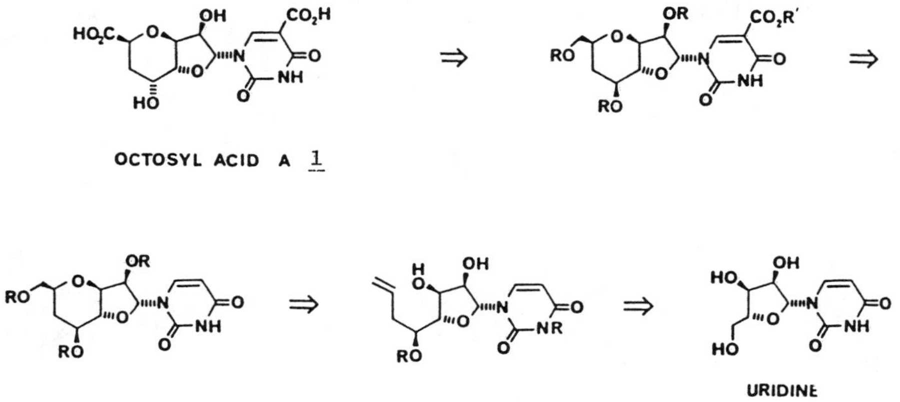

OCTOSYL ACID A **1**

URIDINE

Figure 13. Retrosynthetic analysis and the use of uridine as a precursor.

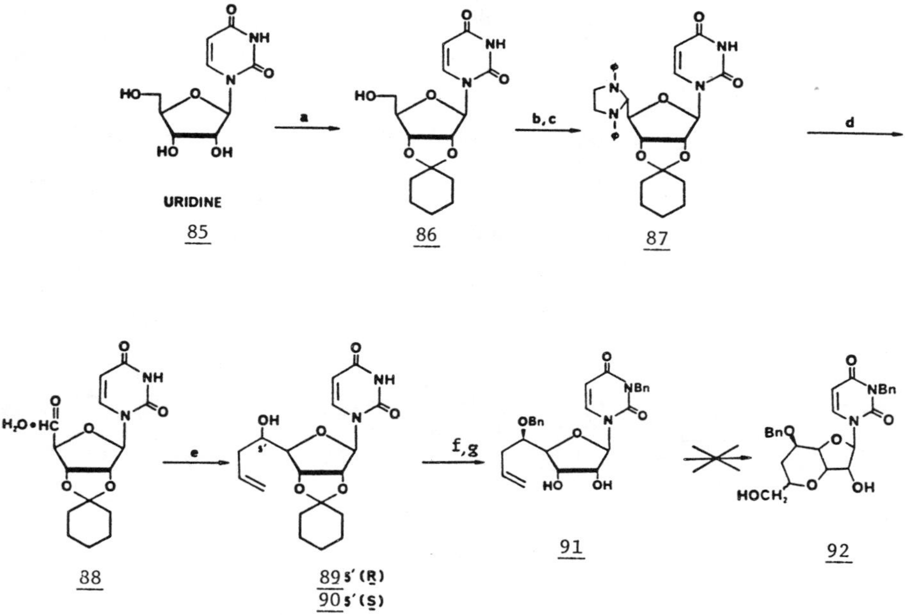

URIDINE
85

86

87

88

89 s'(**R**)
90 s'(**S**)

91

92

a. cyclohexanone, HC(OEt)$_3$, H$_2$SO$_4$, DMF; b. DMSO, DCC, Cl$_2$CHCO$_2$H;
c. PhNHCH$_2$CH$_2$NHPh; d. H+ resins, THF–H$_2$O; e. AllylMgCl, THF,
–78°C; f. BnBr, NaH; g. TFA–H$_2$O, 1:1

Figure 14. The intramolecular epoxide opening approach.

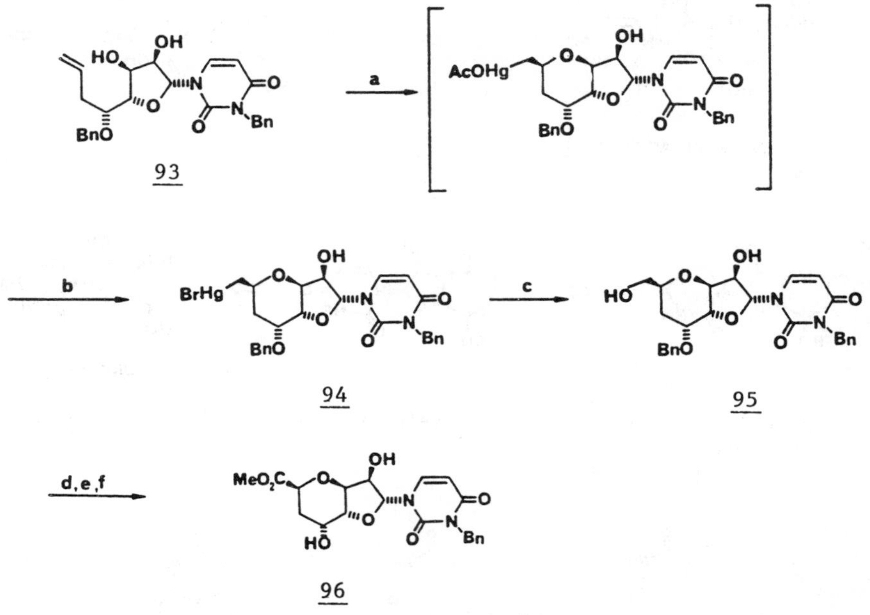

a. $Hg(OAc)_2$, THF;   b. NaBr;   c.. $NaBH_4$, O_2, DMF; d. Pt, O_2, $NaHCO_3$, H_2O, 90°C; e. H^+, MeOH; f . 20% $Pd(OH)_2/C$, H_2,

Figure 15.   The intramolecular oxymercuration approach as a strategy for the octosyl acids.

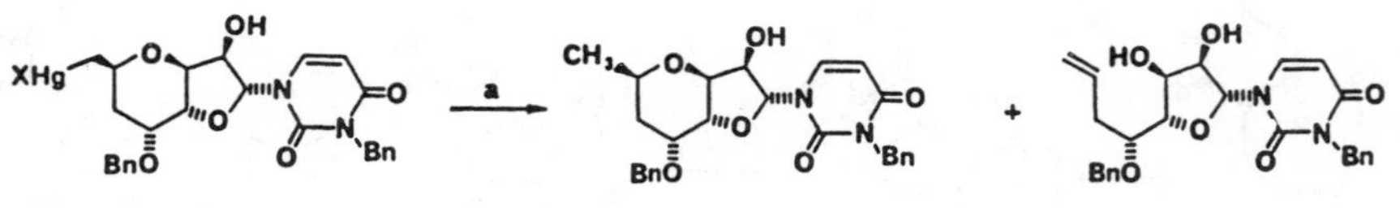

	97	93
X = OAc	50	50
X = Br	75	25
X = Cl	0	100

a. $NaBH_4$, DMF

Figure 16.   Effect of the halide in the oxymercuration-reduction model reactions.

the alkylmecurial compound under conditions of reduction with sodium
borohydride in dimethylformamide (26), as a measure of the formation
of cyclized product. Interestingly, the C-8' methyl derivative 97
was not always the major product isolated. It would seem that in
the case of the alkylmecurial chloride the radical generated at C-8'
was too unstable to be captured by a hydrogen radical, and the ring
must have reopened to the starting diol 93.

Since the cyclohexylidene protective group was found to be too
acid stable, the isopropylidene protective group was tested and
found to be more compatible with this approach. The known 98 (28)
was then subjected to a 3-carbon extension using allyl magnesium
bromide to afford two C-5' epimers in a ratio of 16:1 and in a
combined yield of 77% (29) (Figure 17). The absolute
stereochemistry of the newly formed asymmetric center at C-5' could
not be unambiguously assigned at this point, but the high field
^{1}H-NMR spectrum of more advanced intermediates later unequivocally
established the configuration of the major isomer.
    This stereoselection can be explained by assuming an attack of
the Grignard reagent via a non-cyclic Cram model (30).

With an efficient synthesis of the Grignard product 99 in hand,
the next challenge was a protective group problem. The most
compatible was found to be the benzyloxymethyl (BOM) group.
Interestingly, the imine nitrogen of uridine derivative 99 could be
selectively protected (31) to give 101, but with stronger
conditions, the dibenzyloxymethyl derivative 102 was accessible
(Figure 18).

The problem of selective hydrolysis of the acetonide was
studied next. The most practical method consisted of a treatment
with acetic acid in aqueous tetrahydrofuran at 65°C which gave a
mixture of diol 103 (major) and triol 104 (minor) in 77 and 15%
yields respectively. Triol 104 could be converted to 103 by a
sequence involving formation of a 2'3'-orthoester, conversion to the
corresponding di-BOM derivative, then mild hydrolysis.
    The next step in the synthesis was the formation of the
tetrahydropyran ring. It was found that mercuric acetate was
slightly superior to mercuric trifluoroacetate in this series.The
^{1}H-NMR spectrum of the crude bicyclic product showed the presence of
two diastereomers 106 and 107 in an 11:1 ratio respectively. The
pure 7'(S) isomer 106 was hydrogenolyzed to the triol 108, and this
was oxidized then esterified to give 109. The last major task was
to introduce a carboxyl group at C-5. To this end, we returned to
the triol 108 (Figure 19).

Two traditional methods have been used to introduce a carboxyl
substituent at C-5 of uracil. One involves bromination at C-5,
metalation via a lithium-halogen exchange, and finally quenching
with carbon dioxide (32) to give the C-5 carboxylic acid. The

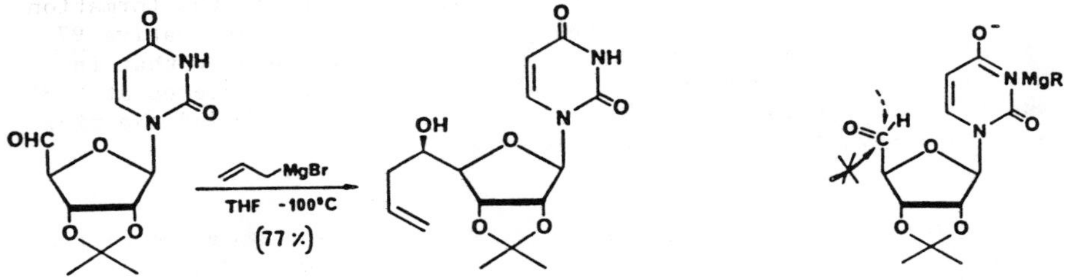

98 99 100

Figure 17. Stereoselective chain elongation and establishment of the correct stereochemistry.

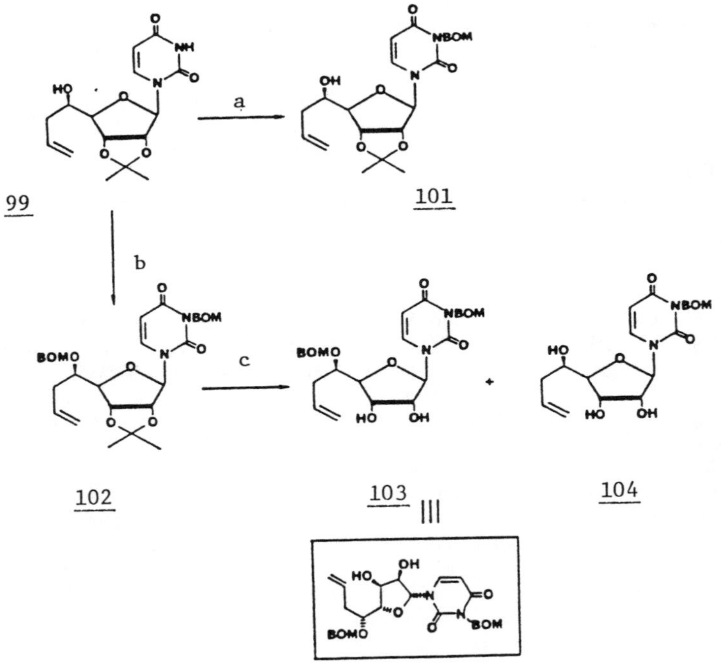

a. BOMCl, DBU, DMF, 0°; b. BOMCl, iPr_2NET, DMAP, THF, 70°; c. AcOH-THF-H_2O, 65°, 72 h.

Figure 18. Selection of the benzyloxymethyl protective group.

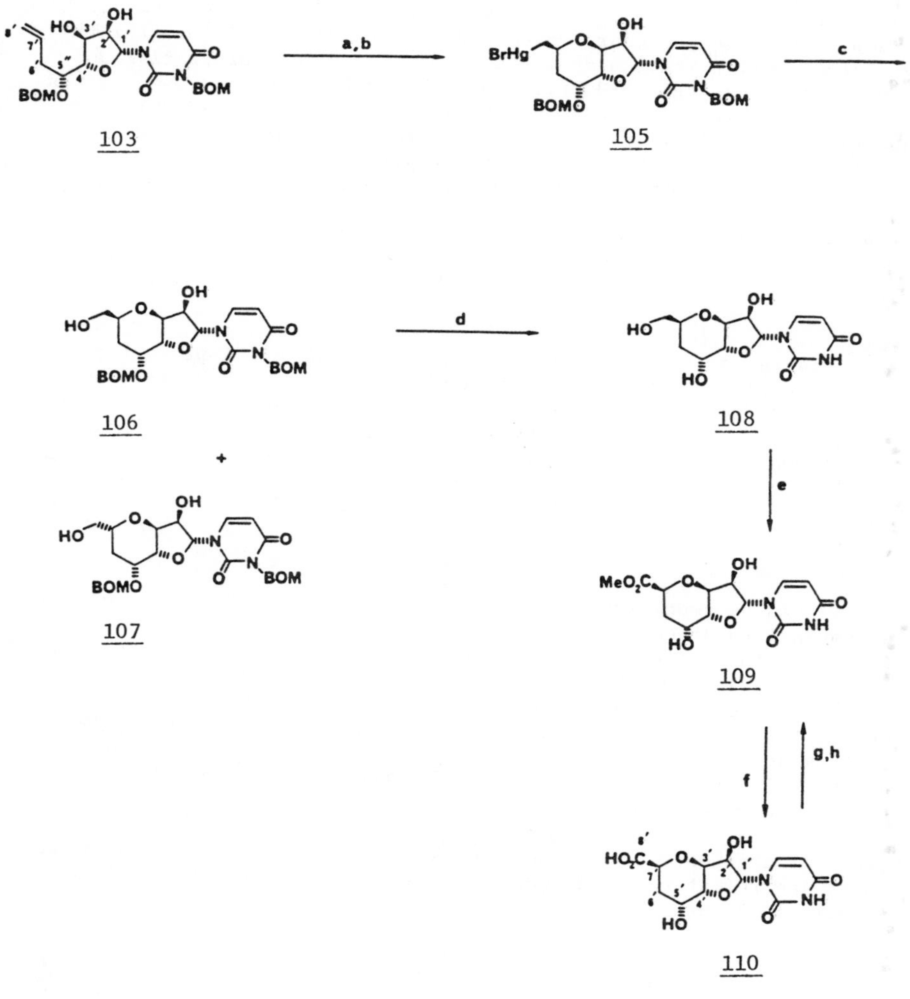

a. $Hg(OAc)_2$, THF; b. NaBr; c. $NaBH_4$, O_2, DMF; d. 20% $Pd(OH)_2/C$, H_2, MeOH; e. Pt, O_2, $NaHCO_3$, H_2O; 90°C; f. H^+, MeOH; g. LiOH, H_2O; h. H^+ resin, H_2O.

Figure 19. Elaboration of the bicyclic ring system of octosyl acid A.

second method relies on mercury-palladium chemistry (33). This
method would involve mercuration at C-5, palladium-mercury exchange,
and alkenyl-palladium exchange to give the C-5 vinyl derivative
which would still have to be subjected to an oxidative cleavage. We
chose to adapt a procedure developed by Miyasaka et al (34) for the
efficient derivatization at the C-5 position of dihydrouracils and
then re-introducing the double bond. The dihydro derivative 111 was
prepared by hydrogenation using rhodium-on-alumina in essentially
quantitative yield (Figure 20). Silylation, formation of the
lithium enolate, and quenching with ethyl chloroformate gave a 1:1
mixture of C-5 ethoxycarbonyl diastereomers 112. The double bond
was re-introduced via a selenide addition-elimination sequence (35)
to give 113 in an overall yield of 88%. With the uracil moiety
successfully functionalized and protected, the final step in the
synthesis of octosyl acid A was to oxidize C-8' selectively.

This was accomplished by first removing the silyl protecting
groups and platinum-catalyzed oxidation which gave the half-ester
half-acid derivative 115. Saponification, followed by acidification
gave octosyl acid A. Dissolution of this acid in acetone, removing
a small amount of insoluble matter and precipitation gave octosyl
acid A as a white powder, mp 285-288° (dec.); $[\alpha]_D$ +9.8° (c 0.5,
1N NaOH); reported (4), mp 290-295° (dec); $[\alpha]$, +13.3° (NaOH)
(Figure 21).

As can be seen, there is some discrepancy in the melting points
and optical rotations with the reported data (4). The small
difference in melting points, is negligible, particularly since
octosyl acid A slowly decomposes at these high temperature.
The elemental analysis of our synthetic octosyl acid A, was
correct for a non-hydrated compound. Secondly, its 400 MHz ^{1}H-NMR
spectrum showed no trace of any other isomer. Danishefsky, and
Hungate (8) had reported $[\alpha]_D$ +9.1° for their synthetic octosyl
acid A in good agreement with our results. In the light of the
spectroscopic and analytical data obtained, we contend that our
sample of synthetic octosyl acid A is pure and that the constants we
report are reliable. A ^{1}H NMR spectrum of synthetic octosyl acid A
is shown in Figure 22.

Synthesis of Octosyl Acid C Dimethyl Acetal

After the successful synthesis of octosyl acid A, our attention was
directed towards the synthesis of octosyl acid C,3. As shown in
Figure 23, octosyl acid C contains a cis-fused bicyclic
oxoperhydrofuropyran. Therefore, a synthetic strategy based on
oxidation at C-5' and epimerization at C-4' in the original octosyl
acid A was devised.

The bicyclic intermediate 106 used in the synthesis of octosyl
acid A became the starting point for the synthesis of octosyl acid C
(Figure 24). Disilylation of 106 followed by hydrogenolysis
liberated the C-5' hydroxyl group which was oxidized to afford the

corresponding C-5' keto derivative 118. Treatment with 1,8-diazobicyclo[5.4.0]undecene in refluxing toluene effected epimerization at C-4' to afford the cis-fused bicyclic system 119. (Figure 25).

With the establishment of the basic carbohydrate skeletal system of octosyl acid C, our attention was then directed toward funcionalizing the uracil moiety. Protection of the keto group as its dimethyl acetal gave 120. Hydrogenation to the dihydro derivative 121 and carbomethoxylation via the enolate as previously described afforded 122 in 70% overall yield. Finally, deprotection, oxidation and esterification gave octosyl acid C dimethyl ester dimethyl acetal 125 in 60% yield. Saponification followed by treatment with a cation-exchange resin gave octosyl acid C dimethyl acetal 126. Various attempts to hydrolyze the dimethyl acetal group were unsuccessful, resulting in decomposition. The ^{1}H NMR spectrum of synthetic octosyl acid C dimethyl acetal is shown in Figure 26.

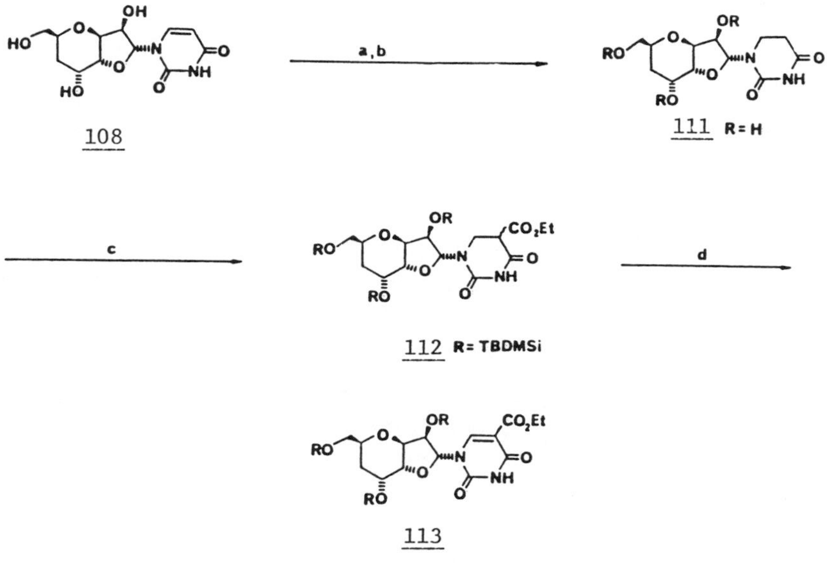

108

111 R = H

112 R = TBDMSi

113

a. 5% Rh on alumina, H_2, MeOH;   b. TBDMSiCl, Et_3N, DMAP, DMF;
c. LDA, $ClCO_2Et$, THF, -78°C;   d. PhSeCl.pyr., CH_2Cl_2, H_2O_2

Figure 20.   Carbethoxylation at the C-5 position.

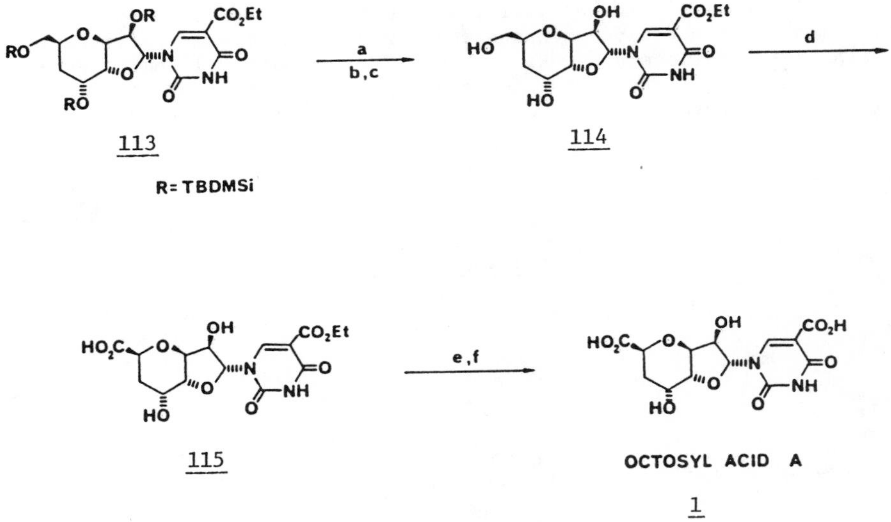

a. Bu₄NF, THF;   b. TMSCl, pyr.;   c. H+ resins, MeOH;   d. Pt, O₂,
NaHCO₃, H₂O, 90°C;   e. LiOH, H₂O;   f. H⁺ resins, H₂O;

Figure 21.   Completion of the synthesis of octosyl acid A.

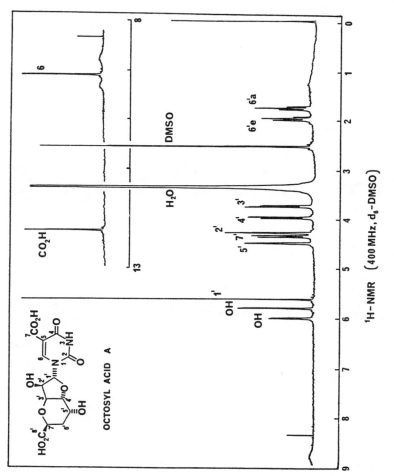

Figure 22. 400 MHz nmr spectrum of synthetic octosyl acid.

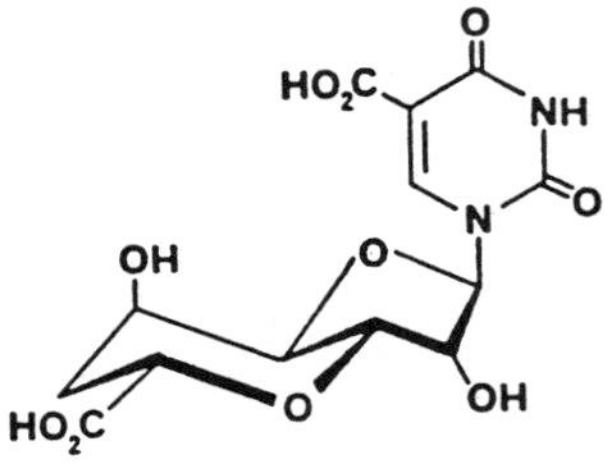

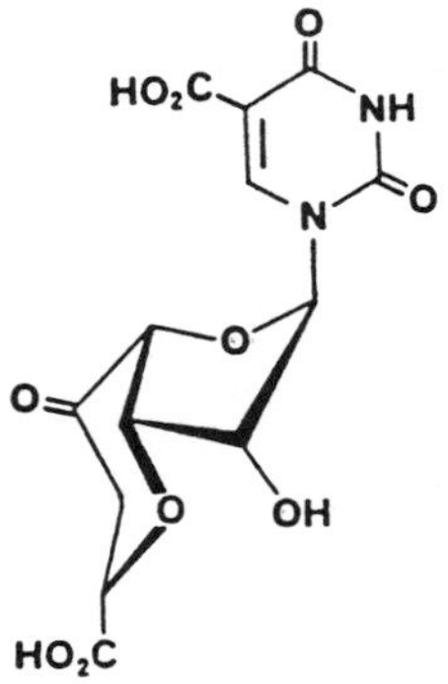

OCTOSYL ACID A, 1

OCTOSYL ACID C, 3

|||

|||

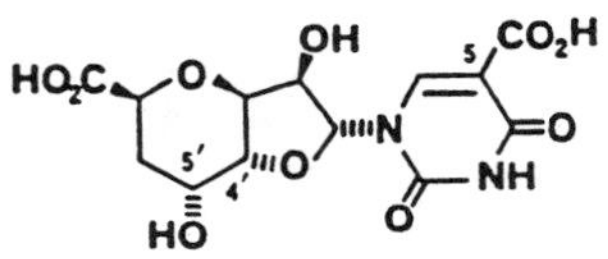

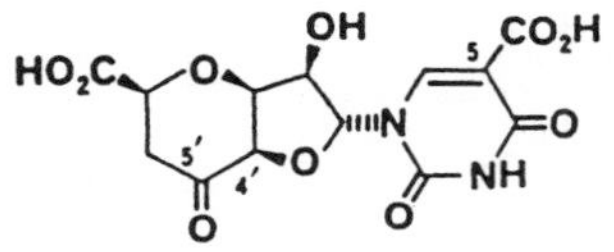

Figure 23. Structures and perspective drawings of octosyl acids A and C.

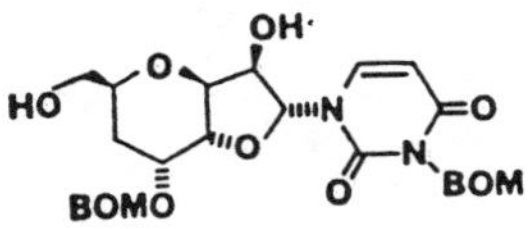

106

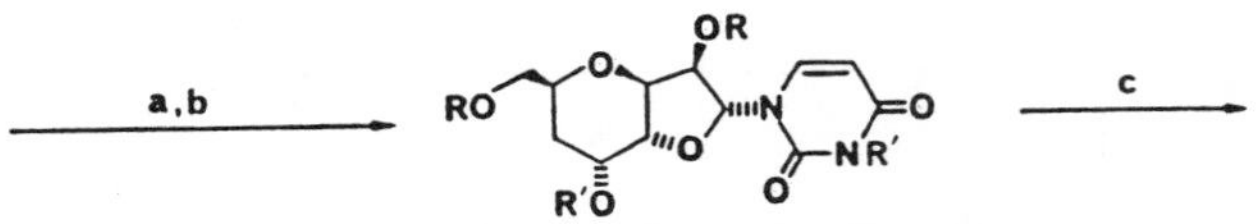

116 R= TBDMSi R'= BOM

117 R= TBDMSi R'= H

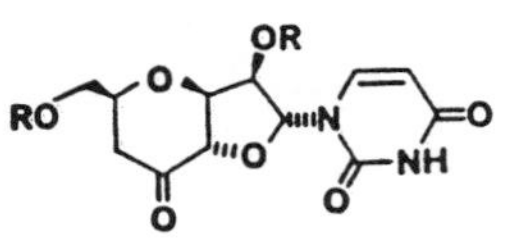

118

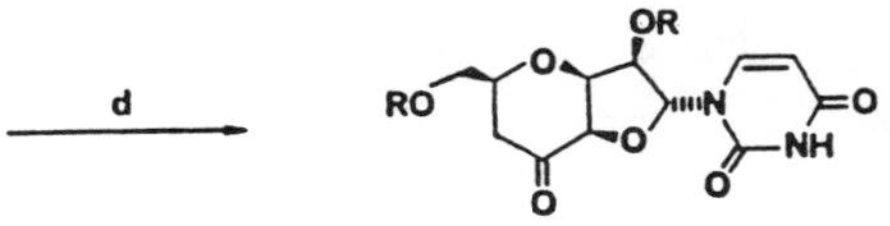

119

a. TBDMSiCl, Et_3N, DMAP, DMF; b. 20% $Pd(OH)_2/C$, H_2, EtOAc; c. PCC, molecular sieves, CH_2Cl_2; d. DBU, $PhCH_3$, reflux.

Figure 24. Key steps in the transformation of octosyl acid A intermediates en route to octosyl acid C.

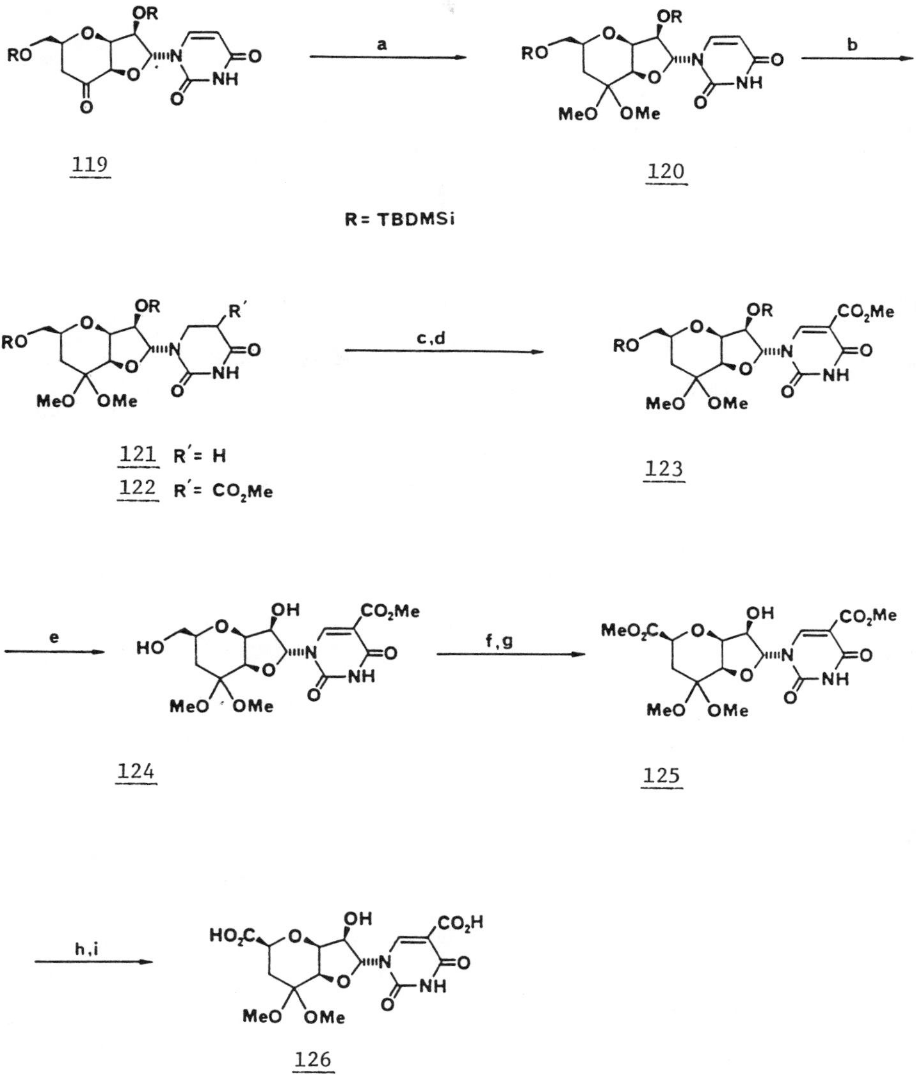

119
120
R= TBDMSi
121 R'= H
122 R'= CO2Me
123
124
125
126
a
b
c,d
e
f,g
h,i

a. $HC(OMe)_3$, PPTS; b. 5% Rh on alumina, H_2, EtOAc; c. LDA, $ClCO_2Me$, THF, -78°C; d. PhSeCl.pyr., CH_2Cl_2; H_2O_2; e. Bu_4NF, THF; f. Pt, O_2, $NaHCO_3$, H_2O, 90°C; g. H^+, MeOH; h. LiOH, H_2O; i. H^+ resins, H_2O.

Figure 25. Final steps in the synthesis of octosyl acid C dimethyl acetal.

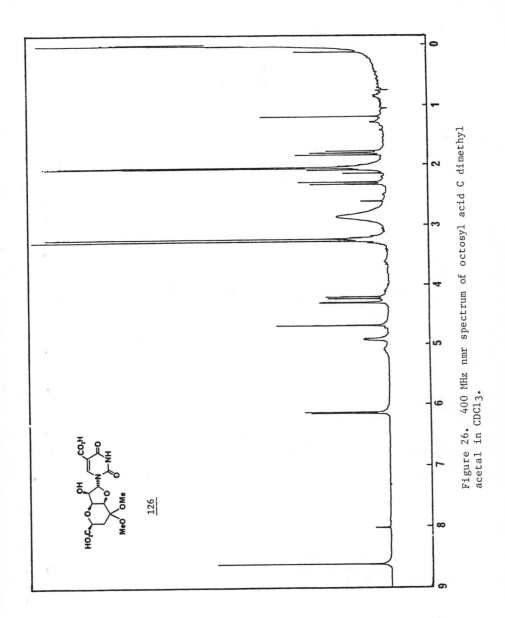

Figure 26. 400 MHz nmr spectrum of octosyl acid C dimethyl acetal in CDCl₃.

Acknowledgment

We thank NSERCC and FCAR for financial assistance and the Shionogi Co., Japan for a sabbatical leave (Tamio Sugawara)

Literature Cited

1. Isono, K.; Suzuki, S. Heterocycles, 1979, 13, 333 and references cited therein.
2. Suhadolnik, R.J. in "Nucleosides as Biological Probes", Wiley, New York, 1979, p. 295.
3. Sunadolnik, R.J. in "Progress in Nucleic Acid Research and Molecular Biology", Vol. 22, p. 193.
4. Isono, K.; Crain, P.F.; McCloskey, J.A. J. Am. Chem. Soc., 1975, 97, 943.
5. a) Azuma, T.; Isono, K.; Crain, P.F.; McCloskey, J.A. Tetrahedron Lett. 1976, 1687. b) Azuma, T.; Isono, K. Chem. Pharm. Bull. 1977, 25, 3347.
6. a) Sakata, K.; Sakurai, A.; Tamura, S. Tetrahedron Lett. 1974, 4327. b) Sakata, K.; Sakurai, A.; Tamura, S. ibid. 1975, 3191.
7. a) Sakata, K.; Uzawa, J. Agric. Biol. Chem. 1977, 41, 413. b) Sakata, K.; Uzawa, J.; Sakurai, A. Org. Magn. Reson. 1977, 10, 230. c) Shibuya, K.; Tanaka, M.; Nanbata, T.; Isono, K.; Suzuki, S. Agric. Biol. Chem. 1972, 36, 1229. d) Isono, K. Suzuki, S.; Tanaka, M.; Nanbata, T.; Shibuya, T. ibid. 1972, 36, 1571.
8. Danishefsky, S.; Hungate, R. J. Am. Chem. Soc., 1986, 108, 2486.
9. Hanessian, S.; Kloss, J.; Sugawara, T. J. Am. Chem. Soc., 1986, 108, 2758.
10. Anzai, K.; Saita, T. Bull. Chem. Soc. Japan, 1977, 50, 169; J. Chem. Soc. Chem. Commun., 1976, 681.
11. Anzai, K.; Saita, T. Nucleic Acids Res., Special Publication No. 2, 1977, 87.
12. Kim, K.S.; Szarek, W.A. Can. J. Chem., 1981, 59, 878; Carbohydr. Res., 1982, 100, 169.
13. Hanessian, S.; Sato, K.; Liak, T.J.; Danh, N.; Dixit, D.; Cheney, B.V. J. Am. Chem. Soc., 1984, 106, 6114.
14. Hanessian, S.; Dixit, D.M.; Liak, T.J. Pure Appl. Chem. 1981, 53, 129.
15. See for example, M.L. Wolfrom, P. McWain and A. Thompson, J. Org. Chem., 1962, 27, 3549; D. Horton, Pure & Appl. Chem., 1975, 42, 301.
16. Hanessian, S.; Dixit, D.M.; Sugawara, T. unpublished results.
17. a) Herr, R.R.; Slomp, G. J. Am. Chem. Soc., 1967, 89, 2444. b) Schroeder, W.; Bannister, B.; Hoeksema, H. J. Am. Chem. Soc., 1967, 89, 2448.
18. Danishefsky, S.; Hungate, R., subsequent lecture.
19. For a review, see: David, S.; Hanessian, S. Tetrahedron, 1985, 41, 643.
20. a) Damodaran, N.P.; Jones, G.H.; Moffatt, J.G. J. Am. Chem. Soc., 1971, 93, 3812. b) Jones, G.H.; Moffatt, J.G. in "Methods in Carbohydrate Chemistry", Vol VI, p. 315, 1972.
21. DeMole, E.; Enggist, P. Helv. Chim. Acta, 1971, 54, 456.

22. Tamaru, Y.; Mizutani, M.; Furukawa, Y.; Kawamura, S.; Yoshida,
    Z.; Yanagi, K.; Minobe, M. J. Am. Chem. Soc., 1984, 106, 1079.
23. a) Clive, D.L.J.; Chittattu, G.; Wong, C.K. Can. J. Chem.,
    1980, 45, 2921, 4063. b) Nicolaou, K.C.; Lysenko, Z.
    Tetrahedron Lett., 1977, 1257. c) Nicolaou, K.C.; Claremon,
    D.A.; Barnette, W.E.; Seitz, S.P. J. Am. Chem. Soc., 1979, 101,
    3704.
24. Spezial, V.; Amat, M.; Lattes, A. J. Heterocyclic Chem., 1976,
    13, 349.
25. Pougny, J.-R.; Nassr, M.A.M.; Sinay, P.: J. Chem. Soc. Chem.
    Commun., 1981, 375.
26. Hill, C.L.; Whitesides, G.M. J. Am. Chem. Soc., 1974, 96, 870.
27. a) Heyns, K.; Paulsen, H. in "Newer Methods of Preparative
    Organic Chemistry", Foerst, W., ed.; Academic Press, New York,
    1963, Vol. II, p. 303. b) Heyns, K.; Blazejewicz, L.
    Tetrahedron, 1960, 9, 67.
28. Corey, E.J.; Samuelsson, B. J. Org. Chem., 1984, 49, 4735.
29. For a related addition of an allyl Grignard to a methyl
    glycoside see: Danishefsky, S.; DeNinno, M. Tetrahedron Lett.,
    1985, 26, 823.
30. Cram, D.L.; Elhafez, F.A.A. J. Am. Chem. Soc., 1942, 74, 5828.
31. Su, T.-L.; Harada, K.; Watanabe, K.A. J. Nucleosides
    Nucleotides, 1984, 3, 513.
32. Pichat, L.; Massé, B.; Deschamps, J.; Dufay; P. Bull. Soc.
    Chim. France, 1971, 2102.
33. a) Ruth, J.L.; Bergstrom, D.E. J. Org. Chem., 1978, 43, 2870.
    b) Bergstrom, D.E.; J. Nucleosides Nucleotides, 1982, 1, 1.
34. Hayakawa, H.; Tanaka, H.; Miyasaka, T.; Tetrahedron, 1985, 41,
    1675.
35. Liotta, D.; Barnum, C.; Puleo, R.; Zima, G.; Bayle, C.; Kesar,
    H.S. III; J. Org. Chem., 1981, 46, 2920; and references cited
    therein.

RECEIVED May 31, 1988

Chapter 5

Use of Unstabilized Carbohydrate Ylides for the Synthesis of Long-Chain Carbohydrates

John A. Secrist III[1], Keith D. Barnes[2], and Shang-Ren Wu[3]

Department of Chemistry, The Ohio State University, Columbus, OH 43210

The challenge embodied in the synthesis of carbohydrates containing more than six or seven carbon atoms was first taken up by Fischer (1), and has continued to the present day. The list of natural products that contain long-chain complex carbohydrates continues to grow, and now includes hikizimycin (2), the related tunicamycins, streptovirudins, and mycospocidin (3-7), sinefungin and the related factor C (8), the octosyl acids and the ezomycins (8), the sialic acids (9), lincomycin (10,11), 3-deoxy-D-manno-2-octulosonic acid (KDO) (12), celesticetin (13), and apramycin and oxyapramycin, two broad-spectrum antibiotics from the nebramycin complex (14,15). Most of these compounds are biologically active, and some of them have very interesting activities indeed. Hikizimycin is an anthelmintic agent and inhibits protein synthesis by preventing the peptide bond-forming reaction. The tunicamycins, the streptovirudins, and mycospocidin are very similar compounds that are potent inhibitors of glycosylation. The tunicamycins have been better studied, and they also have been found to have some antiviral properties, though the compounds are quite toxic. The mechanism of action of these compounds appears to be the inhibition of the transfer of N-acetylglucosamine from DP-N-acetylglucosamine to dolichyl phosphate, with the compounds apparently acting as bisubstrate inhibitors of the enzymic reaction. Sinefungin and factor C are decosyl nucleosides with an amino acid at the terminus of the chain extension, and they are both potent inhibitors of a wide variety of biological methyl transfer reactions, acting as competitive inhibitors of the methyltransferase enzymes. Sinefungin has been found to have both antiviral activity and anti-parasitic activity. The ezomycins, anhydrooctosyl uronic acid nucleosides, have modest antifungal activity, though little is known about their mechanism of action. Lincomycin, which contains the octosyl sugar lincosamine, is a clinically used antibacterial agent. The sialic acids, which are N- and O-acylated derivatives of neuraminic acid, are components of glycoproteins found in many types of cells. KDO is a component of the lipopolysaccharides of gram-negative bacteria. Future research will undoubtedly uncover more compounds that have long-chain carbohydrate constituents.

[1]Current address: Organic Chemistry Research Department, Southern Research Institute, Birmingham, AL 35255–5305

[2]Current address: Agricultural Chemical Group, Chemical Research and Development, FMC Corporation, Princeton, NJ 08540

[3]Current address: Syracuse Research Laboratory, Allied-Signal, Inc., Solvay, NY 12309

0097–6156/89/0386–0093$06.00/0
© 1989 American Chemical Society

Some years ago we became interested in the synthesis of certain of these longer chain carbohydrates, in particular those with chain lengths greater than eight carbons. At that time there was very little literature on the subject. Several papers described specific reactions that resulted in the formation of long chain carbohydrates, but these reactions offered no hope of selectivity or of general application. In two cases dimerization reactions were involved (16,17), and in the third the long-chain carbohydrate was a minor byproduct from a commercial process (18). During the time that the chemistry described in this chapter was developed, Paulsen and coworkers were also developing an approach to longer chain carbo-hydrates involving the condensation of the dianion of a monounpro-tected dithiane-substituted carbohydrate with suitably blocked carbo-hydrate aldehydes (19,20). More recently, other elegant and useful approaches to specific compounds such as hikizimycin and tunicamycin have been developed (21-24), and some of that work is touched on in other chapters in this volume. Routes that lead to some of the other natural products mentioned above have also been developed (25-49).

Our goal was to develop a versatile synthetic method for the construction of long-chain carbohydrates, a method that might be employed to prepare, for example, ten to twelve carbon sugars from readily available starting materials. More importantly, a method was desired that would allow specification of the absolute configurations of as many carbons as possible in the long-chain carbohydrate produced. Application of the method to the synthesis of specific targets might then be pursued.

A method that appeared to offer the desired versatility involved application of the Wittig reaction. Previously, the Wittig reaction had been used mainly to condense aldehydo or keto sugars with simple stabilized or unstabilized ylides to prepare a variety of chain-extended or chain-branched carbohydrates (50,51). A stabilized phosphorane had also been generated in a carbohydrate framework and had been found to condense with certain aromatic aldehydes (52,53). Systems such as these, where the carbon of the phosphorane is insulated from the remainder of the carbohydrate, did not appear useful for the attainment of the goal of this research program, though very recently carbohydrate-derived stabilized ylides have been employed to prepare long-chain carbohydrates (54). In order to embrace the desired scope, the ylide would need to be an intrinsic part of the carbohydrate. It appeared, therefore, that an unstabilized carbohydrate ylide would need to be generated and successfully condensed with various aldehydes, including carbohydrate aldehydes. Thus, the initially envisioned route to long chain carbohydrates could be schematically described as shown below.

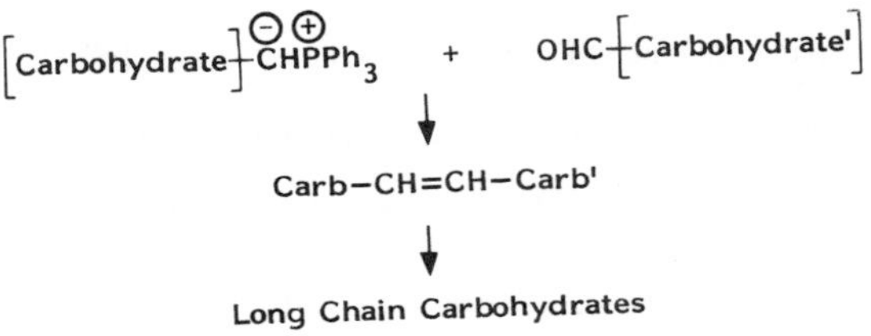

$$\left[\text{Carbohydrate}\right]\overset{\ominus}{\text{CH}}\overset{\oplus}{\text{PPh}}_3 \quad + \quad \text{OHC}\left[\text{Carbohydrate}'\right]$$

$$\downarrow$$

$$\text{Carb}-\text{CH}=\text{CH}-\text{Carb}'$$

$$\downarrow$$

$$\textbf{Long Chain Carbohydrates}$$

Clearly, care would need to be exercised in selecting stable protecting groups for both reactants. By appropriate construction of the two partners, a wide variety of chain-extended carbohydrates would become available. The double bond produced by the condensation reaction would, of course, be amenable to suitable manipulation in order to incorporate additional needed or desired functionality. Such manipulation would be facilitated if the proposed reaction produced only one configuration about the double bond, though separation of the two isomers would not be a major stumbling block.

Analysis of a potential carbohydrate phosphorane from a typical carbohydrate immediately pointed up the major problem that any such compound would have, namely that an oxygen-containing leaving group generally would be present β to the ylide carbon. Such a group might be expelled readily from an unstabilized ylide, producing a vinyl phosphonium salt. An initial examination of the literature was not too encouraging with regard to this question. The phosphonium salts I and II had been prepared and were reported not to generate ylides with any stability, though the conditions for the generation of the ylides were harsh relative to those actually needed for the formation of such ylides (55). Presumably vinyl phosphonium salt formation was the outcome in their systems. Precedent for a reversible β-elimination, however, can be found in the ylide generated from tetrahydrofurfuryl-triphenylphosphonium bromide (56). All of this information made it clear that great care would need to be taken in the development of conditions for our scheme.

I   II

For the initial experiments, a ribose-derived system was chosen, and the phosphonium salt III was synthesized by standard methods from methyl 2,3-O-isopropylidene-β-D-ribofuranoside (57). Displacement of a leaving group at C-5 proved to be more difficult than anticipated, and a variety of conditions were examined in order to maximize our yield. Sulfolane was found to be a particularly useful solvent for such displacements with quaternization, and it was routinely employed for phosphonium salt formation in other systems unless some other factor influenced the solvent choice. In line with the considerations of the previous paragraphs, formation of the ylide derived from III under very mild conditions (as low a temperature as possible) appeared to be desirable. The solubility of III in solvents such as THF, however, was very low, and it was not possible to obtain a homogeneous solution even at ice temperatures. Such a problem was found with all of the phosphonium salts that were prepared. A solvent mixture containing THF and hexamethylphosphoramide (HMPA) was found to be effective in overcoming this obstacle. By adding increasing amounts of HMPA, it proved possible to keep the phosphonium salts in solution at reasonably low temperatures, allowing generation and utilization of the ylides. The ylide derived from III was generated with n-BuLi at -50 °C and then quenched with benzaldehyde, affording two products in 79% yield.   These products proved to be the E- and Z-isomers IVa and IVb, with the configuration at C-4 inverted. Both aromatic and aliphatic aldehydes gave good yields of chain-extended

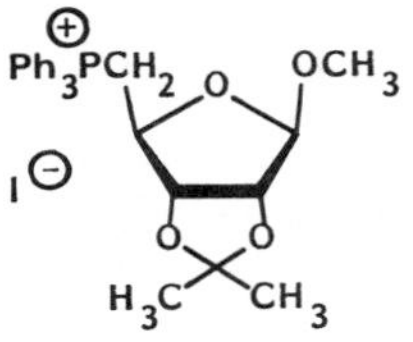

III

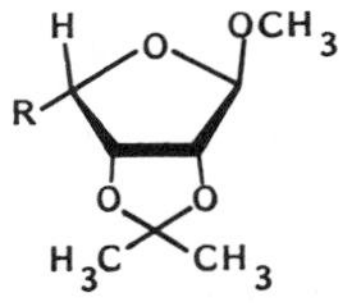

IV

a) R = CH=CHPh(E)
b) R = CH=CHPh(Z)

carbohydrates, in all cases with the α-D-lyxo configuration rather than the original β-D-ribo. Several conclusions were drawn from this research. First, the furan ring of ylide V, once formed, was clearly opening to form the alkoxy vinylphosphonium salt VI, which reclosed to form the thermodynamic product (VII), in this case having the α-D-lyxo configuration. The success of the procedure, then, depended upon the attachment of the β-oxygen to the carbohydrate through another set of bonds, thus enabling an intramolecular attack to reform an ylide that could react with the carbonyl component. With ylide V, the inversion took place within seconds.

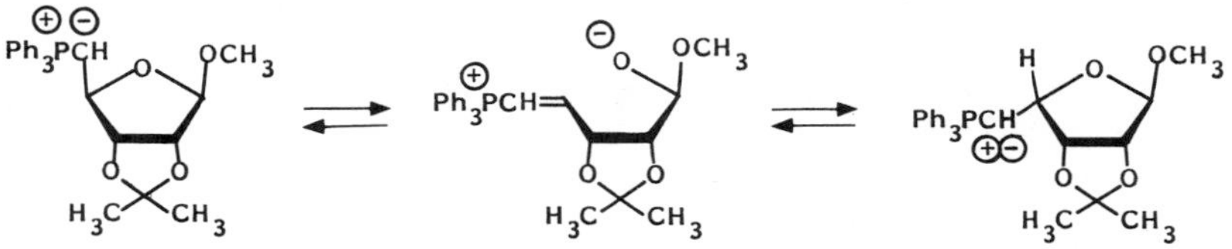

V

VI

VII

After these results had established the feasibility of generating and utilizing a carbohydrate phosphorane, the two systems that had been reported earlier were examined in order to determine if similar conditions would allow them to undergo the Wittig reaction. The ylide derived from phosphonium salt I condensed with both benzaldehyde and 4-chlorobenzaldehyde to produce good yields of olefinic products VIIIa and VIIIb. The ylide derived from phosphonium salt II also was successfully condensed with benzaldehyde, but the yield of IX was only 30%, presumably because of its extremely poor solubility even in an HMPA-THF solvent mixture. Both of these systems supported the tenet that it was possible to use unstabilized carbohydrate phosphoranes if the conditions are proper and if the β-oxygen is attached to the carbohydrate through another set of bonds.

With these results in hand, examination of the generality of the method as well as its application to the condensation of unstabilized carbohydrate ylides with carbohydrate aldehydes was undertaken. At

VIII   IX

a) R = CH=CHPh (77:1, Z/E)
b) R = CH=CHpClPh (76:2, Z/E)

this stage it seemed appropriate to have some long-range goal in mind, and hikosamine (actually a blocked derivative thereof), the undecose portion of the nucleoside antibiotic hikizimycin, was chosen as a suitable target. Retrosynthetic analysis on the structure of hikosamine suggested a natural cleavage point between carbons 6 and 7. That is, it seemed particularly attractive to construct two pieces such that the Wittig reaction would form that bond, as Scheme 1 (next page) shows. Another obvious question concerned when to introduce the nitrogen at C-4. Because such nitrogens are typically introduced by the displacement of a leaving group of the opposite configuration, it seemed initially reasonable to suggest the condensation of either a D-galacto ylide with a D-arabino aldehyde or of a D-arabino ylide with a D-galacto aldehyde. Inversion of C-4 of the D-galacto portion would allow introduction of the nitrogen in the proper configuration. Possible utilization of all four types of molecules was explored.

The phosphonium salt X derived from 1,2:3,4-di-O-isopropylidene-α-D-galactopyranose was readily prepared (58), and the ylide derived from it was condensed effectively with both aliphatic and aromatic aldehydes. The products all had the α-D-galacto configuration, and the aliphatic aldehydes afforded only the Z-isomers. The less constrained permethylated phosphonium salt XI was also prepared, and

X   XI

the ylide derived from it was found to condense with benzaldehyde to produce a good yield of an E/Z mixture (Z predominating) of the olefin again with exclusively the α-D-galacto configuration (58). In both examples the D-galacto product is presumably the thermodynamic isomer. Thus, there is no evidence in these systems as to whether the ylide is opening and reclosing to the same ylide, or whether it is reacting without ring opening. In fact, the ribose-derived system presented in the preceding paragraph is the only example where a change of configuration was found.

Condensation of the ylide derived from X with carbohydrate aldehyde X proceeded smoothly to afford only the Z-isomer XIII. Proof of the olefinic configuration was garnered by photochemical isomeri-

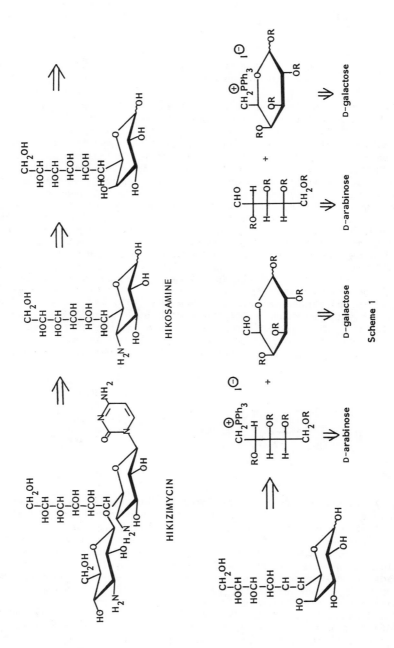

HIKIZIMYCIN

HIKOSAMINE

Scheme 1

zation to the E-isomer, which was readily distinguishable from the Z-isomer by comparison of their proton NMR spectra. The configuration at what was the C-4 position of the ribo aldehyde was also shown to be

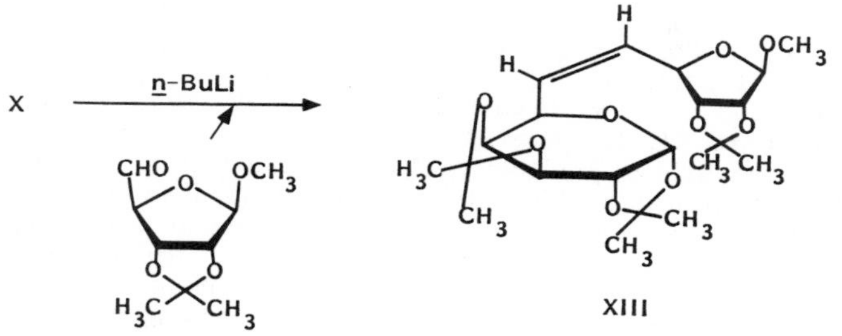

XII   XIII

unchanged by the reaction conditions. A condensation of the same ylide with aldehydes XIV and XV provided similar results, again with no isomerization adjacent to the aldehydes.

As an example of a D-arabino phosphonium salt, XVIa was prepared. Simple Wittig reactions with the ylide derived from XVIa did not proceed as smoothly as with the other ylides, and for that reason salt XVIb, which proved to be much better behaved (results not shown), was also prepared.

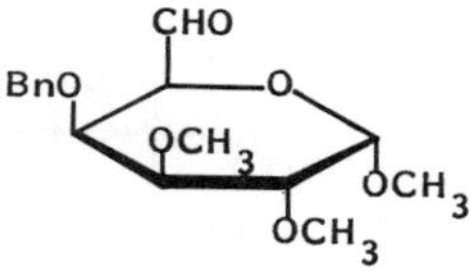

XIV (Bn = benzyl) 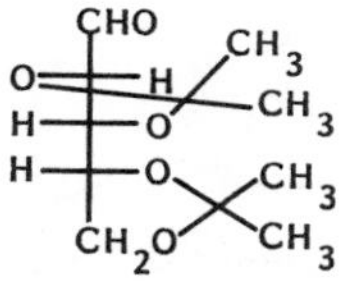   XV

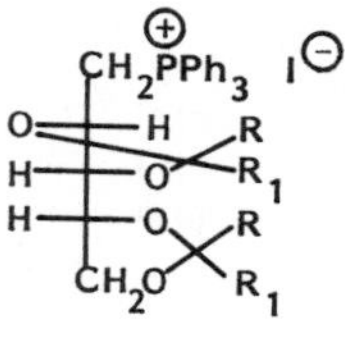

XVI   XVII

a) $R = R_1 = CH_3$
b) $R, R_1 = (CH_2)_5$

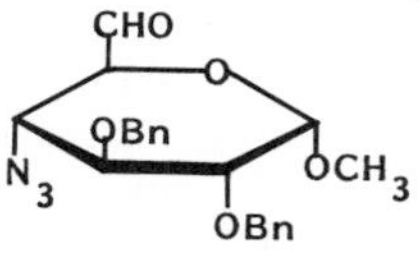

Once it had been established that either D-galacto or D-arabino ylides could be successfully condensed with carbohydrate aldehydes while maintaining their configurational integrity, the route leading to hikosamine that appeared most straightforward was selected. It seemed best not to have a nitrogen functionality in the ylide partner if it could be avoided, and it appeared desirable to carry out as much manipulation as possible prior to the Wittig reaction. With these considerations in mind, aldehyde XVII was chosen as a promising partner to the ylide derived from salt XVIb. Synthesis of this aldehyde was achieved by standard methods (59). Because of its slight instability, XVII was stored as the N,N-diphenylimidazolidine derivative, with the aldehyde being liberated just prior to use. In order to test its suitability in the proposed scheme, aldehyde XVII was initially condensed with the ylide derived from X to afford a 65% yield of dodecose XVIII. Employing the ylide derived from XVIb and aldehyde XVII as Wittig partners allowed formation of a 50% yield of Z-isomer XIXa, which was proven to have the indicated configurations. From this structure only two stereogenic centers needed to be established, those derived from the double bond. In order to obtain the required configurations for hikosamine, a trans addition across the Z-olefin was needed. Various approaches were examined, none of which proved satisfactory. It was found, however, that osmylation of the Z-isomer proceeded smoothly to produce only one of the two possible isomers that could result from cis addition. If this observation held for the E-isomer, a route to a hikosamine derivative might be developed. After reduction of XIXa to XIXb with LiAlH$_4$, isomerization to XIXc was achieved by irradiation in cyclohexane solution in the presence of diphenyl disulfide. In order to avoid significant loss of material, the isomerization was stopped at about a 3/2 (Z/E) ratio, and the olefins were separated. After N-acetylation, osmylation of XIXd again produced only one isomer, which

XVIII

XIX

a) Q = N$_3$, Z isomer
b) Q = NH$_2$, Z isomer
c) Q = NH$_2$, E isomer
d) Q = NHAc, E isomer

proved to be the one derived from hikosamine. Removal of the cyclohexylidene and benzyl blocking groups and peracetylation produced methyl peracetyl α-hikosaminide identical to that prepared from naturally-occurring material (59). Research carried out in other laboratories since the completion of this work allows prediction of the success of the above osmylation reaction and provides a framework for applying osmylations to a variety of systems (37,60-66).

During the period of this research, the application of our approach to a tunicamycin derivative was undertaken. Though the work was not pursued far enough to obtain tunicamine or tunicaminyluracil, the scope of the general reaction was expanded somewhat, and some of these efforts are worth recounting briefly. Based upon results reported herein, the brief retrosynthetic analysis presented below seemed to provide a reasonable approach to the undecose moiety of the tunicamycins. A key question was what to use as group M. A phthalimide group was selected in order to determine whether such a nitrogen-containing group would be acceptable in the phosphorane partner. Salt XX was prepared by standard means as shown (next page),

tunicamycin (R = branched fatty acid side chain)

tunicamine   ᴅ–galacto   ᴅ–ribo

but its solubility was so poor that it was not possible to examine it properly. Salt XXI, in which the silyloxy group could be selectively manipulated to incorporate a nitrogen functionality later was then prepared. Condensation of the ylide from XXI with aldehyde XVII went smoothly to produce XXII in 55% yield. The silyl protecting group was readily removed with tetra-n-butylammonium fluoride, allowing for selective manipulation at that site.

An important question with regard to the scope of this method for complex carbohydrate synthesis was whether an unstabilized carbohydrate phosphorane might be compatible with a suitably blocked nucleoside aldehyde. Condensation of the ylide derived from XXI with uridine derivative XXIII proceeded under the usual conditions to give a 25% yield of XXIV, from which the N-benzoyl group could be readily removed with methanolic ammonia. The yield from this condensation reaction was not optimized, but its success clearly demonstrated the feasibility of such a transformation, if design considerations dictate the use of a nucleoside directly.

To summarize, unstabilized carbohydrate phosphoranes are useful entities for the synthesis of long-chain carbohydrates by a Wittig process as long as the proper experimental conditions are employed and as long as the β-oxygen is attached by another set of bonds to

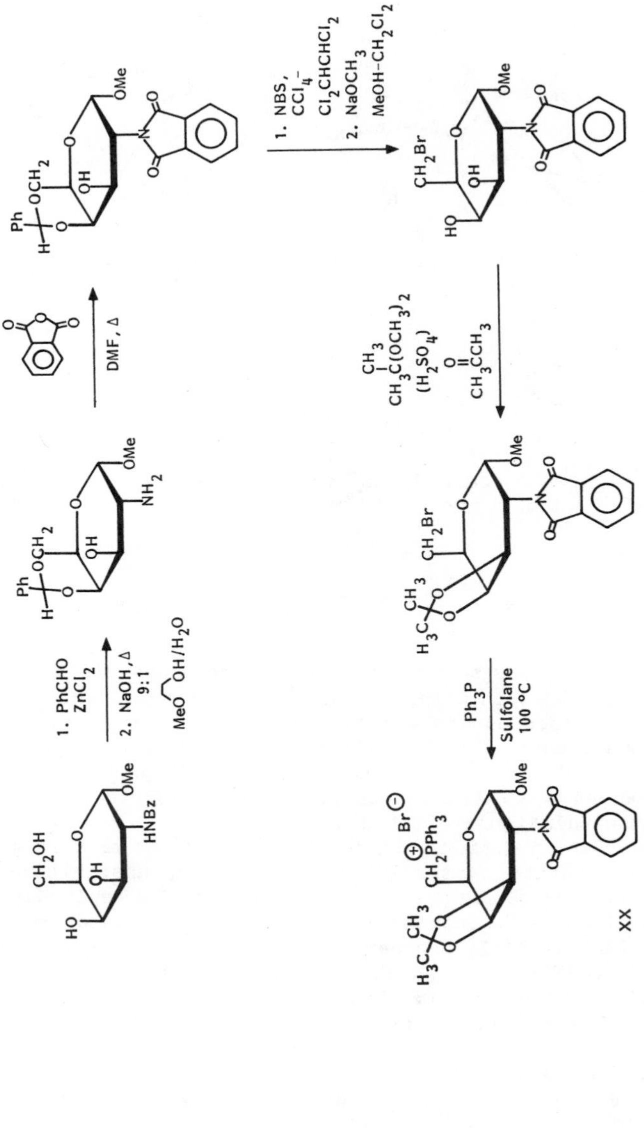

XX

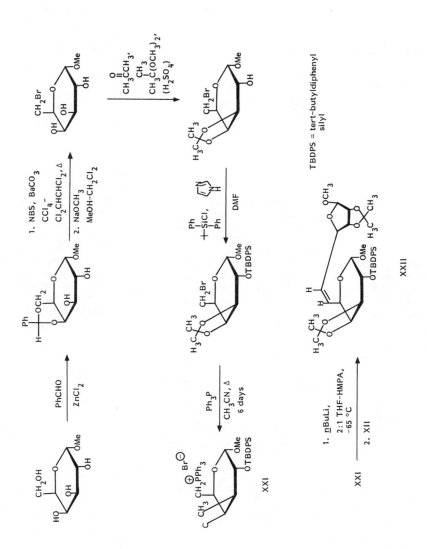

the carbohydrate. This process allows the rapid assembly of eight to ten stereogenic centers with defined configurations from readily available precursors and allows manipulation of the newly formed double bond to generate two more such centers, depending upon the target in question. Thus from the available five and six carbon carbohydrates, a considerable array of ten- to twelve-carbon complex carbohydrates can be constructed. This research has also shown the compatibility of a variety of blocking groups with the procedure Beyond the applications noted herein, the concept of generating a usable nucleophilic center   to an oxygen functionality is one that has some generality (67).

Literature Cited

1. Fischer, E.; Passmore, F. Chem. Ber., 1890, 23, 2226-39.
2. Vuilhorgne, M.; Ennifar, S.; Das, B. C.; Paschal, J. W.; Nagarajan, R.; Hagaman, E. W.; Wenkert, E. J. Org. Chem., 1977, 42, 3289-91.
3. Tkacz, J. S. Antibiotics (N. Y.), 1983, 6, 255-78.
4. Eckardt, K. J. Nat. Prod., 1983, 46, 544-50.
5. Tamura, G., Ed., Tunicamycin; Japan Sci. Soc. Press, Tokyo, Japan, 1982.
6. Tkacz, J. S.; Wong, A. Fed. Proc., 1978, 37, 1766.
7. Nakamura, S.; Arai, M.; Karasawa, K.; Yonehara, H. J. Antibiotics, 1957, 10A, 248-53.
8. Suhadolnik, R. J. Nucleosides as Biological Probes; Wiley: New York, 1979; pp. 19-23.
9. Schauer, R. Adv. Carbohydr. Chem. Biochem., 1982, 40, 131-234.
10. Phillips, I. J. Antimicrob. Agents Chemother., 1981, 7 (Suppl. A), 11-18.
11. Eble, T. E. In Kirk-Othmer Encycl. Chem. Technol.; 3rd ed., Grayson, M.; Eckroth, D., Eds.; Wiley: New York, 1978; Vol. 2, pp. 930-36.

12. Unger, F. M. Advan. Carbohydr. Chem. Biochem., 1981, 38, 323-88.
13. Hoeksma, H. J. Am. Chem. Soc., 1968, 90, 755-7.
14. O'Connor, S.; Lam, L. K. T.; Jones, N. D.; Chaney, M. O. J. Org. Chem., 1976, 41, 2087-92.
15. Koch, K. F.; Merkel, K. E.; O'Connor, S. C.; Occolowitz, J. L.; Paschal, J. W.; Dorman, D. E. J. Org. Chem., 1978, 43, 1430-4.
16. Roy, R. B.; Chilton, W. S. J. Org. Chem., 1971, 36, 3242-43.
17. Kornilov, V. I.; dyk Shung, L.; Zhdanov, Yu. A. J. Gen. Chem. USSR, 1971, 202-04.
18. Wolfrom, M. L.; Binkley, W. W.; Spencer, C. C.; Lew, B. W. J. Am. Chem. Soc., 1951, 73, 3357-58.
19. Paulsen, H.; Roden, K.; Sinnwell, V.; Koebernick, W. Angew. Chem., Int. Ed. Engl., 1976, 15, 439-40.
20. Paulsen, H.; Roden, K.; Sinnwell, V.; Luger, P. Liebigs Ann. Chem., 1981, 2009-27.
21. Danishefsky, S.; Barbachyn, M. J. Am. Chem. Soc., 1985, 107, 7761-2.
22. Danishefsky, S.; Maring, C. J. Am. Chem. Soc., 1985, 107, 7762-4.
23. Suami, T.; Sasai, H.; Matsuno, K.; Suzuki, N. Carbohydr. Res., 1985, 143, 85-96.
24. Sasai, H.; Matsuno, K.; Suami, T. J. Carbohydr. Chem., 1985, 4, 99-112.
25. Danishefsky, S.; Hungate, R. J. Am. Chem. Soc., 1986, 108, 2486-7.
26. Hanessian, S.; Kloss, J.; Sugawara, T. J. Am. Chem. Soc., 1986, 108, 2758-9.
27. Geze, M.; Blanchard, P.; Fourrey, J. L.; Robert-Gero, M. J. Am. Chem. Soc., 1983, 105, 7638-40.
28. Danishefsky, S.; DeNinno, M. P. J. Org. Chem., 1986, 51, 2617-8.
29. Auge, C.; David, S.; Gautheron, C.; Veyrieres, A. Tetrahedron Lett., 1985, 26, 2439-40.
30. Auge, C.; David, S.; Gautheron, C. Tetrahedron Lett., 1984, 25, 4663-4.
31. Mirzayanova, M. N.; Davydova, L. P.; Samokhvalov, G. I. Zh. Obshch. Khim., 1970, 40, 693-7; Chem. Abstr., 1970, 73, 25797.
32. How, M. J.; Halford, M. D. A.; Stacey, M.; Vickers, E. Carbohydr. Res., 1969, 11, 313-20.
33. Kuhn, R.; Baschang, G. Liebigs Ann. Chem., 1962, 659, 156-63.
34. Carroll, P. M.; Cornforth, J. W. Biochim. Biophys. Acta, 1960, 39, 161-2.
35. Cornforth, J. W.; Firth, M. E.; Gottschalk, A. Biochem. J., 1958, 68, 57-61.
36. Larson, E. R.; Danishefsky, S. J. Am. Chem. Soc., 1983, 105, 6715-16.
37. Danishefsky, S.; Larson, E.; Springer, J. J. Am. Chem. Soc., 1985, 107, 1274-80.
38. David, S. M.; Fischer, J. C. Carbohydr. Res., 1976, 50, 239-46.
39. Magerlein, B. J. Tetrahedron Lett., 1970, 33-6.
40. Saeki, H.; Ohki, E. Chem. Pharm. Bull., 1970, 18, 789-802.
41. Howarth, G. B.; Szarek, W. A.; Jones, J. K. N. J. Chem. Soc. D, 1969, 1339-40.
42. Danishefsky, S. J.; Pearson, W. H.; Segmuller, B. E. J. Am. Chem. Soc., 1985, 107, 1280-85.

43. Zhdanov, Yu. A.; Kornilov, V. I.; Turik, S. V. Bioorg. Khim., 1983, 9, 104-11; Chem. Abstr., 1983, 98, 161085.
44. Perry, M. B.; Williams, D. T. Methods Carbohydr. Chem., 1976, 7, 44-8.
45. Charon, D.; Sarfati, R. S.; Strobach, D. R.; Szabo, L. Eur. J. Biochem., 1969, 364-9.
46. Hershberger, C.; Davis, M.; Binkley, S. B. J. Biol. Chem., 1968, 243, 1585-8.
47. Hershberger, C.; Binkley, S. B. J. Biol. Chem., 1968, 243, 1578-84.
48. Ghalambor, M. A.; Levine, E. M.; Heath, E. C. J. Biol. Chem., 1966, 241, 3207-15.
49. Tatsuka, K.; Akimoto, K.; Takahashi, H.; Hamatsu, T.; Annaka, M.; Kinoshita, M. Bull. Chem. Soc., 1984, 57, 529-38.
50. Zhdanov, Yu. A.; Alexeev, Yu. E.; Alexeeva, V. G. Advan. Carbohydr. Chem. Biochem., 1972, 27, 227-99.
51. Tronchet, J. M. J.; Bonenfant, A.-P. Helv. Chim. Acta, 1980, 63, 1644-53, and other papers in this series.
52. Zhdanov, Yu. A.; Uzlova, L. A. Zh. Obshch. Khim., 1972, 42, 759-62; Chem. Abstr., 1972, 77, 140426.
53. Zhdanov, Yu. A.; Polenov, V. A. Carbohydr. Res., 1971, 16, 466-8.
54. Jarosz, S.; Mootoo, D.; Fraser-Reid, B. Carbohydr. Res., 1986, 147, 59-68.
55. Bohlmann, F.; Herbst, P. Chem. Ber., 1959, 92, 1319-28.
56. Schweizer, E. E.; Creasy, W. S.; Light, K. K.; Shaffer, E. T. J. Org. Chem., 1969, 34, 212-8.
57. Secrist, J. A., III; Wu, S.-R. J. Org. Chem., 1977, 42, 4084-8.
58. Secrist, J. A., III; Wu, S.-R. J. Org. Chem., 1979, 44, 1434-8.
59. Secrist, J. A., III; Barnes, K. D. J. Org. Chem., 1980, 45, 4526-8.
60. Brimacombe, J. S.; Hanna, R.; Kabir, A. K. M. S. J. Chem. Soc., Perkin Trans. 1, 1986, 823-8.
61. Brimacombe, J. S.; Hanna, R.; Kabir, A. K. M. S.; Bennett, F.; Taylor, I. D. J. Chem. Socl, Perkin Trans. 1, 1986, 815-21.
62. Vedejs, E.; McClure, C. K. J. Am. Chem. Soc., 1986, 108, 1094-6.
63. Brimacombe, J. S.; Kabir, A. K. M. S.; Taylor, I. D. Carbohydr. Res., 1985, 140, C9-C12.
64. Annunziata, R.; Cinquini, M.; Cozzi, F.; Raimondi, L. J. Chem. Soc., Chem. Commun., 1985, 403-5.
65. Cha, J. K.; Christ, W. J.; Kishi, Y. Tetrahedron, 1984, 40, 2247-55.
66. Stork, G.; Kahn, M. Tetrahedron Lett., 1983, 24, 3951-4.
67. Unpublished material from this presentation was abstracted from: Barnes, K. D., Ph.D. Thesis, Ohio State University, 198; Wu, S.-R., Ph.D. Thesis, Ohio State University, 1979.

RECEIVED October 14, 1988

Chapter 6

Synthesis of Chiral Pyrrolidines from Carbohydrates

J. Grant Buchanan, Alan R. Edgar, Brian D. Hewitt,
Veerappa B. Jigajinni, Gurdial Singh, and Richard H. Wightman

Department of Chemistry, Heriot-Watt University, Riccarton,
Edinburgh EH14 4AS, United Kingdom

We have extended our work on a new synthesis of the antiprotozoal antibiotic anisomycin to the necine bases of
the pyrrolizidine alkaloids, in particular retronecine and
crotanecine. The key intermediate, (2R,3S,4R)-2-(alkoxy-
carbonylmethyl)-3,4-isopropylidenedioxypyrrolidine, has
been prepared by three distinct routes from D-ribose and
D-erythrose, using reactions of high stereoselectivity.
A new approach to anisomycin from D-erythrose using
Wittig methodology is outlined.

We were first attracted to chiral pyrrolidines by the possibility of
applying methods used in C-nucleoside synthesis (1) to the synthesis
of the antiprotozoal antibiotic anisomycin (1) from D-ribose (2).
The approach, which differs from other recent syntheses (3,4,5), is
outlined in Scheme 1. Three points may be noted: (i) in the
Grignard addition to 2,3-0-isopropylidene-D-ribose (2) the D-allo
configuration in (3) is in accordance with the Felkin-Anh model (6)
and is to be expected from our earlier work (1); (ii) methane-
sulfonylation of the oxime (4) serves not only to dehydrate the
oxime but to introduce a leaving group for ring closure at the next
step; (iii) the intramolecular displacement to form the pyrrolidine
ring [(5) → (6)] proceeds cleanly and with complete inversion of
configuration (3,5,7).

The Geissman-Waiss lactone (7)(8) is a well-known precursor of
(+)- retronecine (8) (8-11), one of the most common necine bases
derived from the pyrrolizidine alkaloids. We envisaged that the
pyrrolidine ester (9) could be converted into the lactone (7), representing a formal synthesis of (+)-retronecine (8) (12). In
addition, (9) should be capable of transformation into the related
necine base crotanecine (10) (13).

Scheme 2 illustrates a synthesis of the benzyloxycarbonyl
derivative (11) of the ester (9a) using chemistry analogous to that
for anisomycin (Scheme 1). In the formation of the D-allo-triol (12),

NOTE: This chapter is dedicated to Professor Luis F. Leloir on the occasion of his 80th birthday.

0097-6156/89/0386-0107$06.00/0
© 1989 American Chemical Society

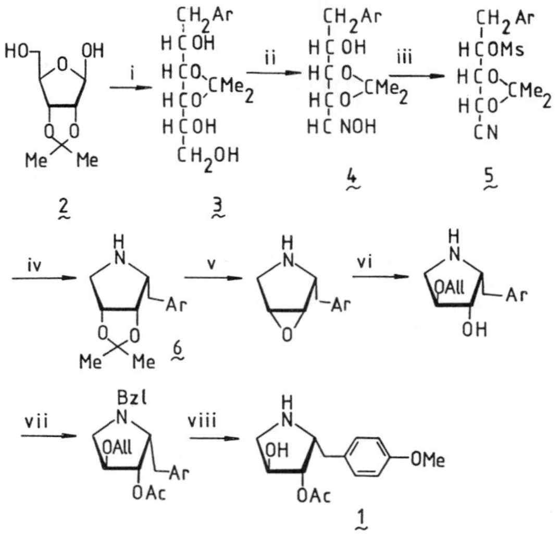

$Ar = \underline{p}\text{-}MeOC_6H_4\text{-}$ $Ms = MeSO_2\text{-}$ $All = CH_2\text{:}CHCH_2\text{-}$ $Bzl = PhCH_2\text{-}$

Reagents: i, $ArCH_2MgCl\text{-}THF$;   ii, $NaIO_4$, then $HONH_3Cl\text{-}C_5H_5N$;
              iii, $MsCl\text{-}C_5H_5N$;   iv, $LiAlH_4$;   v, $HBr\text{-}HOAc$, then KOH;
              vi, $AllOH\text{-}HClO_4$;   vii, $BzlCl$, then $Ac_2O\text{-}C_5H_5N$;
              viii, $PdC\text{-}H^+$, then $PdC\text{-}H_2$.

Scheme 1

diallyl zinc, formed from the Grignard reagent in situ gave higher stereoselectivity compared to the Grignard reagent itself (14). The yield of the final oxidation step was poor and other avenues to esters (9) were explored. The first of these used the Wittig reaction as an important step, as outlined in Scheme 3.

(7)

(8) R = H

(10) R = OH

(9a) R = Me
(9b) R = Et

2,3-O-Isopropylidene-D-erythrose (13) (15), obtained either by acetonation of D-erythrose (16) or by periodate oxidation of 3,4-O-isopropylidene-D-arabinose (15,17), reacted with ethoxycarbonylmethylenetriphenylphosphorane in refluxing benzene (18) to give the E-alkene (14) as the major product (56%) together with the Z-alkene (15) (21%). As expected (18-20) the alkenes (14) and (15) readily cyclized to tetrahydrofurans (16) under very mild basic conditions. Initially the β anomer of (16) was favored [86% from (14) and 100% from (15)]; at equilibrium the α anomer preponderated (82%) (19). Our intention was to convert the alcohols (14) and (15) into the corresponding amines and then to effect cyclization to the pyrrolidine (9b). In the event, this objective was achieved more easily and with complete stereoselectivity.

Treatment of the alcohol (14) with trifluoromethylsulfonic anhydride (triflic anhydride) at −78°C afforded the ester (17) which could be isolated and characterized. We knew from previous experience (2) that sulfonyl esters vicinal to an isopropylidene acetal are relatively stable. The triflate (17) reacted cleanly with potassium azide and 18-crown-6 in dichloromethane at room temperature. The crystalline product [68% overall from (14)] was not the azide (18) but the isomeric Δ^2-triazoline (19). Clearly the initially formed azide (18) had undergone intramolecular 1,3-cycloaddition to the double bond of the unsaturated ester (21-24). The stereochemistry of the triazoline (19), determined by proton nmr spectroscopy, showed that the reaction was stereospecific. There are several known examples of this reaction (24), including one in the carbohydrate series (25). When the triazoline was treated with sodium ethoxide (26) the diazoester (20) was rapidly formed by ring-opening and was isolated in 85% yield. Hydrogenolysis of the diazo group of (20) gave the required pyrrolidine ester (9b) (90%).

The Z-alkene (15) was subjected to the same sequence (Scheme 4). The triflate (21) was easily obtained, but in this case reaction with azide ion gave directly the diazoester (22). Molecular models show that the triazoline corresponding to (19) has severe steric interactions (27) and is more accessible to deprotonation (cf. ref. 23). Stereochemical and mechanistic aspects of the azide cycloadditions

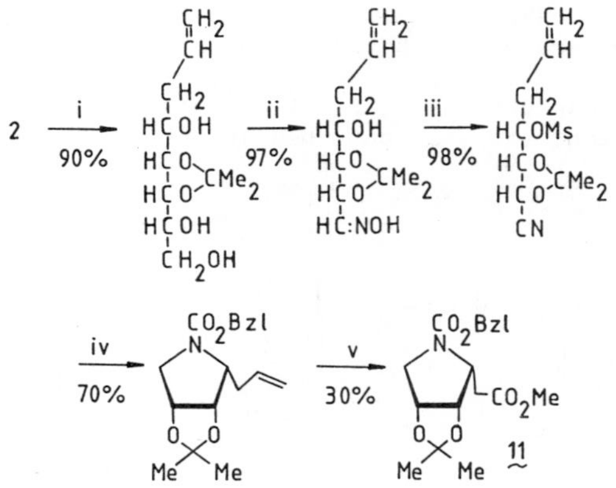

$$CH_2$$
$$\parallel$$
$$CH$$

2 $\xrightarrow{i}$ 90%

CH_2	CH_2	CH_2
CH	CH	CH

i, All$_2$Zn
ii, NaIO$_4$, then HONH$_3$Cl-C$_5$H$_5$N
iii, MsCl-C$_5$H$_5$N

CH_2
HCOH
HCO$\diagdown$CMe$_2$
HCO$\diagup$
HCOH
CH$_2$OH

$\xrightarrow{ii}$ 97%

CH_2
HCOH
HCO$\diagdown$CMe$_2$
HCO$\diagup$
HC:NOH

$\xrightarrow{iii}$ 98%

CH_2
HCOMs
HCO$\diagdown$CMe$_2$
HCO$\diagup$
CN

$\xrightarrow{iv}$ 70%

CO$_2$Bzl
N
(pyrrolidine ring with vinyl, O-O-CMe$_2$)
Me Me

$\xrightarrow{v}$ 30%

CO$_2$Bzl
N
(pyrrolidine ring with CO$_2$Me, O-O-CMe$_2$)
Me Me 11

Reagents: i, All$_2$Zn; ii, NaIO$_4$, then HONH$_3$Cl-C$_5$H$_5$N;
iii, MsCl-C$_5$H$_5$N; iv, LiAlH$_4$, then BzlOCOCl;
v, NaIO$_4$-KMnO$_4$, then CH$_2$N$_2$.

Scheme 2

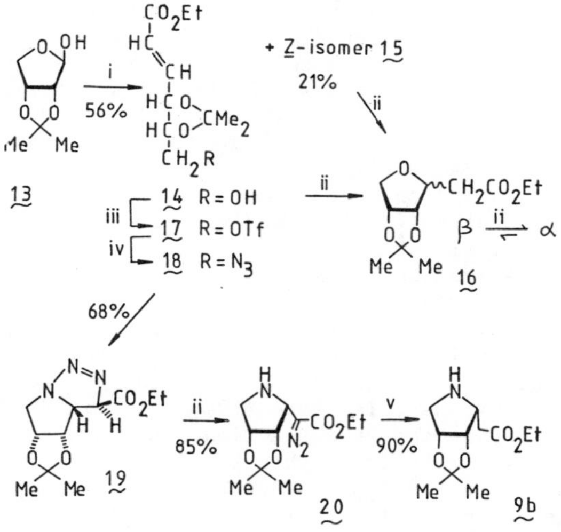

13

$\xrightarrow{i}$ 56%

CO$_2$Et
HC
CH
HCO$\diagdown$CMe$_2$
HCO$\diagup$
CH$_2$R

+ Z-isomer 15
21%

14 R=OH
17 R=OTf
18 R=N$_3$

iii
iv

$\xrightarrow{ii}$

(furanose ring) CH$_2$CO$_2$Et
O-O
Me Me 16 $\beta \rightleftharpoons \alpha$ (ii)

68%

(triazoline ring) CO$_2$Et
N=N
N
H H
O-O
Me Me 19

$\xrightarrow{ii}$ 85%

H
N
(pyrrolidine) CO$_2$Et
N$_2$
O-O
Me Me 20

$\xrightarrow{v}$ 90%

H
N
(pyrrolidine) CO$_2$Et
O-O
Me Me 9b

Reagents: i, Ph$_3$P=CHCO$_2$Et-C$_6$H$_6$, 80°C; ii, NaOEt-EtOH;
iii, Tf$_2$O-C$_5$H$_5$N-CH$_2$Cl$_2$, -78°C; iv, KN$_3$-18 crown 6-
CH$_2$Cl$_2$, RT; v, PdC-H$_2$.

Scheme 3

are discussed in a full paper (Buchanan, J.G.; Edgar, A.R.: Hewitt, B.D. J. Chem. Soc., Perkin Trans. 1, in press). Hydrogenolysis of (22) gave the β-ester (23). A further route to the α-ester (9b) emerged when (22) was heated in boiling toluene to give the expected vinylogous urethane (24) (28). When (24) was treated with sodium cyanoborohydride under acidic conditions reduction occurred at the β-face to give ester (9b). This reduction played a part in another synthesis of (9a) which is now described (Scheme 5).

2,3-O-Isopropylidene-D-erythrose (13) was converted, via the oxime, into the cyanomethanesulfonate (25). In a Blaise reaction (29), the zinc enolate derived from methyl bromoacetate reacted with (25) to give the enamino esters (26). Cyclization was effected with 1,8-diazabicyclo[5.4.0]undec-7-ene (DBU) and the product (27) reduced with cyanoborohydride. The resulting pyrrolidine α-ester (9a) was identified by reaction with benzyl chloroformate to give the amide (11), whose structure had been rigorously assigned. The amide (11), prepared by this method, was used for the subsequent transformations.

The conversion of (11) into the Geissman-Waiss lactone is shown in Scheme 6. Acidic hydrolysis of the isopropylidene group was accompanied by lactone ring formation to give (28) in 82% yield. Deoxygenation by the Barton procedure (30) afforded the lactone (29) (90%) which was easily deprotected to give the Geissman-Waiss lactone as the hydrochloride (7), constituting a formal synthesis of (+)-retronecine (8) (8-10).

The ester (9a) contains the necessary oxygen functionality, of the correct stereochemistry, for a synthesis of crotanecine (10) (Scheme 7). Alkylation of the pyrrolidine ring nitrogen was achieved using ethyl bromoacetate, producing the diester (30) in 85% yield. Attempts to induce Dieckmann cyclization of diester (30) directly under several conditions failed, so it was converted by acidic hydrolysis, into the lactone (31). Protection of the hydroxyl group in (31) was effected as the tert-butyldimethylsilyl ether (32). When treated in toluene at room temperature with potassium ethoxide (32) underwent the Dieckmann condensation. The intermediate ketoester (33) was reduced with borohydride and the resulting diastereomeric mixture acetylated to give the diacetates (34) in 40% yield. Elimination of acetic acid from (34) (DBU) afforded the unsaturated ester (35) (70%). The ester group in (35) has been reduced, by means of diisobutylaluminium hydride, to give the protected crotanecine (36), but we have experienced great difficulty in isolating crotanecine in substance after fluoride ion deprotection. (Buchanan, J.G.; Jigajinni, V.B.; Singh, G.; Wightman, R.H. J. Chem. Soc., Perkin Trans 1, in press).

At this stage in our work, Benn and his colleagues (31) described a synthesis of crotanecine from (2S,4R)-4-hydroxyproline (37) in which the silyl ether (32) is an intermediate. The subsequent reactions are similar to our own projected synthesis, involving both (33) and (34).

We have also investigated an alternative route to anisomycin (1) from 2,3-O-isopropylidene-D-erythrose (13) using a Wittig reaction as the first step. It was argued that p-nitrophenylmethylenetriphenyl-phosphorane (38) (32) would be an ideal reagent for construction of the carbon skeleton of anisomycin (Scheme 8). It was envisaged that the p-nitro group in the alkene products [(39) and (40)] would enable

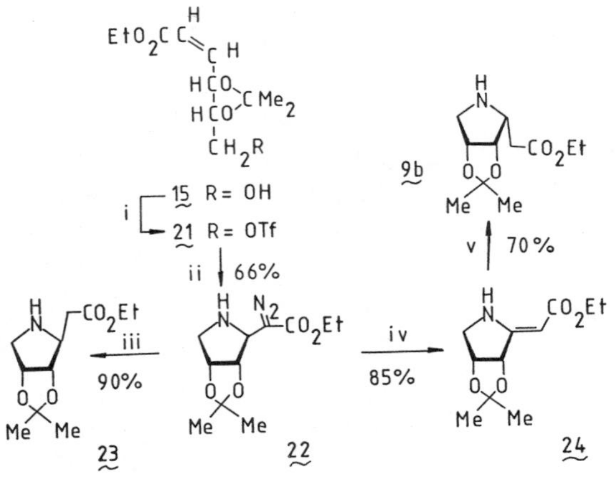

Scheme 4

Reagents: i, $Tf_2O-C_5H_5N--CH_2Cl_2$, $-78°C$;   ii, KN_3-18 crown 6-
CH_2Cl_2, RT;   iii, $PdC-H_2$;   iv, PhMe, $110°C$;
v, $NaBH_3CN$, EtOH, H^+.

Scheme 5

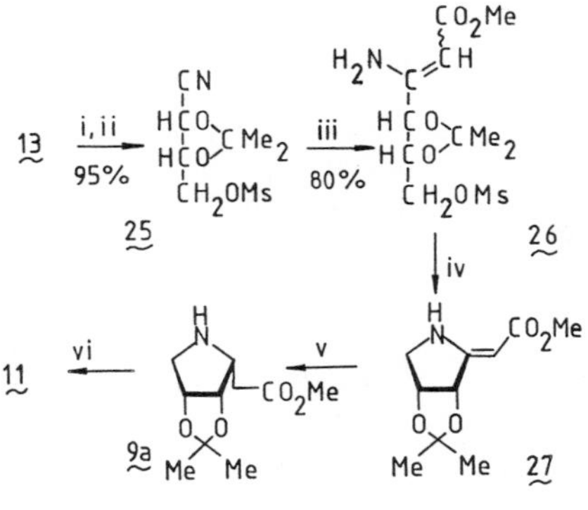

Reagents.   i, $HONH_3Cl-C_5H_5N$, RT;   ii, $MsCl$(12 equiv.)$-C_5H_5N-23°C$;
iii, Zn, $BrCH_2CO_2Me$ (5 equiv.)$-THF$, reflux;   iv, DBU
(3 equiv.)$-CH_2Cl_2$, RT;   v, $NaBH_3CN$, MeOH, HCl;
vi, $BzlOCOCl-Et_3N-CH_2Cl_2$.

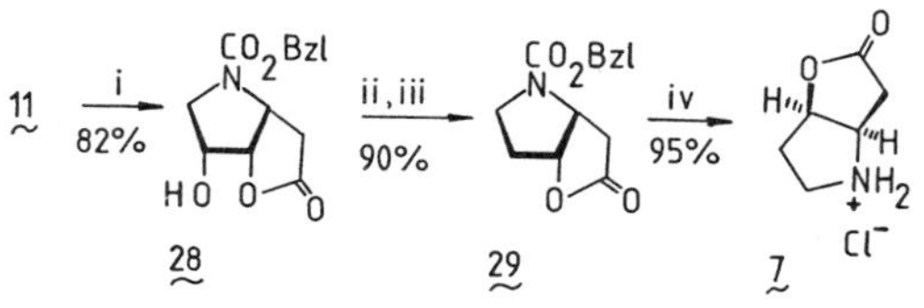

Reagents: i, 80% aq. CF_3CO_2H, RT; ii, 1,1'-thiocarbonyl-diimidazole-C_5H_5N-THF, reflux, iii, Bu_3SnH (2.2 equiv.)-C_6H_6, reflux; iv. PdC-H_2

Scheme 6

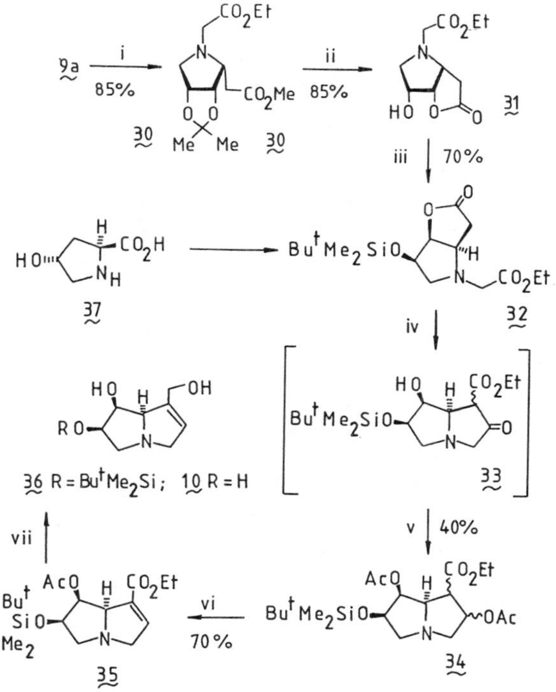

Reagents: i, $BrCH_2CO_2Et$-NEt_3-THF; ii, 80% aq. TFA, RT; iii, t-$BuMe_2SiCl$-imidazole-DMF; iv, KOEt-PhMe, RT, then HOAc; v, $NaBH_4$-EtOH, then Ac_2O-C_5H_5N; vi, DBU-CH_2Cl_2, RT; vii, DIBAL-hexane-CH_2Cl_2, -78°C.

Scheme 7

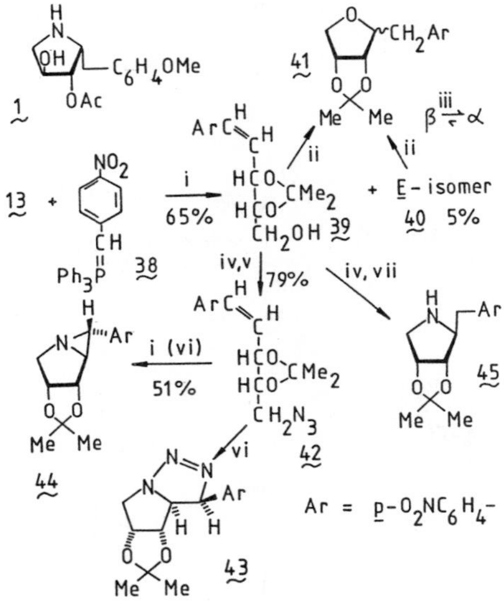

Reagents: i, C_6H_6, 80°C; ii, NaOMe-MeOH; iii, KOt-Bu-
t-BuOH; iv, $Tf_2O-C_5H_5N-CH_2Cl_2$, -78°C;
v, KN_3-18 crown 6-CH_2Cl_2, RT; vi, CH_2Cl_2, RT;
vii, NH_3 (liquid)-CH_2Cl_2, RT.

Scheme 8

conjugate addition to take place (33) and permit the formation of a pyrrolidine ring. Reaction of (38) with (13) in boiling benzene gave mainly the Z-isomer (39) (65%) together with some E-isomer (40) (5%). When each was treated with sodium methoxide in methanol ring closure to the tetrahydrofurans (41) occurred, but much more slowly than in the analogous esters (14) and (15) (Scheme 3). The β-isomer of (41) was the sole product from (39) and the major product (5:1) from (40). The two isomers of (41) could be equilibrated using potassium tert-butoxide to give a mixture favouring the α-isomer (3:1), in agreement with the ester series (19).

When the triflate of the Z-alkene (39) was treated with azide ion, the corresponding azide (42) could be isolated in 79% yield. Clearly the 1,3-cycloaddition occurs less readily than in the ester series (Schemes 3,4). Attempts to convert the azide (42) into the Δ^2-triazoline (43) were unsatisfactory. When (42) was heated in benzene solution the aziridine (44) was the major product (51%). The structure was determined by Dr K.J. McCullough by X-ray crystallography. At room temperature, dissolved in dichloromethane, the azide (42) decomposed slowly (~50% after 7 days) to give low yields of aziridine (44) and triazoline (43).

The triflate of (39) has been converted into the pyrrolidine (45) directly by treatment with ammonia. The β-configuration, inappropriate for a synthesis of anisomycin, was expected, assuming that displacement of the triflate group is the first stage in the reaction. We have so far been unable to rearrange the β-isomer (45) into the required α-series. This work is being continued.

Acknowledgments

We thank Professor M.H. Benn and Dr V. Yadav (Calgary), Professor K. Narasaka (Tokyo), and Dr C.C.J. Culvenor (Parkville, Victoria) for the provision of samples and spectroscopic data. Financial support from S.E.R.C. and from the Nuffield Foundation (One-Year Science Research Fellowship to R.H.W.) is gratefully acknowledged. We thank Dr I.H. Sadler and his staff for n.m.r. spectra obtained on the S.E.R.C. high-field facility at the University of Edinburgh.

Literature Cited

1. Buchanan, J.G.; Dunn, A.D.; Edgar, A.R. J. Chem. Soc., Perkin Trans. 1 1975, 1191-1200.
2. Buchanan, J.G.; MacLean, K.A.; Wightman, R.H.; Paulsen, H. J. Chem. Soc., Perkin Trans. 1 1985, 1463-1470.
3. Verheyden, J.P.H.; Richardson, A.C.; Bhatt, R.S., Grant, B.D.; Fitch, W.L.; Moffatt, J.G. Pure Appl. Chem. 1978, 50, 1363-1383.
4. Schumacher, D.P.; Hall, S.S. J.Am. Chem. Soc. 1982, 104, 6076-6080.
5. Iida, H.; Yamazaki, N.; Kibayashi, C.J. Org. Chem. 1986, 51, 1069-1073.
6. Bartlett, P.A. Tetrahedron 1980, 36, 2-72.
7. Gateau, A., Sepulchre, A.-M.; Gaudemer, A.; Gero, S.D. Carbohydr. Res. 1972, 24, 474 478.
8. Geissman, T.A.; Waiss, A.C., Jr. J. Org. Chem. 1962, 27, 139-142.

9. Benn, M., Rueger, H. Heterocycles 1982, 19, 23–25; 1983, 20, 1331–1334.
10. Narasaka, K.; Sakakura, T.; Uchimaru, T.; Guedin-Vuong, D. J. Am. Chem. Soc. 1984, 106, 2954–2961.
11. Gurjar, M.K.; Patil, V.J. Indian J. Chem. 1985, 24B 1282–1283.
12. Buchanan, J.G.; Singh, G.; Wightman, R.H. J. Chem. Soc., Chem. Commun. 1984, 1299–1300.
13. Atal, C.K., Kapur, K.K.; Culvenor, C.C.J.; Smith, L.W. Tetrahedron Lett. 1966, 537–544.
14. Fronza, G.; Fuganti, C.; Grasselli, P.; Pedrocchi-Fantoni, G. Zirotti, C. Tetrahedron Lett. 1982, 23, 4143–4146.
15. Ballou, C.E. J. Am. Chem. Soc. 1957, 79, 165–166.
16. Barker, R.; MacDonald, D.L. J. Am. Chem. Soc. 1960, 82, 2301–2303.
17. Gelas, J.; Horton, D. Carbohydr. Res. 1975, 45, 181–195.
18. Collins, P.M.; Overend, W.G.; Shing, T. J. Chem. Soc., Chem. Commun. 1981, 1139–1140; 1982, 297–298.
19. Ohrui, H.; Jones, G.H., Moffatt, J.G.; Maddox, M.L.; Christensen A.T., Byram, S.K. J. Am. Chem. Soc. 1975, 97, 4602–4613.
20. Buchanan, J.G.; Edgar, A.R.; Power, M.J.; Theaker, P.D. Carbohydr. Res. 1974, 38, C22–C24.
21. Logothetis, A.L. J. Am. Chem. Soc. 1965, 87, 749–754.
22. Padwa, A. Angew. Chem., Int. Ed. Engl. 1976, 15, 123–136.
23. Sundberg, R.J.; Pearce, B.C. J. Org. Chem. 1982, 47, 725–730.
24. Kadaba, P.K.; Stanovnik, B.; Tišler, M. Adv. Heterocycl. Chem. 1984, 37, 217–349.
25. Farrington, A.; Hough, L. Carbohydr. Res. 1974, 38, 107–115.
26. Huisgen, R.; Szeimies, G., Möbius, L. Chem. Ber. 1966, 99, 475–490.
27. Huisgen, R. Angew. Chem., Int. Ed. Engl. 1963, 2, 633–645.
28. Szeimies, G.; Huisgen, R. Chem. Ber. 1966, 99, 491–503.
29. Hannick, S.M.; Kishi, Y. J. Org. Chem. 1983, 48, 3833–3835.
30. Barton, D.H.R.; McCombie, S.W. J. Chem. Soc., Perkin Trans. 1 1975, 1574–1585.
31. Yadav, V.K., Rueger, H.; Benn M.H. Heterocycles 1984, 22, 2735–2738.
32. McDonald, R.N.; Campbell, T.W. J. Org. Chem. 1959, 24, 1969–1975.
33. Dale, W.J.; Strobel, C.W. J. Am. Chem. Soc. 1954, 76, 6172–6174.

RECEIVED May 31, 1988

Chapter 7

Dimethyl(methylthio)sulfonium Triflate as a Promoter for Creating Glycosidic Linkages in Oligosaccharide Synthesis

Fredrik Andersson, Winnie Birberg, Peter Fügedi[1],
Per J. Garegg[2], Mina Nashed[3], and Åke Pilotti

Department of Organic Chemistry, Arrhenius Laboratory, University
of Stockholm, S–106 91 Stockholm, Sweden

Dimethyl(methylthio)sulfonium triflate has been evaluated as a promoter in glycoside synthesis employing thioglycosides as glycosyl donors. Those having a participating group in the 2-position produce 1,2-_trans_-glycosides with high stereoselectivity. A nonparticipating group in the 2-position usually gives a preponderance of 1,2-_cis_-glycosides. The addition of tetraalkylammonium bromide to the latter reaction mixtures (nonparticipating benzyl group in the 2-position) transforms the reaction into a halide-assisted one with enhanced selectivity for 1,2-_cis_-glycosidation. These novel methods makes it possible to construct oligosaccharides with the 1-position protected as a stable alkylthio (arylthio) glycoside. When required, the thioglycoside is activated with the foregoing thiophilic reagent, and the oligosaccharide is joined to the desired acceptor.

Oligosaccharides, either free or covalently linked to such other substances as lipids or proteins are involved in a host of biological interaction processes in which they act as carriers of biological information. Their efficient synthesis is therefore a matter of some importance. Two main problems have to be dealt with. One is that of devising a protective-group strategy that allows chemical reaction at the desired hydroxyl group(s) while leaving the others intact. The second one is that of stereospecific glycosylation. This paper will address the second one of these two topics. The subject of oligosaccharide synthesis has been extensively reviewed (1–12).

[1]Current address: Institute of Biochemistry, L. Kossuth University, P.O. Box 55, H-Debrecen, Hungary
[2]Address correspondence to this author.
[3]Current address: Faculty of Science, Chemistry Department, Alexandria University, Alexandria, Egypt

In order to make a 1,2-cis-glycoside carrying an amino or hydroxyl function at C-2, a nonparticipating substituent at C-2 is mandatory. On the other hand, the synthesis of a 1,2-trans-glycoside usually requires an efficient participating C-2 substituent that directs the incoming nucleophile to approach from the opposite side of the furanose or pyranose ring.

The classical glycosyl donors are glycosyl halides. Those having a nonparticipating C-2 substituent usually lead preponderantly to 1,2-cis-glycosides, and those having a participating C-2 substituent give 1,2-trans-glycosides.

The use of a block strategy, rather than adding one monomer at a time, is highly advantageous in the synthesis of large oligosaccharides. The main reason for this is that most of the protecting-group manipulations in a block synthesis are performed on small rather than on large fragments. One disadvantage of this approach is that it normally requires the conversion of large oligosaccharides into glycosyl halides for use as glycosyl donors in further condensations; this may lead to severe difficulties and result in diminished yields (13).

Thioglycosides have recently been reintroduced as potential glycosyl donors. Their advantage lie in the fact that they are stable to normal protecting-group manipulations encountered in oligosaccharide synthesis, although they may be selectively activated to become glycosyl donors (14-20). This is illustrated in Scheme 1. The branched heptasaccharide 1 corresponds to the glucan fragment responsible for triggering the defense of the soybean to infections by the mould Phytophthora megasperma. In our original synthesis of this heptasaccharide (13), we ran into the problem of having to convert a tetrasaccharide into a glycosyl halide. Extensive experimentation did not permit us to raise the yield in that step above 24%. Gram quantities of the heptasaccharide were required for the continued phytochemical investigations, and our original synthesis was clearly not suitable for that purpose. Recourse was therefore taken to the use of thioglycosides and a novel glycosylation procedure recently developed by Lönn in this laboratory (20,21).

The synthesis was based upon the following key methods: (a) the use of glucosyl bromides carrying participating benzoyl groups (22,23) in the 2-position and silver triflate-promoted glycosylation (24) for making the smaller fragments; (b) the use of regio-selective reductive cleavage of benzylidene acetals (25); and (c) the use of thioglycosides activated by methyl triflate in glycosylation reactions involving large blocks. The synthesis is outlined in Scheme 1 (26).

2,3,4,6-Tetra-O-benzoyl-α-D-glucopyranosyl bromide (27) (1 molar equivalent) was condensed with methyl 4,6-O-benzylidene-1-thio-β-D-glucopyranoside, m.p. 184–185°, $[\alpha]_D$ –49° (c 0.6, $CHCl_3$), in dichloromethane in the presence of silver triflate to yield 55% of the 1,3-β-linked disaccharide 2, $[\alpha]_D$ +6° (c 1.1, $CHCl_3$). This route was preferable to one involving initial

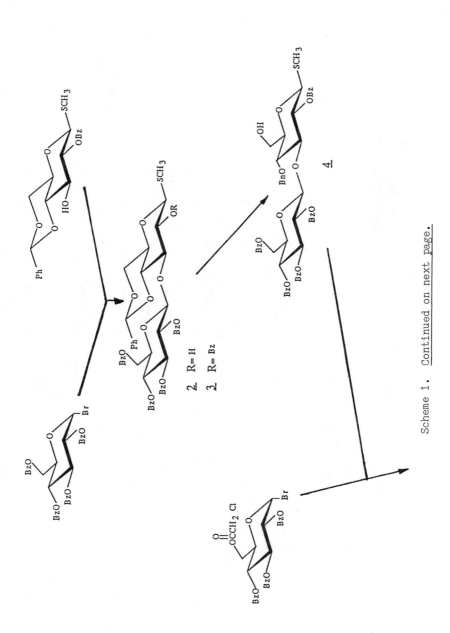

2 R = H
3 R = Bz

Scheme 1. Continued on next page.

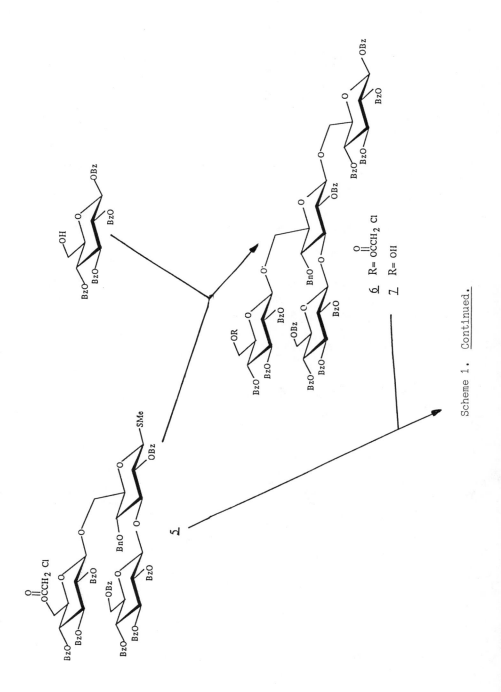

Scheme 1. Continued.

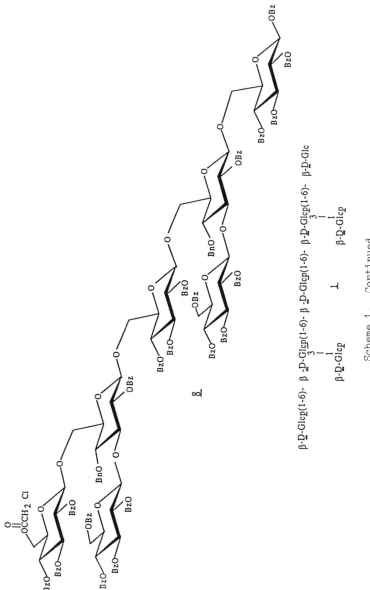

β-D-Glcp(1-6)- β-D-Glcp(1-6)- β-D-Glcp(1-6)- β-D-Glcp(1-6)- β-D-Glcp(1-6)- β-D-Glc
                        3   3
                        | |
                   β-D-Glcp   β-D-Glcp

1

8

Scheme 1. Continued.

protection by partial benzoylation of the 2-hydroxyl group in the glycosyl acceptor. The remaining free hydroxyl group in the 2-position of 2 was benzoylated. Treatment of the product 3, m.p. 228-229°, $[\alpha]_D$ +15° ($\underline{c}$ 1.1, CHCl$_3$), with borane— trimethylamine and aluminum chloride ($\underline{25}$) in dichloromethane afforded 4 having a free hydroxyl group in the 6-position in 90% yield; $[\alpha]_D$ +5° ($\underline{c}$ 1.2, CHCl$_3$).

1,2,3,4-Tetra-$\underline{O}$-benzoyl-β-D-glucopyranose, m.p. 179-181°, $[\alpha]_D$ -15° ($\underline{c}$ 1.7, CHCl$_3$), was chloroacetylated ($\underline{28}$) and then converted into the α bromide in near-quantitative yield by treatment with hydrogen bromide in acetic acid. Condensation of this glycosyl bromide with 4 in dichloromethane using silver triflate as promoter yielded the key compound 5 in 83% yield; $[\alpha]_D$ -6° ($\underline{c}$ 0.9, CHCl$_3$). This was the glycosyl donor in the final condensation. It was also the precursor for the glycosyl acceptor, obtained as follows:

Condensation of 5 with 1,2,3,4-tetra-$\underline{O}$-benzoyl-β-D-gluco-pyranose in dichloromethane using methyl triflate as promoter afforded tetrasaccharide 6 in 83% yield; $[\alpha]_D$ -18° ($\underline{c}$ 1.0, CHCl$_3$). The chloroacetyl group in 6 was removed by treatment with hydrazine dithiocarbonate in acetonitrile to give 7 in 88% yield $[\alpha]_D$ -16° ($\underline{c}$ 1.5, CHCl$_3$).

The final condensation of the glycosyl donor 5 with the acceptor 7 in dichloromethane, again using methyl triflate as promoter, gave the protected heptasaccharide in 93% yield; $[\alpha]_D$ -20° ($\underline{c}$ 1.2, CHCl$_3$). This was then deprotected by hydrogenolysis and debenzoylation to give the heptasaccharide 1, which had optical rotation, n.m.r. spectra, l.c. retention time and phyto-elicitor activity identical to the material previously synthesized (13,29).

The foregoing synthesis clearly demonstrates the utility of thioglycosides for the block synthesis of oligosaccharides. However, in other work in this laboratory it was found that, quite apart from the health hazard involved in the routine handling of methyl triflate as a glycosylation promoter, other disadvantages may be encountered, namely, elimination reactions leading to 2-hydroxy-glycal derivatives as byproducts, and the possibility of competing methylation of the hydroxyl group(s) in the glycosyl acceptor, expected to be more severe with unreactive glycosyl donors. Attention was therefore turned to more-specifically thiophilic reagents.

Regiospecific activation of a sulfur center should require the use of a "soft" Lewis acid. Dimethyl(methylthio)sulfonium tetra-fluoroboronate ($\underline{30}$) reacts rapidly with dimethyl sulfide ($\underline{31}$) and it has many synthetic applications ($\underline{32,33}$). In pilot experiments, it was indeed effective in activating alkyl 1-thioglycosides, but the yields in glycosylation were moderate. Triflates have been found most useful in glycosylation reactions ($\underline{22-24}$, $\underline{34-36}$) and the anion was therefore replaced by triflate. Dimethyl(methylthio)sulfonium triflate (DMTST) ($\underline{37}$) was thereby found to be a most efficient promoter in glycosylation reactions.

The result of a number of model experiments testing the usefulness of DMTST as a promoter in 1,2-trans-glycosylation reactions involving the glycosylation of hydroxyl groups in the 2-,3-,4- and 6-positions of suitable hexopyranoside glycosyl acceptors is shown in Scheme 2 (38). In these experiments, DMTST (4—5 equivalents) was added to a mixture of the thioglycoside (1.2 equivalents) and the glycosyl acceptor (1 equivalent) in dichloromethane containing 4Å molecular sieves at room temperature. Yields of isolated, pure compounds ($\left[\alpha\right]_D$ in CHCl$_3$) are given in Scheme 2. In accordance with previous findings, a participating benzoyl group in the 2-position of the glycosyl donor gives higher yields than does an acetyl group. The use of an N-phthaloyl group in the 2-position affords an efficient synthesis of a protected 2-amino-2-deoxy-β-D-glucosylated product. The yields are uniformly high. This glycosylation procedure is potentially more advantageous than existing ones for the synthesis of 1,2-trans-glycosides.

The usefulness of the method is further illustrated in the following two examples of unpublished syntheses by other workers in this laboratory. In a synthesis of a trisaccharide fragment corresponding to the repeating unit in a fucolipid accumulating in human colonic and liver adenocarcinoma, the thioglycoside 25 was to be condensed with the acceptor 26 (Scheme 3). In this instance, the use of methyl triflate produced a complicated mixture containing considerable quantities of the 1,2-elimination product from 25. In another experiment, 25 was treated with bromine to give the corresponding α-bromide, which was then condensed with 26 in the presence of silver triflate. Only moderate yields of the product 27 was obtained. However, under the foregoing conditions, DMTST gave an acceptable yield of the product 27 (39).

In the other example (Scheme 4), involving the simultaneous glycosylation of two hydroxyl groups, (compound 29) an α-mannosylation involving a glycosyl donor (28) having a non-participating benzyl group in the axial 2-position was attempted. The synthesis illustrates the compatibility of the promoter with the presence of a phosphoric triester group in the glycosyl donor. In this particular example, DMTST was generated in situ in the mixture. The deprotected pentasaccharide provides a model for studies of mannose 6-phosphate-specific receptors and the uptake by fibroblasts (40).

The mannosylation experiment raises the question of the use of DMTST as a general promoter for the synthesis of 1,2-cis--glycosides using thioglycosyl donors with a nonparticipating (such as benzyl) group in the 2-position. Thus the DMTST (4—5 molar equivalents) was added to a mixture of the thioglycoside donor 31 (41) (1.2 equivalents) and the glycosyl acceptors 10 (42), 16 (26), and 34 (43) (1 equivalent) in dichloromethane containing 4Å molecular sieves at room temperature. Yields of isolated product, separated by means of column chromatography on silica gel, are

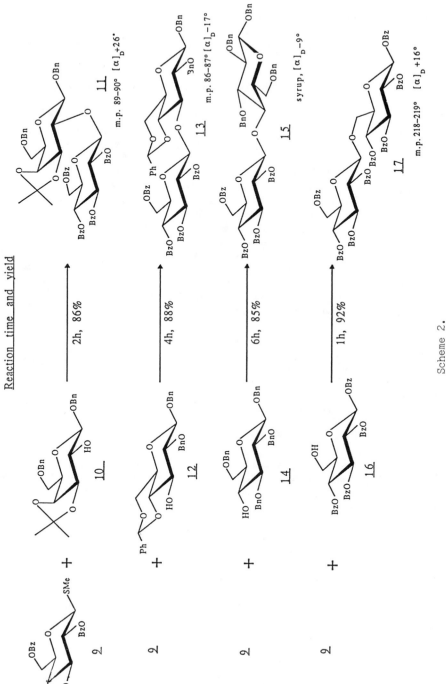

Reaction time and yield

Scheme 2.

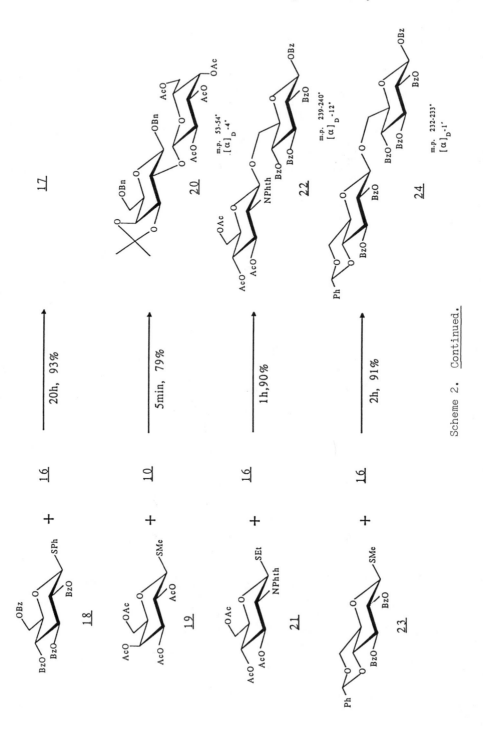

Scheme 2. Continued.

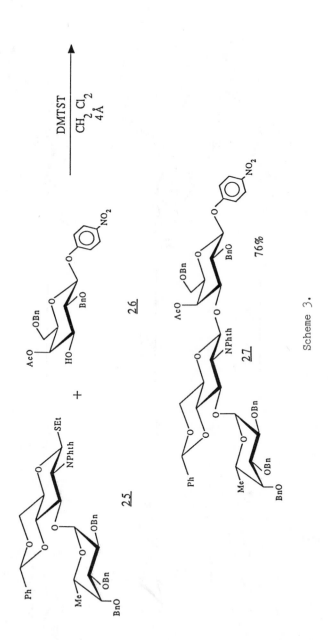

Scheme 3.

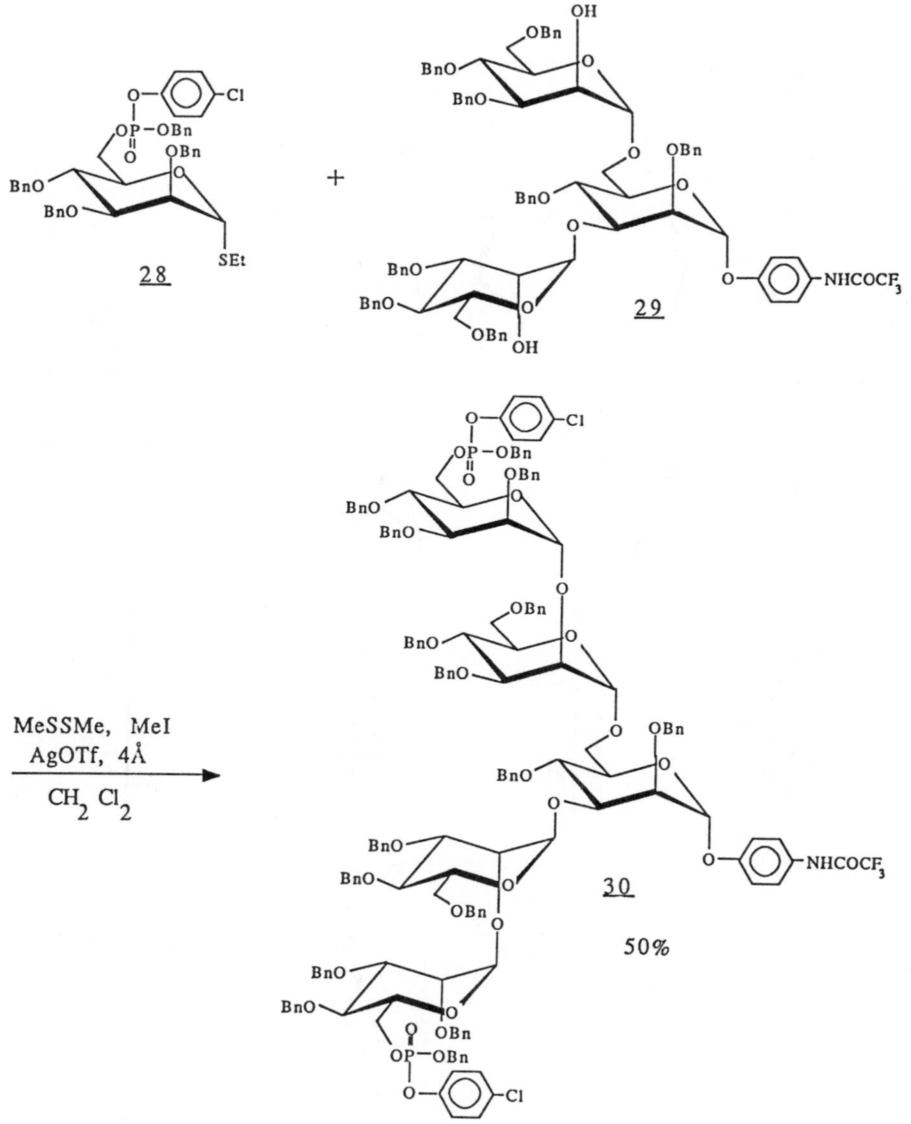

Scheme 4.

shown in Scheme 5. The overall yields are good, but not the stereospecificity, which is similar to that found in this solvent using methyl triflate as promoter (21) and to that normally encountered using silver triflate as promoter in glycosidation reactions with 2-O-benzylglycosyl bromides as donors. A most stereoselective, albeit slow, 1,2-cis-glycosylation procedure is the halide-assisted method (44). In order to see if DMTST could be

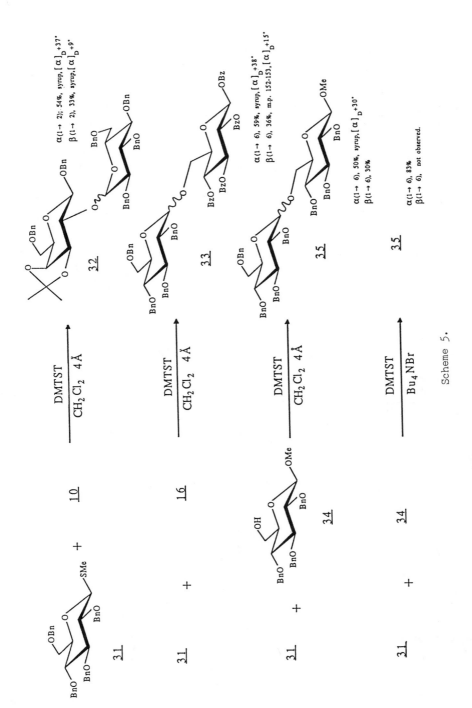

Scheme 5.

used in conjunction with 2-O-benzylated thioglycosides for the in situ generation of a halide-assisted reaction, the condensation of thioglycoside 31 with the acceptor 34 activated by DMTST was repeated in the presence of 5 equivalents of tetrabutylammonium bromide. The reaction was monitored by t.l.c., which showed that the thioglycoside immediately disappeared from the mixture and gave rise to 2,3,4,6-tetra-O-benzyl-α-D-glycopyranosyl bromide, which then effected a slow, but stereospecific, glycosylation of 34, giving an 85% yield of the α-linked product 35. In a control experiment, the same glycosyl bromide was treated with the acceptor 34 under halide-assisted conditions (excluding DMTST). A similar reaction-rate and stereospecificity was observed; the product 35 was isolated in 83% yield (45).

These final model experiments indicate that thioglycosides may be used in the block synthesis of 1,2-cis-linked glycosides with DMTST as promoter. With unreactive acceptors it may be advisable to refrain from transforming the reaction into a halide-assisted one by adding halide ion to the mixture. When the acceptor reactivity is sufficiently high, however, the presence of halide ion may significantly improve the stereospecificity and the yield of the desired product.

ACKNOWLEDGMENTS

This work was supported by the Swedish National Board for Technical Development and by the Swedish Natural Science Research Council.

REFERENCES

1. Kochetkov, N.K.; Chizhov, O.S.; Bochkov, A.F. MTP Internat. Rev. Sci. Org. Chem. Ser. 1, 1973, 7, Carbohydrates, 147.
2. Bochkov, A.F.; Zaikov, G.E. Chemistry of the O-Glycosidic Bond. Formation and Cleavage. Pergamon Press 1979.
3. Wulff, G.; Röhle, G. Angew. Chem. Int. Ed. Engl., 1974, 13,157.
4. Flowers, H. Methods Enzymol., 1978, 50, 93.
5. Lemieux, R.U. Chem. Soc. Revs., 1978, 7, 423.
6. Paulsen, H. Angew. Chem. Int. Ed. Engl., 1982, 21, 155.
7. Overend, W.G., in Pigman, W. and Horton, D. The Carbohydrates, Chemistry and Biochemistry, Second Ed., 1972, 1A, 279.
8. Pazur, J.H., in Pigman, W. and Horton, D. The Carbohydrates, Chemistry and Biochemistry, Second Ed. 1970, IIA, 69.
9. Igarashi, K. Adv. Carbohydr. Chem. Biochem., 1977, 34, 243.
10. Sinaÿ, P. Pure Appl. Chem., 1978, 50, 1437.
11. Paulsen, H. Chem. Soc. Reviews, 1984, 13, 15.
12. Schmidt, R.R. Angew. Chem. Int. Ed. Engl., 1986, 25, 212.
13. Ossowski,P.; Pilotti, Å.; Garegg, P.J.; Lindberg, B. Angew. Chem. Int. Ed. Engl., 1983, 22, 793; J. Biol. Chem., 1984, 259,11337.
14. Ferrier, R.J.; Hay, R.W.; Vethiviyasar, N. Carbohydr. Res., 1973, 27, 55.
15. Van Cleve, J.W. Carbohydr. Res., 1979, 70, 161.

16. Hanessian, S.; Bacquet, C.; Lehong, N. Carbohydr. Res., 1980, 80, c17.

17. Garegg, P.J.; Henrichson, C.; Norberg, T. Carbohydr. Res., 1983, 116, 162.

18. Nicolaou, K.C.; Seitz, S.P.; Papahatjis, D.P. J. Am. Chem. Soc., 1983, 105, 2430.

19. Nicolaou, K.C.; Dolle, R.E.; Papahatjis, D.P.; Randall, J.L., J. Am. Chem. Soc., 1984, 106, 4189.

20. Lönn, H. Carbohydr. Res., 1985, 139, 105, 115.

21. Lönn, H. Chem. Commun. Univ. Stockholm, No 2, 1984.

22. Garegg, P.J.; Norberg, T. Acta Chem. Scand., 1979, B33, 116.

23. Garegg, P.J.; Konradsson, P.; Kvarnström, I.; Norberg, T.; Svensson, S.C.T.; Wigilius, B. Acta Chem. Scand., 1985, B39, 569.

24. Hanessian, S.; Banoub, J. Carbohydr. Res., 1977, 53, C13.

25. Ek, M.; Garegg, P.J.; Hultberg, H.; Oscarson, S. J. Carbohydr. Chem., 1983, 2, 305.

26. Fügedi, P.; Birberg, W.; Garegg, P.J.; Pilotti, Å. Carbohydr. Res., 1987, 164, 297.

27. Ness, R.K.; Fletcher, H.G. Jr; Hudson, C.S. J. Am. Chem. Soc., 1950, 72, 2200.

28. Heran, N.; Utille, J.P.; Vottero, P.J.A. Carbohydr. Res., 1977, 53, 268.

29. Sharp, J.; Albersheim, P.; Ossowski, P.; Pilotti, Å.; Garegg, P.J.; Lindberg, B. J. Biol. Chem., 1984, 259, 11341.

30. Meerwein, H.; Zenner, K.-F, Gipp, R., Justus Liebigs Ann. Chem., 1965, 688, 67.

31. Smallcombe, S.H.; Caserio, M.C. J. Am. Chem. Soc., 1971, 93, 5826.

32. Trost, B.M.; Sato, T. J. Am. Chem. Soc., 1985, 107, 719.

33. Trost, B.M.; Martin, S.J. J. Am. Chem. Soc., 1984, 106, 4263 and references therein.

34. Lemieux, R.U.; Takeda, T.; Chung, B.Y. in El Khadem, H.S. (Ed.) Synthetic Methods for Carbohydrates, ACS Symp. Ser., 1974, 39.

35. Lemieux, R.U.; Ratcliffe, R.M.; Arreguin, B.; de Vivar, A.R.; Castillo, M.J. Carbohydr. Res., 1977, 55, 113.

36. Lemieux, R.U.; Ratcliffe, R.M. Can. J. Chem., 1979, 57, 1244.

37. Ravenscroft, M.; Roberts, R.M.G.; Tillett, J.G. J. Chem. Soc. Perkin Trans. II, 1982, 1569.

38. Fügedi, P.; Garegg, P.J. Carbohydr. Res., 1986, 144, c9.

39. Helland, L.; Ottosson, H.; Lönngren, J., To be published.

40. Ottosson, H.; Lönngren, J., To be published.

41. Weygand, F.; Ziemann, H. Ann., 1962, 657, 179.

42. David, S.; Thieffrey, A.; Veyrières, A. J. Chem. Soc. Perkin Trans. I, 1981, 1796.

43. Keglević, D.; Ljevaković, D. Carbohydr. Res., 1978, 64, 319.

44. Lemieux, R.U.; Hendricks, K.B.; Stick, R.V.; James, K. J. Am. Chem. Soc., 1975 97, 4056.

45. Andersson, F.; Fügedi, P.; Garegg, P. J.; Nashed, M. Tetrahedron Lett., 1986, 27, 3919.

RECEIVED May 31, 1988

Chapter 8

Approaches to Deoxy Oligosaccharides of Antibiotics and Cytostatics by Stereoselective Glycosylations

Joachim Thiem

Organisch-Chemisches Institut, Westfälische Wilhelms-Universität, Orléans-Ring 23, D–4400 Münster, Federal Republic of Germany

Problems associated with stereoselective glycosylations in the 2-deoxy sugar series are outlined and exemplified. General solutions are provided for the 2-deoxy-α-glycosides employing the $\underline{N}$-iodosuccinimide glycosylation. The 2-deoxy-β-glycosides are available via certain 2-bromo-2-deoxyglycosyl bromides accessible from simple isopropylidene derivatives by dibromomethyl methyl ether reactions. Syntheses of four different E-D-C trisaccharides of the various aureolic acids are reported that make extensive use of these procedures. A novel approach for the preparation of the C-B-A trisaccharide glycoside of class II anthracyclines is described. Here the combined azide --$\underline{N}$-iodosuccinimide procedure is applied to construct a precursor of the aminodeoxy sugar unit. After subsequent $\underline{N}$-iodosuccinimide glycosylations, the precursor trisaccharide is converted straightforwardly into the target molecule.

The basic concepts of glycosylation have been known for more than eighty years. Nevertheless, the selective formation of a full acetal constitutes and remains one of the major challenges in carbohydrate chemistry. Within the current decade, a number of attractive approaches for the glycosylation of simple alcohols and also more-complex aglycons (including sugar derivatives) have been developed (e.g. $\underline{1},\underline{2}$). In all cases a high stereoselec-

0097–6156/89/0386–0131$06.00/0

tivity is desired, and as has been abundantly shown, the
simple transfer of a procedure worked out for a certain
sugar series does not necessarily apply to another, isomeric series. All of the presently used and effective
glycosylation procedures in the normal sugar series in
general make use of a neighboring group at the position
next to the anomeric center, be it by a real anchimeric
assistance or by the operation of steric influence.

The present contribution centers on stereoselective
syntheses of mono- and in particular oligo-saccharides of
the 2-deoxy- and the 2,6-dideoxy series as well as some
branched-chain species. These are the principal sugar
portions in a large number of important natural glycosides, such as the cardiac glycosides (3), the orthosomy-
cins (4), the tetronic acids (5), the aureolic acids (6,
see below), and the anthracyclines (7, see below), to
name just a few.

Nature's approach to the 6-deoxy sugars is known to
involve oxidation of a nucleotide-sugar glycoside with
oxidoreductase and subsequent reduction with NADPH (8).
The introduction of methylbranches was shown (9) to proceed at the stage of the keto sugar intermediates under
the action of "active methionine" [S-(5'-adenosyl)methio-
nine]. There is little known about the formation of the
2-deoxyfunction. The transfer of cytidine diphosphate
into 2'-deoxycytidine 5'-diphosphate is catalyzed by CDP
reductase in Escherichia coli (10). Also described is the
formation of 2-deoxy-D-erythro-pentofuranose 5'-phosphate
catalyzed by an aldolase in Lactobacillus plantarum (10).
The detailed mechanism is unknown, and until now there
have been no studies in the 2-deoxyhexose series (H.G.
Floss, personal communication 1986).

Evidently the particular problems in the chemical
synthesis of 2-deoxy sugar glycosides are the missing
neighboring group, and are also associated with the enhanced lability of their glycosyl halides. For instance,
treatment of the α- or the β-glycosyl halide (the former
beeing slightly more stable because of the anomeric
effect) in the 2,6-dideoxy-D-arabino series 1 or 2 with
an alcohol in the presence of a silver promoter is supposed to proceed via the oxocarbenium intermediate 4.
After nucleophilic attack of the alcohol, both of the
protonated precursors are formed, which after release of
the proton give the α- and the β-glycosides 5 and 8.
Mostly the former prevails, probably again by the operation of an anomeric effect. Another frequent stabilization of 4 in this series occurs by direct deprotonation
to give the cyclic enol ether (glycal) 7. These glycals
bearing a leaving group at the allylic position (carbon
3) can easily undergo acid-catalyzed glycosylation with

allylic rearrangement (Ferrier reaction, 11,12) and give
the alkyl α- and β-D-erythro-hex-2-enopyranosides 3 and
6. It is not attractive to refunctionalize such alkenic
sugars as 3 or 6 into the required target glycosides such
as 5 or 8.
      This general overview may be illustrated by an ex-
ample: treatment of 3,4-di-O-acetyl-2,6-dideoxy-α-D- ara-
bino-hexopyranosyl bromide (9) with methyl 4-O-acetyl-
2,6-dideoxy-α-D-lyxo-hexopyranoside (10) under systemati-
cally developed and adjusted conditions gave in 56% optimized yield both the α-(1->3)- and the β-(1->3)-linked
disaccharides 11 and 12 in the ratio of approximately 2:1
(13,14). Apart from this delicate chemistry, at least one
separation step is required. Consequently, stereoselective or even better stereospecific solutions for the
preparation of these oligosaccharides would be certainly
appreciated.
      We have developed such methods, and these are brief-
ly outlined in general. Treatment of a glycal derivative
13 with N-halosuccinimide (X = Br,: 15; X = I, 16) sup-
posedly gives 1,2-halonium ions (perhaps in resonance
with a 2-halo-oxocarbenium ion). These are attacked by
the nucleophile R'OH to give mainly the 1,2-trans-2-halo
glycosides 14. Further reductive cleavage of the 2-halo
substituent concludes an easy, attractive and highly
stereoselective approach to 2,6-dideoxy α-glycosides 15
in the D-arabino, -lyxo, and -ribo series. As demonstrated conclusively in many examples (eg. 17) the use of NIS
(16) is considerably superior to NBS (15) with respect to
the yields of the glycosylation (13->14) as well as the
halide-cleavage step (14->15). The NIS procedure has also
been tried for the synthesis of 2-deoxy-β-glycosides.
Although this could be realized, the method proved
attractive only in particular situations (18-20).
      The most attractive approach for the 2-deoxy-β-
glycosides starts with 2-bromo-2,6-dideoxyhexopyranosyl
bromides (21) such as 16. These are available regio- and
stereo-specifically from readily accessible and simple
precursors by reaction with dibromomethyl methyl ether
(DBE) (22,23). In principle, the formation of related
compounds may be anticipated by the addition of bromine
to glycal precursors. Previous studies (24-27), however,
proved these processes to yield several isomers which
renders this approach of little preparative value.
      Treatment of the halides 16 with an alcohol promoted
by silver salts does yield mainly the β-glycosides 17
(28,29). Obviously the bromo substituent directs the
incoming nucleophile by steric reasons or possibly via a
1,2-bromonium<->bromo-oxocarbenium intermediate. Following a reduction step, the syntheses of β-glycosides 18

in the D-arabino (28,29) and -lyxo series (30) were
realized. Another approach for the D-ribo series was
demonstrated by Wiesner's group (31). Starting from the
thioglycoside 19, mercury-assisted solvolysis is considered to generate a 1,3-acyloxonium intermediate 20 which
is attacked by the alcohol principally from the β-face to
give 21. Even though there remain some inconsistencies
with respect to the anomeric ratios obtained in the
synthesis of mono- or oligo-saccharide glycosides, this
method proved useful in the selective preparation of
cardiac glycosides.
    Both procedures for selective syntheses of 2,6-
dideoxy-α- and β-glycosides will be outlined in the aureolic acid and also the anthracycline oligosaccharide
series.

Aureolic Acid Oligosaccharides

    These tetrahydroanthracenone oligosaccharide anti-
biotics received their name from their characteristic
golden appearance. The most prominent members, chromomycin A₃ (22), olivomycin A (23), and mithramycin (24)
constitute potent cytostatic agents which, even though
they are extremely toxic, enjoy selected clinical application in the treatment of certain tumors (6). The cytostatic activity is supposed to result from a strong and
selective inhibition of the DNA-dependent RNA synthesis
(32,33). In the presence of Mg2+, a complexation of
guanosine-rich DNA fragments was observed, and further
information as to this mechanism has been recently discussed (34).
    Earlier studies proved the structure of the almost
similar tetrahydrocenone aglycons in 22-24 having the
chiral, five-carbon side chain at C-3 as well as those of
the individual monosaccharides (35,36). Attempts were
made to apply Klyne's rule and deduce the structure of
the oligosaccharides and their attachment to the aglycon
but this was not uniformly convincing (37). The complete
sugar sequence and the direction and type of their interglycosidic linkages were assigned by extended n.m.r.
spectrocopy and supported by syntheses (13,14,38,39).
There are only minor deviations between 22 and 23, although they are produced by different streptomyces
strains. In both of these compounds a differently substituted α-(1->3)-linked bis-2,6-dideoxy-D-lyxo-unit B-A
is attached to the phenolic site at C-6. Their E-D-C
trisaccharide shows a terminal 3-C-methyl-branched sugar,
L-olivomycose E, attached by an α-(1->3)-linkage to the
β-(1->3)-bound dimeric olivosyl-olivose D-C. In mithramycin (24), the B-sugar is a 2,6-dideoxy-D-arabino unit
attached to A via a β-(1->3)-linkage. In the E-D-C trisaccharide part, only β-(1->3)-linkages occur. The D-C
unit is likewise similar, and the terminal sugar again is
of the 3-C-methyl-branched type, but this time it is D-
mycarose.

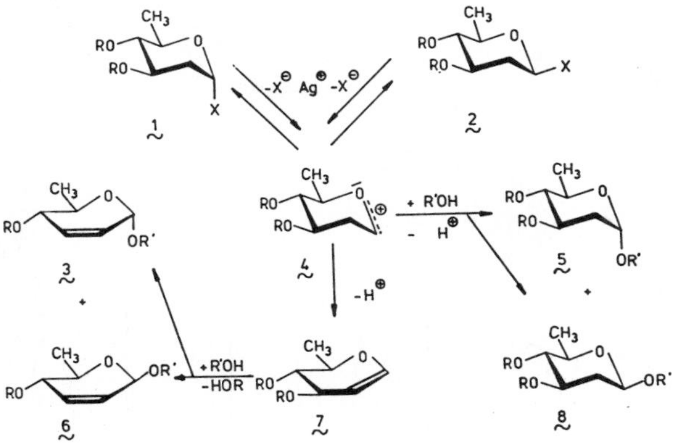
CH3
RO
RO
O
X
1
-X⊖ Ag⊕ -X⊖
2
4
+ R'OH
- H⊕
-H⊕
3
5
6
7
8
OR'
+R'OH
-HOR

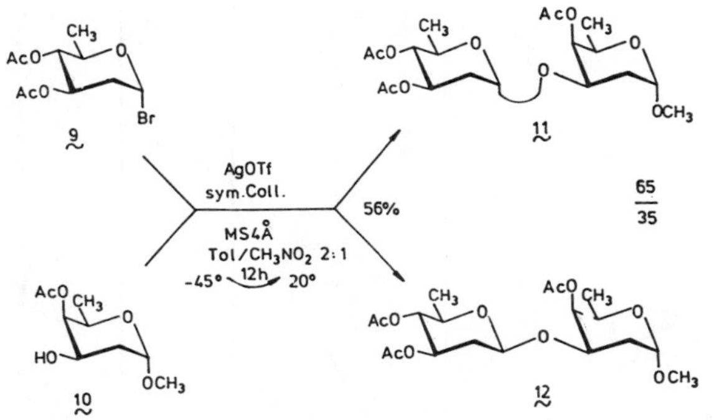
AcO
Br
9
AgOTf
sym.Coll.
MS4Å
Tol/CH3NO2 2:1
-45° 12h 20°
56%
11
65
35
10
HO
OCH3
12

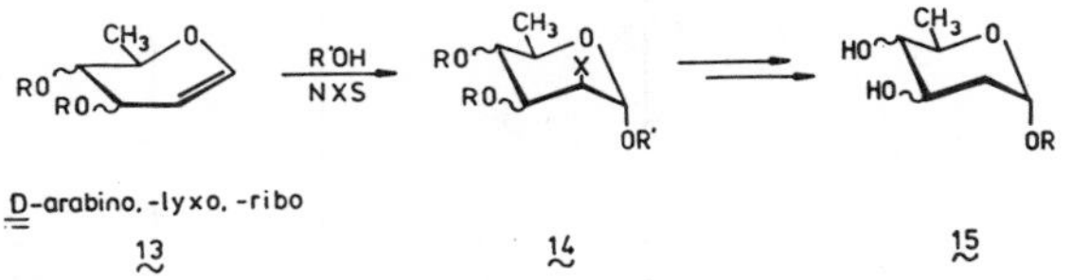
R'OH
NXS
D-arabino, -lyxo, -ribo
13
14
15

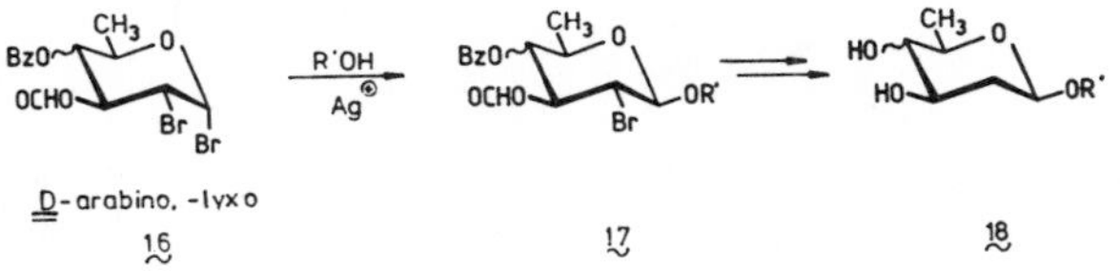
BzO
OCHO
Br
R'OH
Ag⊕
D-arabino, -lyxo
16
17
18

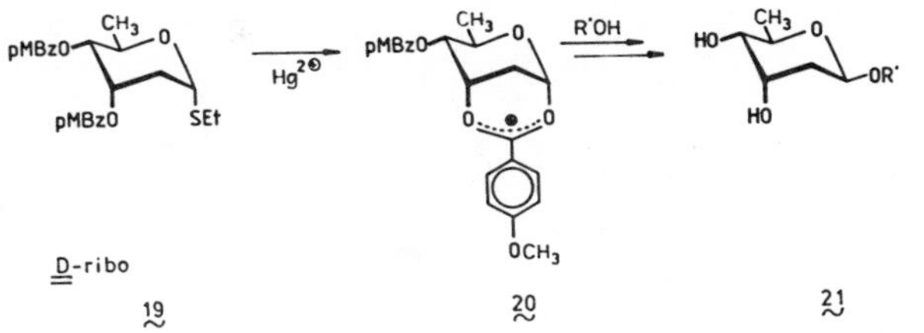
pMBzO
SEt
Hg2⊕
R'OH
OCH3
D-ribo
19
20
21

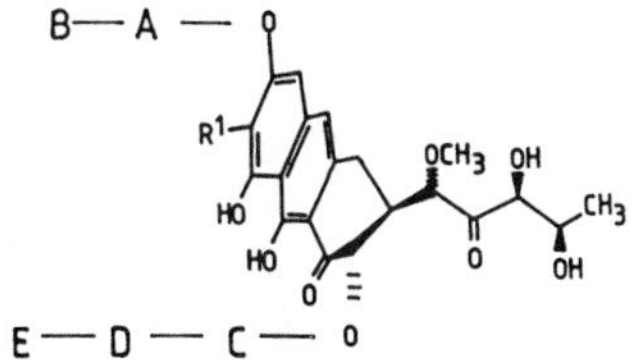

	R^1	B — A	E — D — C
Chromomycin A_3 **22**	CH_3	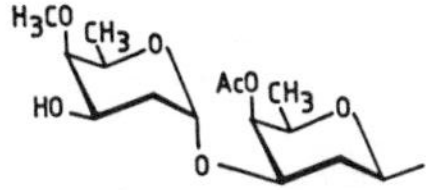	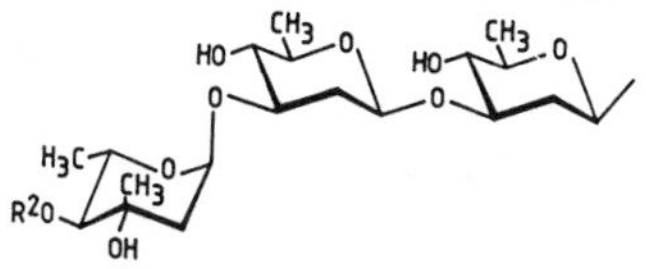 $R^2 = CH_3CO$
Olivomycin A **23**	H	—"—	—"— $R^2 = (CH_3)_2CHCO$
Mithramycin **24**	CH_3	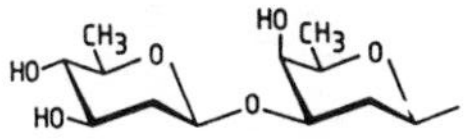	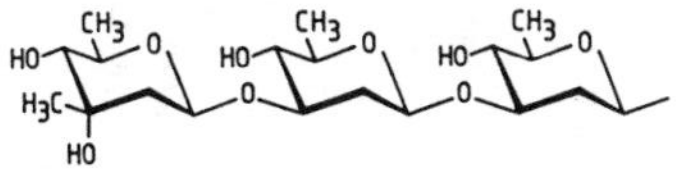

We have previously reported the preparation of the various B-A units (13,40) and have also developed syntheses for the E-D (41) as well as the uniform D-C unit (42). This presentation focuses on a sequential assembling of these complex oligosaccharide units, and discusses novel approaches to a number of E-D-C trisaccharides.

Starting with methyl α-D-mannopyranoside there is ready access to the 2,3-O-isopropylidene rhamnoside 25. Treatment of 25 in dichloromethane with DBE and zinc bromide for 5.5 hours at room temperature gives the 2,3-O-isopropylidene rhamnosyl bromide 26 as a very reactive intermediate which may prove useful in further reactions to give mannosides (29). After another 6.5 hours (12 hours altogether) the 3-O-formylated 2-bromo-2,6-dideoxy-α-D-glucopyranosyl bromide 27 is obtained in high yield. The mechanism of formation may be supposed to proceed via 2,3-formoxonium intermediates (21). Benzylglycosylation of 27 leads virtually exclusively to the crystalline glycoside 28. Treatment with hot methanol containing one drop of concentrated hydrochloric acid achieves selective cleavage of the formyl group and furnishes the crystalline monohydroxy compound 29. The glycosyl bromide 27 and the sugar aglyconic component were condensed under diligently optimized conditions with respect to the solvent mixture (4:1 toluene--nitromethane), the temperature range and rate of increase (-78° over 2 days to room temperature), and the promoter (silver triflate, 43). This resulted in a very good yield (92%) and a favourable α:β ratio of 30:31 = 1:6.5. Fortunately the desired β-(1->3)-linked disaccharide 31 could be fractionally crystallized from the mixture.

After having assembled the D-C precursor the α-(1->3)-attachment of the E unit was to be performed by the NIS method. Therefore a favorable access to the L-olivomycal 37 was needed. We had previously prepared this compound by a five step route from L-arabinose (44). Recently an advantageous preparation of 37 and its epimer L-mycaral (36) by treatment of methyl 2,3-O-benzylidene α-L-rhamnopyranoside (35) with methyllithium, was worked out (45).

Following the selective removal of the 3'-formyloxy function, the disaccharide 32 was obtained and this did not undergo any reaction with the glycal 37 in the presence of NIS. The assumption was at hand that the 2'-bromo substituent might impede the accessibility of the 3'-hydroxy group. However, tributylstannane reduction of 32 gave compound 33, and again, this product showed no reaction in the NIS glycosylation with 37. Amazingly, it was observed that a disaccharide glycal made up from two molecules of 37 could be obtained in modest yield (46). This result reflects a higher nucleophilicity of the tertiary hydroxy group in 37 than that of the secondary one in 33.

Finally, the 4'-benzoate function was also saponified. The unblocked 3',4'-diol component 34 did indeed undergo a smooth NIS glycosylation with the L-olivomycal 37. This occured regiospecifically in overall 64% yield with formation of only the interglycosidic (1"->3')-linkage and also proceeded stereospecifically towards the α-glycoside with respect to the terminal saccharide. The E-D-C derivative 38 was further hydrogenated to the trisaccharide unit 39 of chromomycin A₃ (22) and olivomycin A (23).

Mithramycin (24) bears the terminal sugar unit D-mycarose. By a similar preparation as for the methyl-branched glycals 36 and 37, the D-enantiomer of 36, namely the labile D-mycaral 40 was obtained (45). Its NIS condensation with the 3',4'-diol disaccharide 34 proceeds smoothly, but the reaction turned out to be rather slow with respect to the stability of the D-mycaral. This results in only a modest yield (12%) of the trisaccharide 41 which was hydrogenolyzed to compound 42. This in turn constitutes an isomer of the trisaccharide sequence in 24.

A novel approach for the mithramycin trisaccharide required a conceptional change. Thus, the uniformly β-(1->3)-linked three sugar units were to be assembled first, and only then the branch in the terminal unit was to be generated. Indeed, glycosylation of the 3'-monohydroxy disaccharide 32 with the glycosyl bromide 27, following the previously established procedure proceeded with almost 70% yield. The ratio α:β = 1:10 was again advantageous in favor of the desired derivative 43, which could be crystallized from the mixture. Further subsequent steps (methanol--HCl) released selectively the 3"-formyloxy group, reduced the three bromo functions (Bu₃SnH), and oxidized at C-3" which gave the trisaccharide 3"-uloside 44 in high overall yield. Finally, treatment of 44 with methyllithium gave stereospecifically the trisaccharides 45 and 46 (peculiarly with an additional benzoate group at C-3), which bore exclusively a terminal D-olivomycose unit E.

Previous branching reactions of alkyl α-hexopyranos-3-ulosides gave mainly or exclusively the ribo derivatives (45, 47-49). This is in accordance with expectations, because of partial blocking of the nucleophile from the lower face of the molecule. Quite recent experiments with the corresponding alkyl β-hexopyranos-3-ulosides such as 47 resulted in arabino:ribo ratios of approximately 2:1 (J. Thiem, M. Gerken, B. Schöttmer, J.Weigand, Carbohydr. Res., in press). Owing to the lack of an axial anomeric substituent in this case, nucleophilic attack from below prevails. A comparable ratio with the trisaccharide uloside 44 would have been expected. The exclusive nucleophilic attack from below may be assumed to be governed by secondary bindings of the reagent in the course of its approach to the carbonyl site.

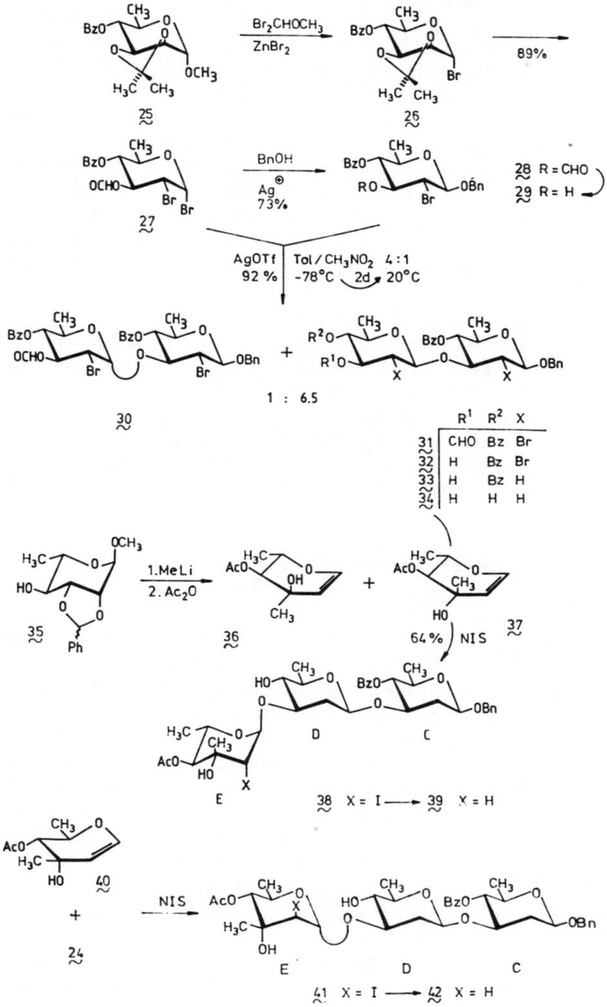

Br2CHOCH3
ZnBr2
89%
BnOH
Ag
73%
28 R = CHO
29 R = H
AgOTf
92 %
Tol / CH3NO2 4 : 1
-78°C 2d 20°C
1 : 6.5
R1 R2 X
31 CHO Bz Br
32 H Bz Br
33 H Bz H
34 H H H
1.MeLi
2. Ac2O
64 %
NIS
38 X = I → 39 X = H
NIS
41 X = I → 42 X = H

Another route to a methyl-branched derivative makes
use of reductive cleavage of spiro epoxides (50). The
realization of this process was tested in the monosaccharide series. Wittig olefination of 47 was used to form
the exocyclic methylene compound 48. This sugar contains
an inherent allyl alcohol fragment, the chiral C-4 alcohol function of which should be ideally suited to determine the chirality of the epoxide to be formed by the
Sharpless method. With tert-butyl hydroperoxide, titanium
tetraisopropoxide and (-)-tartrate (for a "like mode"
process) no reaction occured. After a number of attempts,
the Sharpless method was abandoned and extended back to
the well-established m-chloroperoxybenzoic acid epoxidation. The (3R)-epoxide 49 was obtained stereospecifically
in excellent yield (83%), and this could be readily
reduced to give the D-ribo compound 50. The exclusive
formation of 49 is unexpected and may be associated with
a strong stereochemical induction by the chiral centers
at C-1, C-4, and C-5.

Subsequent to this success, extension to the trisaccharide series was attempted. In this case, Peterson
olefination of the 3"-uloside 44 gave the 3"-exo-methy-
lene derivative 51. Treatment of 51 with m-chloroperoxybenzoic acid under conditions similar to those as elucidated for the monosaccharides gave stereospecifically the
(3"R)-spiroepoxide compound 52. Subsequent reduction
furnished the correct E-D-C trisaccharide sequence 53 of
mithramicin.

Anthracycline Oligosaccharides

Daunorubicin (54) and the less-toxic adriamycin (55)
belong to the class I anthracycline antibiotics
(6,51,52). Owing to their pronounced cytostatic activity
they are valuable chemotherapeutic agents (53-55). Anthracyclines having oligosaccharide chains, such as aclacinomycin (56) (56) or marcellomycin (57), named class II
anthracycline antibiotics, exhibit a more-favorable therapeutic breadth. Among other advantages, the class II
compounds are of similar cytostatic activity as the class
I compounds but show considerable less cumulative cardio-
toxicity (58). A few structure--activity correlations
have been outlined, e.g. the importance of a 3-amino-
2,3,6-trideoxy-L-lyxo derivative attached as the sugar A
unit (58).

Thus, it is of interest to generate oligosaccharides
of the type C-B-A in 56, all of which are linked inter-
glycosidically by α-(1->4)-bonds. We have previously synthesized various derivatives of the terminal C-B disaccharides employing the Ferrier glycosylation approach
(59, 60). At that same time we could combine the azide
functionalization of glycals (61) with the N-iodosuccinimide method (16) and develop a general approach to 3-

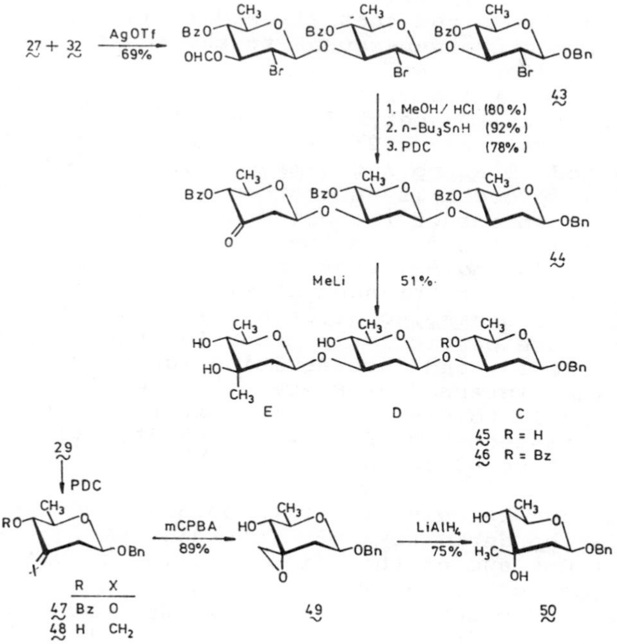
27 + 32
AgOTf
69%
43
1. MeOH/HCl (80%)
2. n-Bu3SnH (92%)
3. PDC (78%)
44
MeLi 51%
E D C
45 R = H
46 R = Bz
29
PDC
mCPBA
89%
LiAlH4
75%
R X
47 Bz O
48 H CH2
49
50

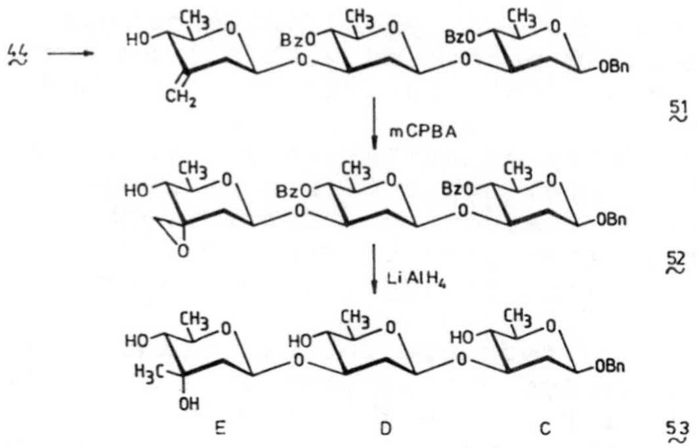
44
51
mCPBA
52
LiAlH4
E D C
53

amino-2,3,6-trideoxy-α-glycosides (62). In essence this
process was further applied to generate other precursors
in this field (63).

Rather early it became evident that the NIS glycosylation of an axial 4-OH group, even in a blocked 3-amino
sugar like daunosamine or rhodosamine, could not be
effected efficiently (64). Consequently, a trisaccharide
synthesis was required that allowed facile inversion of a
precursor structure subsequent to the advantageous use of
NIS glycosylation steps.

The azide--N-iodosuccinimide glycosylation (62)
employed with tri-O-acetyl-D-glucal (57) and benzyl alcohol gave in 92% yield a 2:1 mixture of the 3-azido-2,3-
dideoxy-2-iodo-D-altro and D-manno epimers (62 and 64).
The reaction process is understood to involve primary
nucleophilic attack of the azide under allylic rearrangement to furnish the α- and the β-D-erythro-hex-2-enopyra-
nosyl azides 58 and 59. These are in equilibrium via a
[3.3]-sigmatropic rearrangement with the 3-azido-3-deoxy-
D-ribo (60) or D-arabino glycals (61). The more-reactive
enol ethers are particularly prone to attack by an electrophile like I⁺, and the resulting iodonium (or 2-iodo
oxocarbenium) intermediates are subsequently attacked by
the nucleophile to give predominantly the products of
trans-diaxial geometry. Following deacetylation, the D-
altro derivative 63 could by crystallized from the mixture.

Reductive deiodination with Bu₃SnH smoothly gave the
3-azido-D-ribo derivative 66. Inversion at C-5 and reduction at C-6 and of the azido function should give the
L-lyxo target molecule. The most convenient path for the
C-5 inversion employs stereospecific hydrogenation of an
exocyclic glycal (60, 65). Application of the NIS/Ph₃P
reagent (66, 67) and subsequent acetylation gave the 6-
iodo component 67 in good yield. Elimination of hydrogen
iodide is generally performed with silver fluoride in
anhydrous pyridine (68), but was effected here preferentially by use of DBU (69). This gave the hex-5-enopyrano-
side 68 in virtually quantitative yield. As the final
step, the labile enol ether 68 was hydrogenated in methanol with platinum/charcoal at 30 bar pressure of hydrogen. Acetylation then gave stereospecifically benzyl N-
acetyl-β-L-daunosaminide (70), obtained crystalline in
good yield.

This reaction sequence was not meant at the outset
to add another daunosaminide synthesis to the number
reported in the literature (70). However, with only six
straightforward steps from D-glucose and an overall yield
of approx. 20% it may indeed constitute a rather
favorable alternative. The main purpose for the development of this sequence resides in the advantageous incorporation of the precursor for the unit A into the trisaccharide synthesis.

By treatment of the D-ribo compound 66 with tert-

54 R = H   Daunorubicin
55 R = OH Adriamycin

56 Aclacinomycin

butylchlorodimethylsilane selective blocking of the primary hydroxy group was achieved to give 71. This derivative was to serve as the A unit precursor onto which the
other sugars are attached, and which finally was to be
converted into a product having the L-lyxo configuration.
    As the precursor for the B unit, the selectively
benzylated L-fucal derivative 70 could be obtained crystalline from L-fucal by applying a phase-transfer-cata-
lyzed process (64). By a Ferrier reaction of di-O-acetyl-
L-rhamnal and a subsequent retro enol ether formation
through hydride attack at C-3 (71) the L-amicetal 72 was
obtained (64); this reaction was concurrently also described by others (72).
    Glycosylation of the glycal 70 with the sugar agly-
con 71 in the presence of N-iodosuccinimide proceeded
smoothly and in good yield to give the disaccharide derivative 73. Cleavage of the 4'-acetoxy group yielded the new
disaccharide aglycon 74. This again could be glycosylated
by the NIS procedure to L-amicetal 72 in high yield, and
this gave the trisaccharide 75. Interestingly, there is
no problem in glycosylation of the axial hydroxyl group
at the 4'-position of the terminal L-galacto residue in
74, in contrast to those derivatives having a 3-amino
substituent and otherwise similar structure.
    After having assembled the three sugar units they
had to be transformed into the deoxy- and aminodeoxy-L-
lyxo structures. The iodo functions are readily removed
by the radically-induced reduction with Bu₃SnH to give
76. Treatment of 76 with fluoride in anhydrous tetrahydrofuran yielded quantitatively the derivative 77, which
in turn was selectively iodinated at position 6 to give
78. The dehydrohalogenation with DBU proceeded even
better than in the monosaccharide series and gave the
exocyclic trisaccharide enol ether, compound 79. Hydrogenation under pressure with platinum/charcoal in methanol
and subsequent acetylation accomplished the synthesis of
the C-B-A trisaccharide glycoside 80 in satisfactory
yield. This compound constitutes the carbohydrate
sequence of dihydroaclacinomycin. It may certainly be
further processed into the trisaccharide part of aclacinomycin (56). The main advantage of this novel approach
resides in its versatility for the construction of selectively varied oligomers similar to that in 56.
    Finally, credit should be given to Monneret's group
who were the first (73) to assemble C-B-A trisaccharide
precursors of class II anthracyclines. The fucosyl bromide 82 was obtained from the methyl glycoside 81 following our bromotrimethylsilane procedure (74). The benzyl
α-daunosaminide 84 obtained from the difficultly accessible compound 83 served as the aglycon sugar unit.
Condensation of these components made use of the Helferich conditions and gave the α-(1->4)-linked disaccharide
85 in only 40% yield. The terminal step was again a NIS
glycosylation (16) of the L-amicetal 72, which gave the

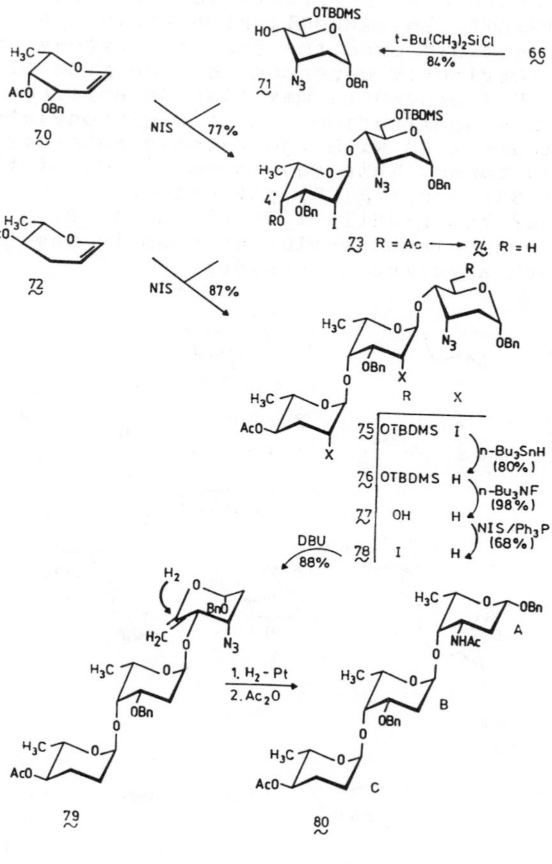
OTBDMS
HO
t-Bu(CH3)2SiCl
84%
66
71
N3
OBn
H3C
AcO
OBn
70
NIS
77%
73 R = Ac → 74 R = H
72
NIS
87%
R
X
75 OTBDMS I
n-Bu3SnH (80%)
76 OTBDMS H
n-Bu3NF (98%)
77 OH H
NIS/Ph3P (68%)
78 I H
DBU
88%
H2
BnO
H2C
1. H2-Pt
2. Ac2O
NHAc
A
B
C
79
80

trisaccharide component 86. Another corresponding preparation of the marcellomycin trisaccharide units was recently published by the same group (75).

A few final words focus on the glycosylation of
anthracyclinones. A number of earlier reports used the
glycosyl halides prepared _in situ_ from 1-acyloxy derivatives of 3-amino-2,3,6-trideoxy-sugar derivatives and
followed the Koenigs-Knorr conditions with amazing stereoselectivities (cf. 52). Recently the direct condensation of the 1-acyloxy compounds to anthracyclinones was
reported to operate with trimethylsilyl trifluoromethane
sulfonate (TMSOTf) in extremly high yield (76). Another
approach successfuly used the Ferrier glycosylation procedure (64). Obviously dependent on the type of anthracyclinone, the NIS procedure may also be applied. As a
particular nice application, the NIS glycosylation of
diacetyl-L-rhamnal 87 with the racemic anthracyclinone 88
should be mentioned. This gave good yields of the diastereomeric (7S,9S) and (7R,9R) glycosides 89 and 90 which
could be separated readily (77, cf. also 78), thus constituting a convenient resolution step in the total synthesis of anthracycline glycosides.

Acknowledgments

Studies of this group have enjoyed continuous support
from the Deutsche Forschungsgemeinschaft and the Fonds
der Chemischen Industrie, which is gratefully acknowledged.

Literature Cited

1. Hanessian, S.; Banoub, J. ACS Symposium Series 1976,
   39, 36.
2. Paulsen, H. Angew. Chem. 1982, 94, 184.
3. Reichstein, T.; Weiss, E. Adv. Carbohydr. Chem. 1962,
   17, 65.
4. Wright, D.E. Tetrahedron 1979, 35, 1207; Ollis, W.D.;
   Smith, C.; Wright, D.E. Tetrahedron 1979, 35, 105.
5. Mallams, A.K.; Puar, M.S., Rossman, R.R.; McPhail,
   A.T.; Macfarlane, R.D.; Stephens, R.L. J. Chem. Soc.,
   Perkin Trans. 1, 1983, 1497.
6. Remers, W.A. The Chemistry of Antitumor Antibiotics,
   Wiley, New York, N.Y. 1979.
7. Kelly, T.R. Annu. Rep. Med. Chem., 1979, 14, 288.
8. Sharon, N. Complex Carbohydrates - Their Chemistry,
   Biosynthesis, and Functions, Addison-Wesley Publ.
   Comp., Reading, Massachusetts 1975.
9. Grisebach, H. Angew. Chem. 1972, 84, 192.
10. Sable, H.Z. Adv. Enzymol. 1966, 28, 391.
11. Ferrier, R.J. Adv. Carbohydr. Chem. 1965, 20, 67.
12. Ferrier, R.J. Adv. Carbohydr. Chem. Biochem. 1969,
    24, 199.
13. Thiem, J.; Schneider, G. Angew. Chem. 1983, 95, 54.
14. Thiem, J.; Schneider, G.; Sinnwell, V. Liebigs Ann.
    Chem.. 1986, 814.
15. Tatsuta, K.; Fujimoto, K.; Kinoshita, M.; Umezawa, S.
    Carbohydr. Res. 1977, 54, 85.
16. Thiem, J.; Karl, H.; Schwentner, J. Synthesis 1978,
    696.
17. Thiem, J.; Elvers, J. Chem. Ber. 1979, 112, 818.
18. Thiem, J.; Ossowski, P.; Ellermann, U. Liebigs Ann.
    Chem. 1981, 2228.
19. Thiem, J.; Ossowski, P. J. Carbohydr. Chem. 1984, 3,
    287.
20. Thiem, J.; Prahst, A.; Lundt, I. Liebigs Ann. Chem.
    1986, 1044.
21. Bock, K.; Pedersen, C.; Thiem, J. Carbohydr. Res.
    1979, 73, 85.
22. Gross, H.; Karsch, U. J. Prakt. Chem. 1965, 29, 315.
23. Bognár, R.; Farkas-Szabó, I.; Farkas, I.; Gross, H.
    Carbohydr. Res. 1967, 5, 241.
24. Lemieux, R.U.; Fraser-Reid, B. Canad. J. Chem. 1965,
    43, 1460.
25. Igarashi, K.; Honma, T.; Imagawa, T. J. Org. Chem.
    1970, 35, 610.
26. Boullanger, P.; Descotes, G. Carbohydr. Res. 1976,
    51, 55.

27. Horton, D.; Priebe, W.; Varela, O. J. Org. Chem.
    1986, 51, 3479.
28. Thiem, J.; Gerken, M.; Bock, K. Liebigs Ann. Chem.
    1983, 462.
29. Thiem, J.; Gerken, M. J. Carbohydr. Chem. 1982/83, 1,
    229.
30. Schöttmer, B. Diplomarbeit, University of Münster
    1985.
31. Wiesner, K.; Tsai, T.Y.R.; Jin, H. Helv. Chim. Acta
    1985, 68, 300.
32. Ward, D.C.; Reich, E.; Goldberg, I.H. Science 1965,
    149, 1259.
33. Behr, W.; Honikel, K.; Hartmann, G. Europ. J. Bio-
    chem. 1969, 9, 82.
34. van Dyke, M.; Dervan, P.B. Biochem. 1983, 22, 2373.
35. Miyamoto, M.; Morita, K.; Kawamatsu, Y.; Naguchi, S.;
    Marumoto, R.; Sasai, M.; Nohara, A.; Nakadaira, Y.;
    Lin, Y.Y.; Nakanishi, K. Tetrahedron 1966, 22, 2761.
36. Berlin, Yu.A.; Kolosov, M.N.; Schemyakin, M.M.
    Tetrahedron Lett. 1966, 1431.
37. Miyamoto, M.; Kawamatsu, Y.; Kawashima, K.; Shino
    hara, M.; Tanaka, K.; Tatsuoka, S.; Nakanishi, K.
    Tetrahedron 1967, 23, 421.
38. Thiem, J.; Meyer, B. J. Chem. Soc., Perkin Trans. 2,
    1979, 1331.
39. Thiem, J.; Meyer, B. Tetrahedron 1981, 37, 551.
40. Thiem, J.; Meyer, B. Chem. Ber. 1980, 113, 3058.
41. Thiem, J.; Elvers, J. Chem. Ber. 1980, 113, 3049.
42. Thiem, J.; Karl, H. Chem. Ber. 1980, 113, 3039.
43. Hanessian, S.; Banoub J. Carbohydr. Res. 1977, 53,
    C 13.
44. Thiem, J.; Elvers, J. Chem. Ber. 1979, 112, 818.
45. Jung, G.; Klemer, A. Chem.Ber. 1981, 114, 740.
46. Thiem, J.; Gerken, M. J. Org. Chem. 1985, 50, 954.
47. Thiem, J.; Elvers, J. Chem. Ber. 1981, 114, 1442.
48. Dyong, I.; Merten, H.; Thiem J. Liebigs Ann. Chem.
    1986, 600.
49. Thiem, J.; Elvers, J. Chem. Ber. 1978, 111, 3514.
50. Yoshimura, J. Advan. Carbohydr. Chem. Biochem. 1984,
    42, 69.
51. Crooke, S.T.; Reich, S.D. (Eds.) Anthracyclines -
    Current Status and Development, Academic Press,
    New York 1980.
52. El Khadem, H.S. (Ed.) Anthracycline Antibiotics,
    Academic Press, New York 1982.
53. Füllenbach, D.; Nagel, G.A.; Seeber, S. (Eds.)
    Adriamycin-Symposium: Ergebnisse und Aspekte, Karger,
    Basel 1981.
54. Arcamone, F. Doxorubicin - Anticancer Antibiotics,
    Academic Press, New York 1981.
55. Jones, S.E. (Ed.) Current Concepts in the Use of
    Doxorubicin Chemotherapy, Farmitalia Carlo Erba,
    Milano 1982.
56. Oki, T. Aclacinomycin A, Chap. 19, p. 323 in lit. 50.

57. Reich, S.D.; Bradner, W.T.; Rose, W.C.; Schurig, J.E.; Madissoo, H.; Johnson, D.F.; DuVernay, V.H.; Crooke, S.T. Marcellomycin, Chap. 20, p. 343 in lit. 50.

58. Doyle, T.W. Anthracycline Oligosaccharides, Chap. 4, p. 27 in Lit. 50.

59. Thiem, J.; Holst, M.; Schwentner, J. Chem. Ber. 1980, 113, 3488.

60. Thiem, J.; Kluge, W.; Schwentner, J. Chem. Ber. 1980, 113, 3497.

61. Heyns, K.; Hohlweg, R. Chem. Ber. 1978, 111, 1632.

62. Heyns, K.; Feldmann, J.; Hadamczyk, D.; Schwentner, J.; Thiem, J. Chem. Ber. 1981, 114 232.

63. Thiem, J.; Springer, D. Carbohydr. Res. 1985, 325.

64. Springer, D. Dissertation, University of Hamburg 1985.

65. Horton, D.; Weckerle, W. Carbohydr. Res. 1975, 44, 227.

66. Appel, R. Angew. Chem. 1975, 87, 863.

67. Hanessian, S.; Lavallee, P. Meth. Carbohydr. Chem. 1976, 7, 49.

68. Helferich, B.; Himmen, E. Ber. Dtsch. Chem. Ges. 1928, 61, 1825.

69. Oediger, H.; Möller, F. Angew. Chem. 1967, 79, 53.

70. Hauser, F.M.; Ellenberger, S.R. Chem. Rev. 1986, 86, 35.

71. Fraser-Reid, B.; Radatus, B.; Tam, S.Y.-K. Meth. Carbohyr. Chem. 1980, 8, 219.

72. Martin, A.; Pais, M.; Monneret, C. Carbohydr. Res. 1983, 113, 21.

73. Martin, A.; Pais, M.; Monneret, C. J. Chem. Soc., Chem. Commun. 1983, 305.

74. Thiem, J.; Meyer, B. Chem. Ber. 1980, 113, 3075.

75. Monneret, C.; Martin, A.; Pais, M. Tetrahedron Lett. 1986, 27, 575.

76. Kimura, Y.; Suzuki, M.; Matsumoto, T.; Abe, R.; Terashima, S. Chem. Lett. 1984, 501.

77. Horton, D.; Priebe, W.; Varela, O. Carbohydr. Res. 1984, 130, C 1.

78. Horton, D.; Priebe, W. Carbohydr. Res. 1985, 136, 391.

RECEIVED May 31, 1988

Chapter 9

Total Synthesis of Cyclodextrins

Yukio Takahashi and Tomoya Ogawa

Riken Institute of Physical and Chemical Research, Wako-shi,
Saitama 351-01, Japan

An approach to the total synthesis of cyclodextrins is
outlined. Intramolecular glycosylative cyclizations
are investigated as the key step, using appropriately
protected malto-oligosyl fluorides, for successful
synthesis of completely protected cyclomalto-hexa-,
hepta-, and -octaoses. Deprotection of these
intermediates afforded the desired, unsubsituted
cyclodextrins.

Cyclodextrins, products of the degradation of starch by an
amylase of Bacillus macerans(1), have been studied in terms of
chemical modifications, mainly for the purpose of developing
efficient enzyme mimics(2). Not only their unique cyclic structures,
but also their ability to form inclusion complexes with suitable
organic molecules, led us to investigate the total synthesis of this
class of molecules(3). We describe here an approach to a total
synthesis of alpha(1), gamma(2), and "iso-alpha" cyclodextrin (3).

0097–6156/89/0386–0150$06.00/0
© 1989 American Chemical Society

Retrosynthetic analysis of cyclodextrins led us to design a linear, key intermediate 4 which could be suitable for a possible intramolecular glycosylation for the synthesis of alpha, gamma, and some isomeric cyclodextrins, according to the kind of protective group at C-6 of the nonreducing glucopyranosyl residue.

[A] Synthesis of cyclomaltohexaose (alpha cyclodextrin)

We first describe a synthesis of alpha cyclodextrin. The synthetic plan is shown in Scheme 1. An immediate precursor for 1 could be the perbenzylated precursor 5, which could be cleaved to give a linear glucohexaosyl fluoride 6. This key intermediate 6 is obtainable from the maltose derivatives 8 and 9 through the intermediacy of the glucohexaose derivative 7.

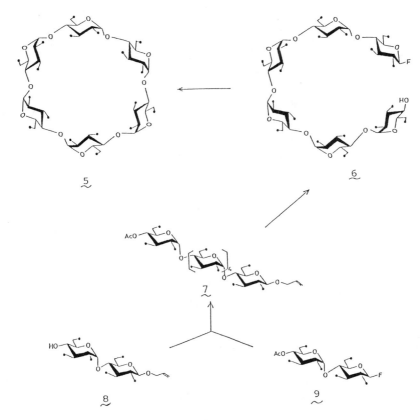

Scheme 1.   Synthesis of cyclomaltohexaose (⏤● = OBn).

Synthetic transformation of β-maltose octaacetate (10) into the glycosyl acceptor 8 and glycosyl donor 9 is shown in Scheme 2.

Compound **10** was converted into allyl glycoside **11** in 73% yield in two steps, (<u>a</u>) Bu$_3$SnOCH$_2$CH=CH$_2$—SnCl$_4$(<u>4</u>), and (<u>b</u>) MeONa—MeOH. Treatment of **11** with dimethoxypropane and TsOH, and then with benzyl bromide—NaH—DMF afforded compound **12** in 51% yield. Solvolysis of compound **12** in MeOH—AcOH, and then monobenzylation by the stannylation—alkylation method(<u>5</u>) gave the desired glycosyl acceptor **8** in 67% yield. Acetylation of compound **8** and then deallylation with PdCl$_2$—AcONa in aq.AcOH(<u>6</u>) afforded a 93% yield of hemiacetal **13**, which was treated with (<u>a</u>) SOCl$_2$—DMF in dichloroethane(<u>7</u>) and (<u>b</u>) AgF—CH$_3$CN(<u>8</u>) to give the desired fluoride **9** in 73% overall yield.

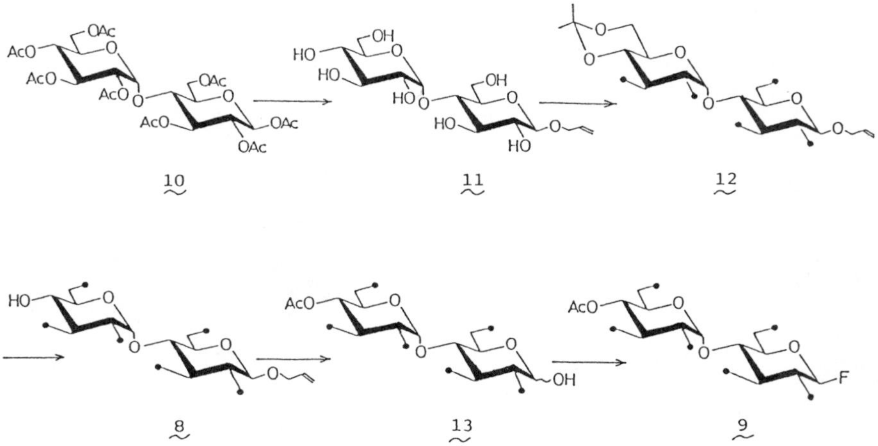

Scheme 2. Synthesis of the glycosyl acceptor and donor.

Having a glucobiosylfluoride **9** and a glycosyl acceptor **8** thus efficiently prepared, glycosylation conditions using both compounds were examined with respect to the kind of solvent and Lewis acid to be employed. In the presence of SnCl$_2$(<u>9</u>), AgOSO$_2$CF$_3$, and 4Å powdered molecular sieves in Et$_2$O, compounds **8** and **9** afforded a 1.8:1 mixture of **14** and **15** in 80% yield as well as the 1,6-anhydro derivative **16** (15%). Other glycosyl donors carrying either chlorine or trichloroacetimidate in place of fluorine gave inferior results in this particular instance.

Further elongation of the glucan chain on the glucotetraose derivative **14** was studied in two ways. First, compound **14** was transformed into a glycosyl acceptor **17**, which was then glycosylated by use of 2 equivalents of the donor **9** under the same conditions as already described to give a 2:1 mixture of **7** and **18** in 65% yield. Second, compound **14** was transformed into the glucotetraosyl donor **19** in 50% overall yield in 3 steps, (<u>a</u>) PdCl$_2$—AcONa—aq.AcOH, (<u>b</u>) SOCl$_2$—DMF, and (<u>c</u>) AgF—CH$_3$CN. Glycosylation of the glycosyl acceptor **8** with 0.9 equivalent of the glycosyl donor **19** afforded a 1.7:1 mixture of **7** and **20** in 55% yield. From the preparative viewpoint, the first route using glucotetraosyl acceptor **17** and

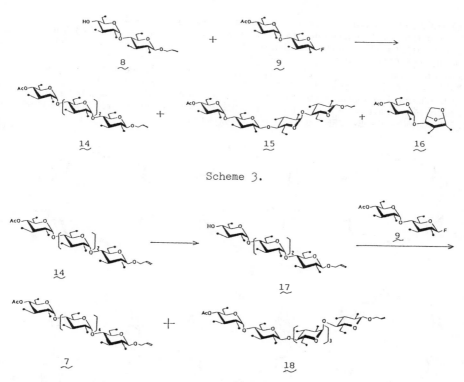

Scheme 3.

Scheme 4. Glucohexaosyl intermediate via the glucobiosyl-donor and glucotetraosyl-acceptor route.

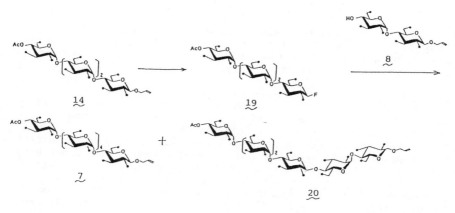

Scheme 5. Glucohexaosyl intermediate via the glucotetraosyl-donor and glucobiosyl-acceptor route.

glucobiosyl donor 9 is more efficient than the second one.
Conversion of the glucohexaosyl derivative 7 into the desired key
intermediate 6 is shown in Scheme 6. Treatment of 7 with (a) NaOMe—
MeOH, (b) (ClCH₂CO)₂O—pyridine afforded in 70% yield the
monochloroacetyl derivative 21, which was deallylated with PdCl₂—
NaOAc—aq.AcOH to give the hemiacetal 22 in 60% yield. The
transformation of 22 into fluoride 23 was achieved in 73% yield in 2
steps: (a) SOCl₂—DMF and (b) AgF—CH₃CN. Compound 23 was deacylated
with NaOMe—MeOH—THF to give 6 in 95% yield.

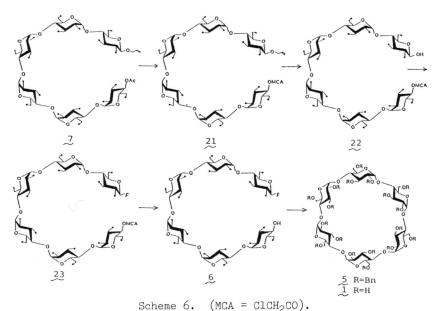

Scheme 6. (MCA = ClCH₂CO).

The crucial intramolecular glycosylation of 6 with SnCl₂—
AgOSO₂CF₃ afforded a 20% yield of compound 5; the latter was
debenzylated with Pd-C—HCO₂H—MeOH(10) to give cyclomaltohexaose
(alpha cyclodextrin) quantitatively.

[B] <u>Synthesis of cyclomaltooctaose (gamma cyclodextrin)</u>
A synthetic approach to gamma cyclodextrin was also examined by
use of the technology developed in [A]. Two routes for the
preparation of a glucooctaosyl intermediate 25 were examined. In a
first approach, the glucohexaosyl acceptor 24, readily obtained from
7 (Scheme 6) was glycosylated with 5.5 equivalents of the donor 9 to
give a 21% yield of the desired intermediate 25 as well as a 5.4%
yield of the isomeric product 26. As a second approach, the
glucotetraosyl fluoride 19 was treated with an equivalent amount of
the glucotetraosyl acceptor 17 to give a 27% yield of a 3:1 mixture
of 25 and the β-anomer.

However, in this reaction, a major product was found to be the
undesired 1,6-anhydro derivative 27, isolated in 30% yield.

As the yield of the desired product 25 by both routes was
comparable and the preparation of glucotetraosyl donor 19 requires 3

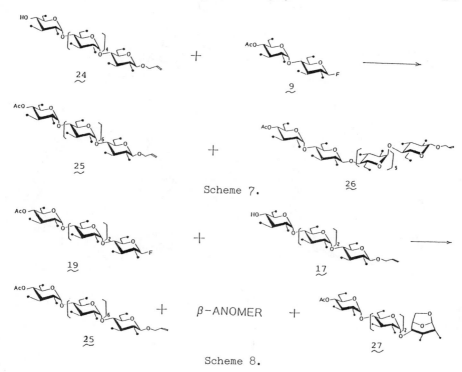

Scheme 7.

Scheme 8.

steps from the intermediate 14 (Scheme 5), it was more practical to use an excess of the glucobiosyl donor 9, as described in the first approach.

The glucooctaosyl derivative 25 was converted into the fluoride 28, in 47% overall yield, in 3 steps (a) PdCl$_2$—AcONa—aq.AcOH, (b) SOCl$_2$—DMF, and (c) AgF. Deacylation of 28 with MeONa—MeOH—THF afforded a 92% yield of the key intermediate 29. Compound 29 was cyclized in the presence of SnCl$_2$—AgOSO$_2$CF$_3$—4Å molecular sieves in Et$_2$O—Cl(CH$_2$)$_2$Cl to give an 8.4% yield of perbenzylated gamma cyclodextrin, which was debenzylated with 10% Pd-C in HCO$_2$H—MeOH—THF—H$_2$O to give an 80% yield of gamma cyclodextrin (2).

[C] Synthesis of iso-alpha cyclodextrin

In the previous section we observed that intramolecular α-(1→4) glycosylative cyclization leading to alpha cyclodextrin was 2.4 times more efficient in terms of the yield obtained than that leading to gamma cyclodextrin. We have examined how regioselective this type of cyclization is when the 4,6-diol derivative 30 is used to give 31 and 32. Product 32 may be regarded as an iso-alpha cyclodextrin derivative.

The synthetic plan is shown in Scheme 10. When used in combination with a glycosyl acceptor 17, the glucobiosyl donor 33 (which carries different kinds of acyl protective groups on O-4' and O-6'), may be suitable for the preparation not only of 30 but also of other intermediates carrying different functional groups at C-6 of the nonreducing-end glucopyranosyl residue of 30.

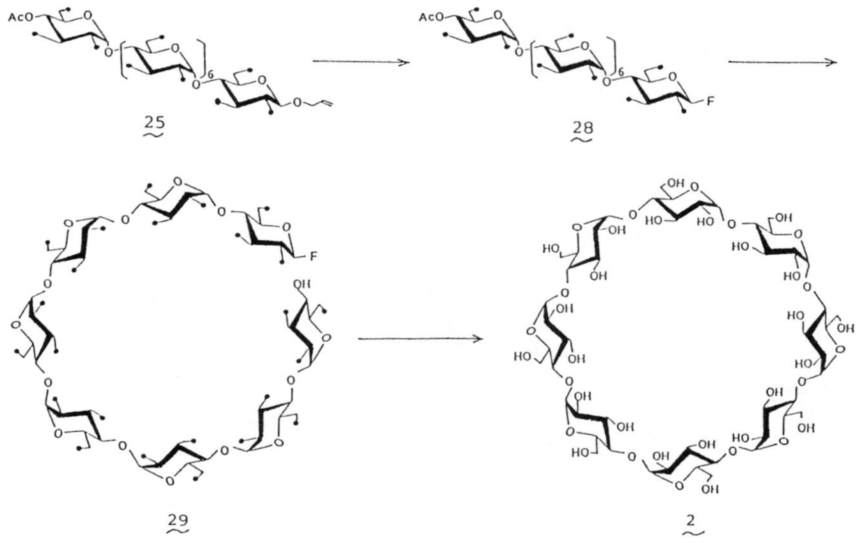

Scheme 9.

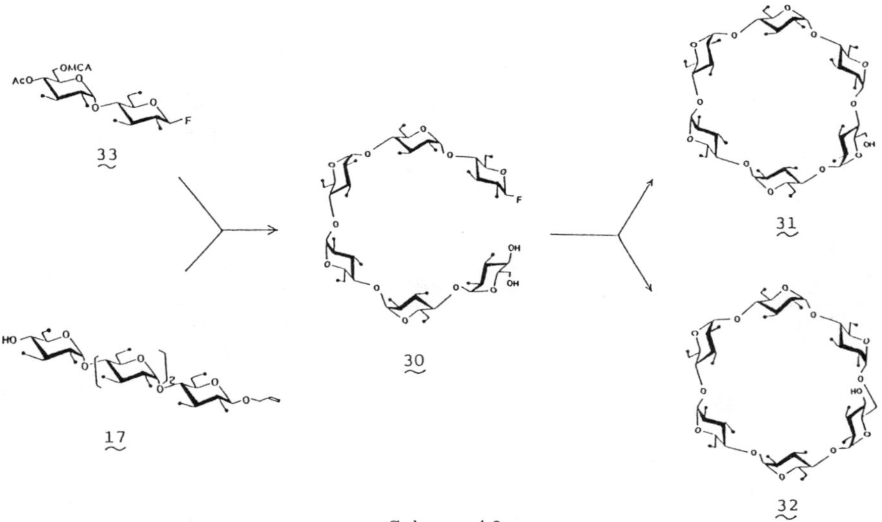

Scheme 10.

Diol **34**, readily obtainable from **12** (Scheme 2), was converted into the fluoride **33** in 5 steps in 35% overall yield. Glycosylation of compound **17** with 4.6 equivalents of the donor **33** afforded the desired glucohexaosyl derivative **35** in 18% yield, together with the β-isomer **36** in 8.2% yield.

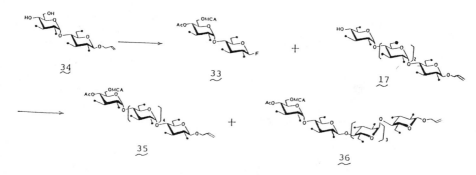

34 → 33 + 17 → 35 + 36

Scheme 11.   Synthesis of a glucohexaosyl intermediate.

The transformation of compound **35** into fluoride **30** was achieved in 50% overall yield <u>via</u> compound **37** in 4 steps: (<u>a</u>) PdCl$_2$—AcONa, (<u>b</u>) SOCl$_2$—DMF, (<u>c</u>) AgF, and (<u>d</u>) MeONa—MeOH—THF. The crucial glycosylative ring-closure was achieved efficiently in the presence of SnCl$_2$—AgOSO$_2$CF$_3$—4Å molecular sieves in Cl(CH$_2$)$_2$Cl in 65% yield to give a 1:2.5 mixture of **31** and **32**. The structure of **31** was readily determined by its transformation into the perbenzyl derivative **5**, which was identified with the same compound obtained previously in Scheme 6. The isomeric product **32** was transformed into the monoacetate **38**, which showed a signal for acetyl methyl at δ 1.901 in its ^{1}H n.m.r. spectrum, and was also debenzylated quantitatively to give "iso-alpha cyclodextrin" **3**, [α]$_D$ +81° (<u>c</u> 0.05, H$_2$O).

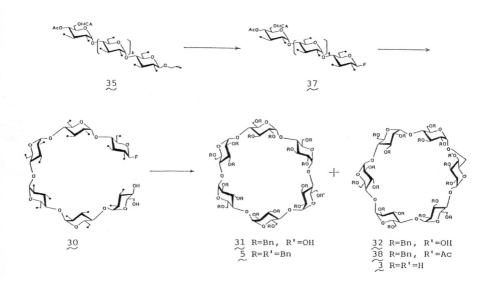

35 → 37

30 → 31 R=Bn, R'=OH ; 5 R=R'=Bn + 32 R=Bn, R'=OH ; 38 R=Bn, R'=Ac ; 3 R=R'=H

Scheme 12.   Glycosylative ring closure.

In conclusion, by employing three key intermediates **6, 29,** and **30,** intramolecular glycosylative ring closures were executed to afford alpha, gamma, and "iso-alpha" cyclodextrins, respectively.(11)

Acknowledgment
We thank Dr. J. Uzawa and Mrs. T. Chijimatsu for recording and measuring the n.m.r. spectra and Dr. H. Honma and his staff for the elemental analyses. We also thank Ms. A. Takahashi and Ms. K. Moriwaki for their technical assistance.

References
1. French, D. Adv. Carbohydr. Chem. Biochem. 1957, 12, 189-60; Saenger, W. Angew. Chem., Int. Ed. Engl. 1980, 19, 344-62.
2. Breslow, R. Science 1982, 218, 532-37; Tabushi, I. Acc. Chem. Res. 1982, 15, 66-72; Croft, A. P.; Bartsch, R. A. Tetrahedron 1983, 39, 1417-74.
3. Ogawa, T.; Takahashi, Y. Carbohydr. Res. 1985, 138, C5-C9.
4. Ogawa, T.; Matsui, M. Carbohydr. Res. 1976, 51, C13-C18.
5. Ogawa, T.; Matsui, M. Carbohydr. Res. 1977, 56, C1-C6; 1978, 62, C1-C4; Tetrahedron 1981, 37, 2363-69; Veyrières, A. J. Chem. Soc., Perkin Trans I 1981, 1626-29.
6. Ogawa, T.; Nakabayashi, S. Carbohydr. Res. 1981, 93, C1-C5; Ogawa, T.; Nakabayashi, S.; Kitajima, T. ibid. 1983, 144, 225-36.
7. Newman, M. S.; Sujeeth, P. K. J. Org. Chem. 1978, 43, 4367-69.
8. Helferich, B.; Gootz, R. Ber. 1929, 62, 2505-07; Hall, L. D.; Manville, J. F.; Bhacca, N. S. Can. J. Chem. 1969, 47, 1-17.
9. Mukaiyama, T.; Murai, Y.; Shoda, S. Chem. Lett. 1981, 431-32.
10. Amin, B. El; Anantharamaiah, G. M.; Royer, G. P.; Means, C. E. J. Org. Chem. 1979, 44, 3442-44; Rao , V. S.; Perlin, A. S. Carbohydr. Res. 1980, 83, 175-77.
11. For further discussion and experimental details, refer to; Takahashi, T.; Ogawa, T. Carbohydr. Res. 1987, 164, 277-96.

RECEIVED October 14, 1988

TOTAL SYNTHESIS OF CARBOHYDRATES

Chapter 10

Total Synthesis of ±-*N*-Acetylneuraminic Acid

A New Strategy for the Synthesis of Glycosides of Sialic Acids

Samuel J. Danishefsky[1], Michael P. DeNinno[1], James E. Audia[1], and Gayle Schulte[2]

[1]Department of Chemistry, Yale University, New Haven, CT 06511
[2]Yale Chemical Instrumentation Center, Yale University, New Haven, CT 06511

A total synthesis of the title compound has been achieved. One of the key strategic elements involved the use of a furan ring as a surrogate for a carboxylic acid. This logic has been applied, on a model basis, to the synthesis of sialo and KDO conjugates.

Gottschalk was the first to encounter N-acetylneuraminic acid (Neu5Ac) during the course of his examination of the action of influenza viruses on various mucins (1). It is a member of a more widely occurring group of compounds known as sialic acids, which are N- and sometimes O- acylated derivatives of neuraminic acid (2,3). The sialic acids usually occur on the non reducing end of several oligosaccharide families. Most notable is the appearance of sialic acid residues in glycoproteins and glycolipids (gangliosides). The identification and evaluation of the full range of functions of sialic acids is an ongoing pursuit. In a general way they have been implicated in determining aggregation phenomena, viscosity and agglutination. The extent of sialylation of glycoconjugates apparently has implications in governing their recognizability by various biological receptors, including those involved in immunological surveillance (4).

Our interest in the synthesis of Neu5Ac involved a confluence of several considerations (5). The program we envisioned would test issues pertinent to the use of the diene-aldehyde cyclocondensation reaction in the synthesis of polyoxygenated natural products, including the higher order monosaccharides (6). Also, the need to deal with attainment of the required side chain configurations at C_7 and C_8 would pose some fresh challenges to another capability, i.e., that of transmitting chirality from a pyranose ring to stereogenic centers emerging on a side chain (7). Moreover, there evolved the possibility that the chemical logic which was basic to the synthesis of Neu5Ac (i.e., use of the furyl function, *vide infra*) might find application in the construction of sialoconjugates. Three important examples of such conjugates are described in Figure 1.

0097–6156/89/0386–0160$06.00/0

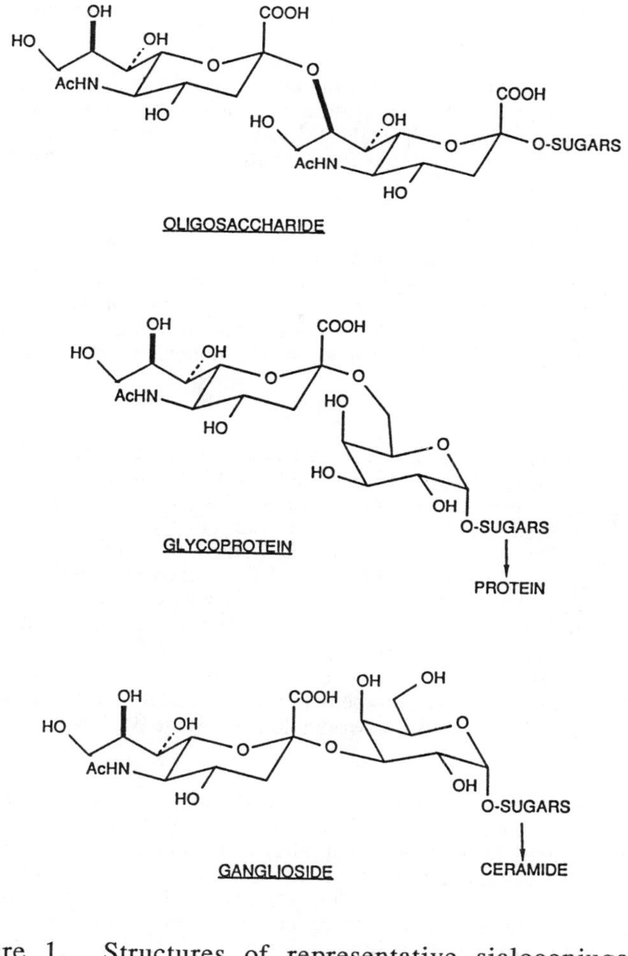

Figure 1.   Structures of representative sialoconjugates.

We considered application of the logic inherent in the Lewis acid catalyzed diene-aldehyde cyclocondensation to the synthesis of Neu5Ac. In principle, an erythrose aldehyde of the type **2** might react with an appropriate version of diene **3** in which a C_1 carboxy equivalent has been incorporated. At this juncture we had very little information as to the facial selectivity of suitably protected versions of **2** (P = protecting group) and as to the ratio of cis : trans dihydropyrone derivatives which might arise from such a cyclocondensation reaction. Also uncertain was the usefulness of the construct which relies on a diene of the type **3**. Smooth access to such a system might not be a simple matter. Even more serious was the question of the magnitude and even direction of regiochemical preference which might be expected from such a diene with a typical aldehyde in the cyclocondensation reaction.

Furthermore, it would be useful if the projected dihydropyrone of the type **4** could be converted to a glycoside (cf. **5**) during the steps required to reach Neu5Ac . The prospects of generating such a glycoside from a system such as **4**, and of maintaining a C_1 carboxyl-containing function during the course of the program, were not very promising .

It seemed that several concerns about the viability of the strategy based on diene **3** could be simultaneously addressed via the use of a furyl diene. Through this conceptual variation, the electronic nature of the C_1 fragment was changed in a fundamental way (cf. **3** and **6**). This strategic reformulation was seen to have potentially favorable ramifications for all of the uncertainties expressed above relevant to **3** (see Figure 2)

The total synthesis of racemic KDO provided the opportunity to test some of these concepts in the field (8). While much simpler than Neu5Ac, KDO, and glycosides thereof, are potentially very important. They are present as substructural units in the cellular system of Gram negative bacteria (9,10). Of importance to us at the moment is that during the course of the KDO effort, the synthesis of diene **7** was achieved from 2-acetylfuran. Lewis acid (BF_3) catalyzed cyclocondensation reaction of **8** with α–phenyselenopropionaldehyde was realized, though the cis : trans ratio in methylene chloride at -78° C was a disappointing 2.4 : 1. The synthesis also established several other feasibility demonstrations. Simple glycosides containing the C_1 furyl fragment could indeed be readily generated and maintained. Moreover, oxidative fragmentation of the furan with retrieval of the C_1 carboxy group at a strategic point was workable (see Figure 3).

We were thus ready to apply these findings to the more difficult Neu5Ac problem. At the planning stage, we did not want to add the question of the facial selectivity of an aldehyde such as **2** to an already formidable list of uncertainties. Therefore, in this testing stage, we first examined the cyclocondensation of the achiral aldehyde **8** with diene **7**. In so doing we would be obliged to address later the issue of communicating relative stereochemistry from a pyranose ring to its side chain. As noted at the outset, the solution to such a problem was of some interest to us.

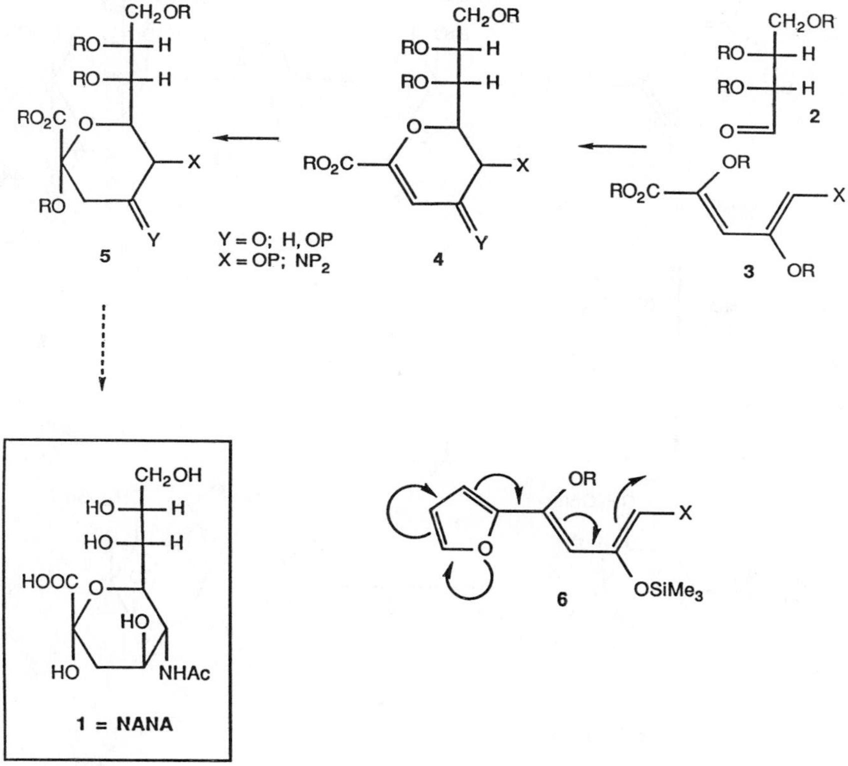

Figure 2. The concept of the furan surrogate.

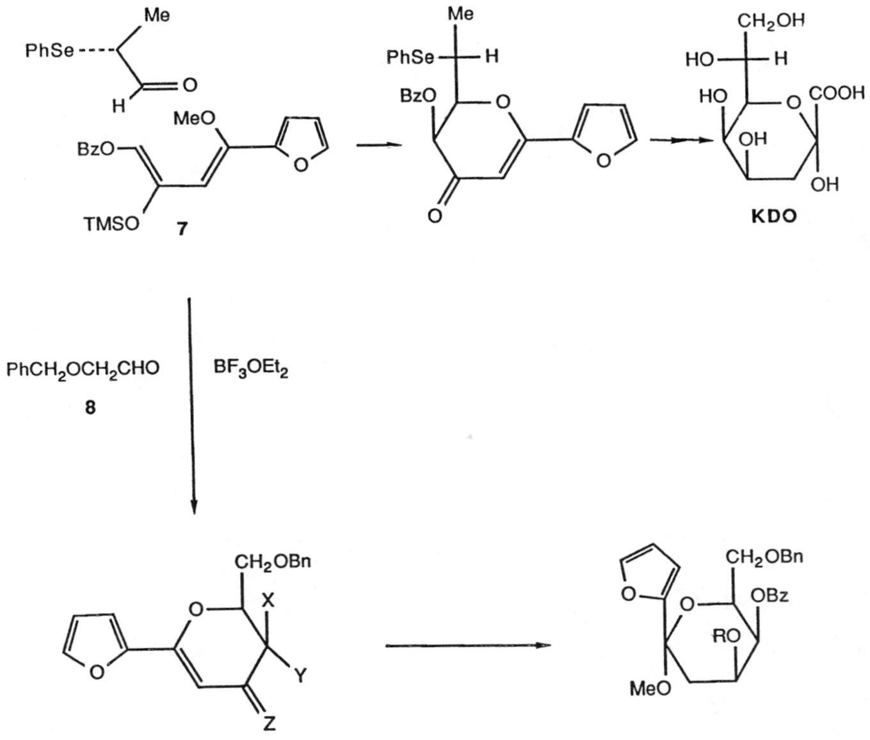

9 $X = OBz$; $Y = H$; $Z = O$

9a $X = H$; $Y = OBz$; $Z = O$

10 $X = OBz$; $Y = H$; $Z = \beta OH,\ \alpha H$

11 $R = H$

12 $R = TBS$

Figure 3. The synthesis of a furyl glycoside.

Of course, with the drastic simplification inherent in the use of aldehyde **8** as the "real world" equivalent of the hypothetical threose **2**, came the postponement of the goal of the synthesis of the natural enantiomer of Neu5Ac. The only way to strive for enantiospecificity would involve replacing **7** with another diene equipped with a chiral auxilliary (11). At this exploratory stage, we preferred to accept a downscaling of goals, confident that lessons learned in the synthesis of racemic Neu5Ac could be applied to the preparation of enantiomer **1** (synthesis of (-) Neu5Ac has been accomplished; DeNinno, M. P., Yale University, unpublished data).

Results

The Total Synthesis of Racemic Neu5Ac. An important practical advance was attained when it was found that cyclocondensation of **7 + 8** in toluene as solvent at -78° C afforded ca. a 5.1 : 1 yield of **9** : **9a** in 78-80% yield. Reduction of the mixture with sodium borohydride in the presence of cerium (III) chloride trihydrate gave an 80% yield of **10**, isolated as a single entity by flash chromatography. As projected, the vinylogous keteneacetal-like character of the C_2-C_3 bond, enabled the smooth addition of methanol under catalysis by camphorsulfonic acid in MeOH-toluene. Compound **11** was obtained in 82% yield. The alcohol function at C_4 was converted (95%) to its TBS ether, **12**, via reaction with tert-butyldimethylsilyltriflate and 2,6-lutidine. The advantages of the unique silyl protection of the C_4 hydroxyl group were to be exploited later in the synthesis, as a device to distinguish the adjacent alcohol at C_5. The early steps of the synthesis are summarized in Figure 3.

The next phase focused on the goal of elaboration of the side chain in the desired sense. The primary alcohol function at C_7 was unveiled by hydrogenolysis $(Pd(OH)_2/EtOAc-MeOH)$. Oxidation of the resultant compound **13** with chromic oxide· pyridine afforded aldehyde **14**, which was now to be elongated through some variation of a Horner-Emmons type of reaction. Shortly before these investigations were launched, Still had demonstrated the use of phosphonate **15** as a device to achieve the two-carbon extension of an aldehyde to a Z-enoate (12). Happily, application of the Still method to compound **14** afforded the desired **16**, mp 120-121° C, in 80% yield as a 20 : 1 mixture of Z : E enoates.

One way in which the Z-α,β-unsaturated carbonyl functionality could be exploited would be via its incorporation into lactone **17**. It could be predicted with some confidence that external reagents would attack the bicyclic lactonic system from its convex face. Such an α attack by osmium tetroxide would provide the correct 7,8-erythro diol stereochemistry required to reach Neu5Ac. This anticipation turned out to be well founded.

The method by which lactone **17** was obtained was not without its own implications for the synthesis. Treatment of **16** with dry tetra n-butylammonium fluoride in acetonitrile achieved desilylation. Not unexpectedly, this process triggered migration of the C_5 benzoyl group to the newly unveiled C_4 alcohol. The C_5 alcohol thereby liberated underwent lactonization to the desired **17** (61% yield from **16**). Indeed, reaction of **17** with stoichiometric osmium tetroxide in pyridine-THF afforded a single diol formulated as **18** in 97% yield (see Figure 4).

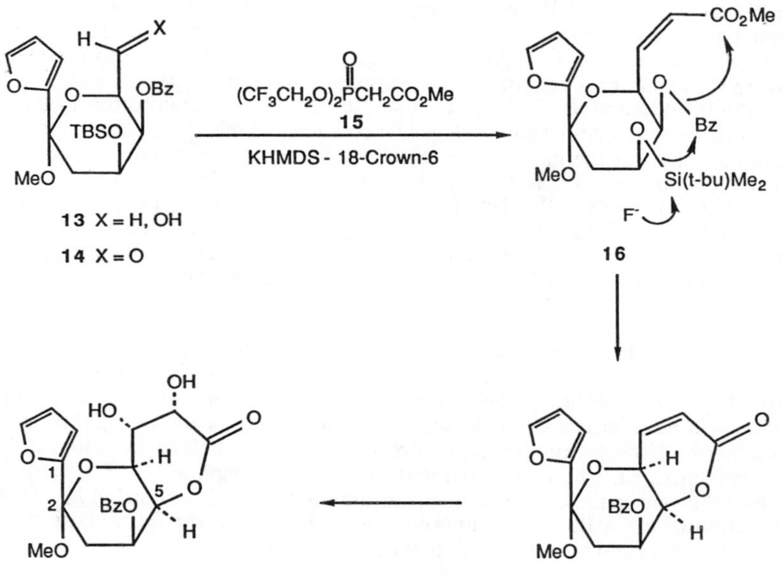

Figure 4.   The lactone option for stereochemical control at carbons 7 and 8.

The next stage would involve conversion of **18** to a tetra acylated pentaol with a uniquely exposed axial alcohol at C_5. Early difficulties were encountered in attempts to cleave the lactone to its corresponding hydroxy ester. Difficulties were also experienced in manipulating the highly polar compounds arising from attempted reductive opening of **18**.

Lack of success during these preliminary skirmishes occasioned us to examine a more direct, if less predictable, possibility, i.e., the osmylation of Z-enoate **16** itself. Obviously, the capacity to foresee the diastereofacial outcome of such a reaction on a substrate containing an acyclic double bond is less than is the case in a conformationally restricted cyclic setting such as lactone **17**. Our hope with compound **16** was that it would assume an antiperiplanar arrangement (see **16a**) and that such a rotamer would be attacked from its less hindered α-face. At the time the experiment was performed, there was virtually no direct analogy to guide us in this matter. An X-ray crystallographic determination of a conceptually related compound, i.e., aldehyde **19**, showed it to assume a nearly antiperiplanar conformation. The α,β-unsaturated ester could well mimic the aldehyde in its conformational preference. It was our conjecture that the antiperiplanar conformation, wherein the pyranosidal oxygen atom is orthogonal to the π-orbitals, would provide the least depletion of the already electronically deficient enoate double bond (*13*).

In the event, osmylation of compound **16** in pyridine, followed by the usual workup, afforded a 92% yield of an 18 : 1 mixture of two diols. That the major one, mp 136-139° C, has the stereochemistry implied in **20** need not be debated in light of a single crystal determination of its 7,8-bis-3,5-dinitrobenzoate derivative **21**, mp 114.5-115.5° C (see Figure 5).

It should be noted that the success of a prediction does not, per se, validate the underlying reasoning. We have investigated the matter in greater detail. The issues of ground state conformation and vectoral sense of attack on compounds such as **16** are treated elsewhere (DeNinno, M. P.; Danishefsky, S. J.; Schulte, G. *J. Am. Chem. Soc.*, manuscript submitted). Suffice it to be said here that the ground state conformation of compound **16**, as determined by X-ray crystallography, reveal it to be of the antiperiplanar type ($O-C_6-C_7-C_8$ dihedral angle = 162°). Thus, there is no experimental reason to doubt the model of α-attack on conformer **16a**, though the matter cannot be considered to be established.

While there is ample room for debate at the theoretical level, the importance of this straightforward route to compound **20** could hardly be missed. Attention could now be directed to the necessary functional group adjustments. Reduction of the ester (lithium triethoxyborohydride) followed by benzoylation (benzoyl chloride, DMAP) afforded an 80% yield of **22**. Treatment of this compound with ruthenium dioxide in the presence of sodium metaperiodate, followed by diazomethane, provided ester **23** in 90% yield.

The overall strategy for installation of the C_5 equatorial acetamido group was foreshadowed by the manner in which lactone **17** was constructed (*vide supra*). Thus, treatment of **23** with HF in methanol led to desilylation and partial benzoyl

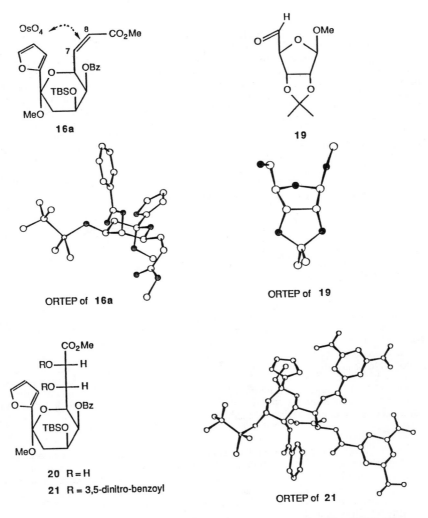

Figure 5. Stereospecific osmylation of the acyclic Z-C_7-C_8 olefin.

migration from C_5. The latter process was furthered by treatment of a methylene chloride solution of the C_4 alcohol with potassium carbonate. It was noted that during this treatment there was a further tendency for benzoyl migration from C_7 to occur. The relative thermodynamic stability of the C_5-C_7 diol monobenzoate permutants has not been determined. Given the $C_7 \rightarrow C_5$ benzoyl migration problem, and given the need to avoid adventitious debenzoylation, the 65% yield realized in the transformation of **23** to **24** was acceptable.

The uniquely distinguished axial alcohol at C_5 was activated by triflation and the resultant triflate was displaced by the action of tetra n-butylammonium azide in benzene at room temperature. Compound **25** was thus available in 86% yield from alcohol **24**. Hydrogenolysis ((H_2/Pd(OH)$_2$C) followed by acetylation afforded compound **26** in 96% yield. The richly detailed NMR spectrum (490 MHz) of **26** was identical in every respect with that of the L-antipode derived from reaction of Neu5Ac, first with acidic methanol and then with benzoyl chloride under the influence of DMAP. The infrared and mass spectra, as well as the chromatographic characteristics of the two materials, were identical.

Finally, saponification of ±**26** followed by acid treatment (Dowex resin) afforded racemic Neu5Ac, whose NMR spectrum (490 MHz) and chromatographic properties were identical with those of authentic Neu5Ac. An efficient and stereoselective total synthesis of ±Neu5Ac had been accomplished as shown in Figure 6.

A New Approach to Sialic Acid Glycosides. One of the key steps in our synthesis of Neu5Ac was the addition of methanol to **10**. The furan presumably stabilizes the cation formed by protonation of the β-carbon of the glycal. Indeed, as discussed at the outset, the de facto nucleophilic character of the furan was one feature which recommended its use.

The possibility presented itself that carboxyglycosides of the KDO or sialoconjugate type might be synthesized from a C-furyl substituted glycosylating agent such as **27** (14-16). If all went well, **27** would react with a unique alcohol of an otherwise protected sugar, to produce **28**. Oxidation of the furan would give rise to **29**. The thought was that, at least in certain respects, this sequence might have advantages over the rather more conventional use of system **30** (see Figure 7).

Of course, to put such an idea to use in the construction of sialoconjugates, the non-trivial matter of synthesizing the C-furyl sialylating agent would have to be addressed. In principle, this could be accomplished by total synthesis, using methodology described above. However, implementation of such a conception would have to await the realization of our ideas for an enantiocontrolled version of the above synthesis. The prospect of using a racemic sialyl donor was not attractive. In principle, the possibility of installing the furan via modification of Neu5Ac itself could be entertained.

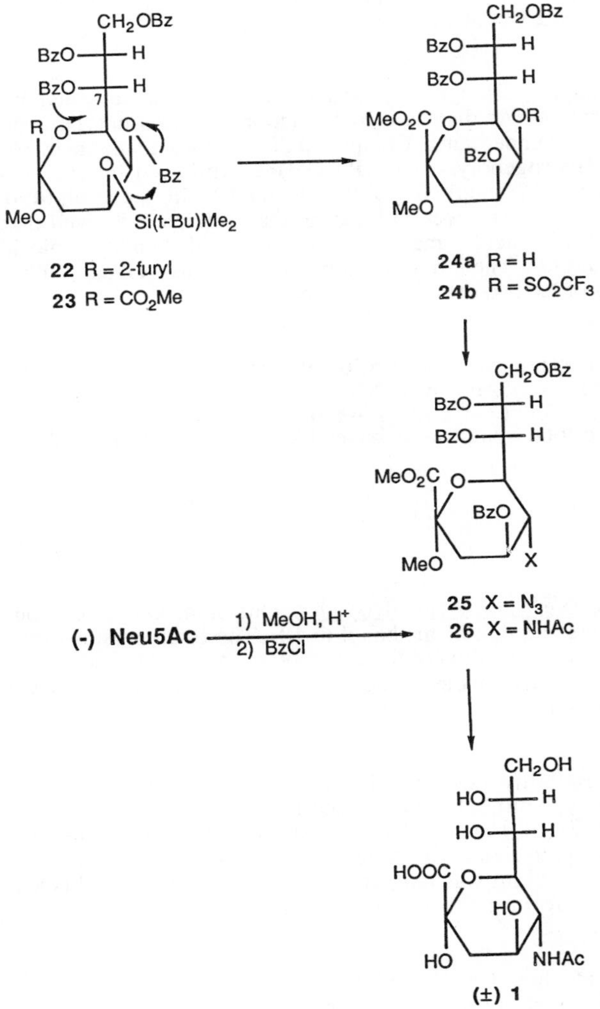

Figure 6. Completion of the synthesis of racemic Neu5Ac.

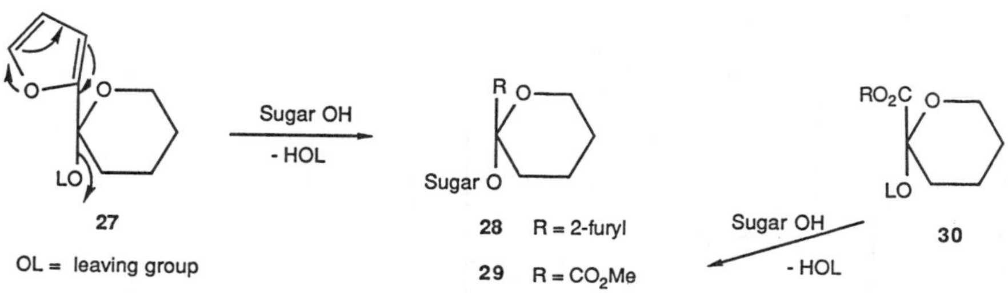

Figure 7. Furan promotion of glycosylation.

Before embarking on either of these routes, it seemed appropriate to investigate the feasibility of the central hypothesis in more accessible model systems. To this end, lactones **31** and **32**, prepared as shown from D-glucal and D-galactal respectively, were converted to the corresponding C_1-furyl substituted methyl glycosides **33** and **34**. We note that in each case the immediate furyllithium addition product existed substantially in open chain form. It was the mixed acetalization process, occasioned by the action of methanol and trimethylorthoformate in the presence of camphorsulfonic acid, which brought about the ring closure.

It was pleasing to find that simple acid-catalyzed exchange between furyl methyl glycoside **33** and the galactose equivalent donor **35** proceeded quite smoothly in benzene under the influence of camphorsulfonic acid. A 60% yield of disaccharides **36** and **37** was obtained. Oxidative cleavage of the mixture followed by esterification gave a 3 : 1 mixture of **38** and **39**. That the major product corresponds to the one with an equatorial glycosidic bond (which is the case with the natural sialoconjugate) was established by NMR analysis in comparison with sialoglycosides of known configuration (19). These results are summarized in Figure 8.

A still more promising result was achieved through the use of compound **40** as the nucleophile (17). The coupling of **40** with **33** and **34** moderated by trimethylsilyltriflate occurred in methylene chloride at -78° C to afford substantially a single isomer in each case in 65-70% yield. The products **36** and **41** were converted, by oxidation-esterification as described above, into **38** and **41**, respectively (see Figure 9).

Thus the furan surrogate provides a promising route to complex conjugates bearing an anomeric C-carboxyl group. It will be recognized that the model systems obtained therefrom correspond, in the relative configurations of the two hexose rings, to that found in complex KDO conjugates. Presumably, with accessibility to the antipodal furyl substituted sialylating agents, the relative configurations found in sialoconjugates could be fashioned. Ongoing studies in our laboratory seek to develop and exploit these themes more fully.

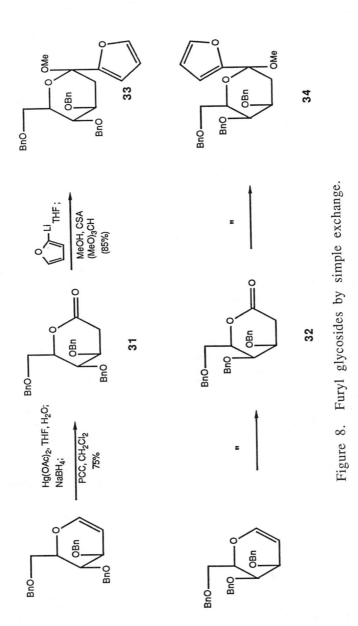

Figure 8. Furyl glycosides by simple exchange.

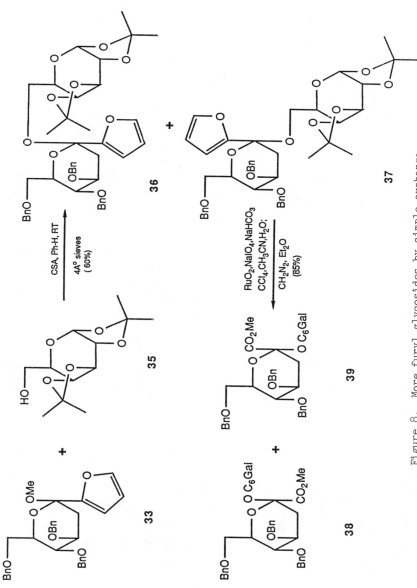

Figure 8. More furyl glycosides by simple exchange.

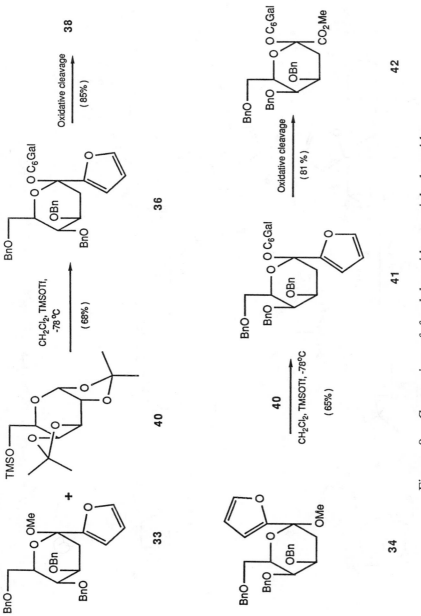

Figure 9. Conversion of furylglycosides to sialoglycoside models.

Acknowledgments
This work was supported by PHS Grant AI 16943. A fellowship from the Corn Refiners Association, Inc. to M.P.D. and a PHS Postdoctoral Fellowship (Grant 1 F32 GM 11104) to J.E.A. are gratefully acknowledged. NMR spectra were obtained through the auspices of the Northeast Regional NSF/NMR Facility at Yale University, which is supported by NSF/Chemistry Division Grant CHE 7916210.

Literature Cited

1. Blix, G.; Gottschalk, A.; Klenk, E. Nature 1957, 179, 1088.
2. Schauer, R. Advances Carbohydr. Chem. Biochem. 1982, 40, 131.
3. Schauer, R. Sialic Acids; Springer-Verlag: Wien, New York, 1982.
4. Sedlacek, H. H.; Weise, M.; Lemmer, A.; Seiler, F. R. Cancer Immunol. Immunother. 1979, 6, 47.
5. Danishefsky, S. J.; DeNinno, M. P. J. Org. Chem. 1986, 51, 2615.
6. Danishefsky, S. J.; DeNinno, M. P. Angew. Chemie Int. Ed. Eng. 1987, 26, 15.
7. Danishefsky, S. J.; DeNinno, M. P.; Phillips, G. B.; Zelle, R. E.; Lartey, P. A. Tetrahedron 1986, 47, 2809.
8. Danishefsky, S. J.; Pearson, W. H.; Segmuller, B. E. J. Am. Chem. Soc. 1985, 107, 1280.
9. Unger, F. M. Adv. Carbohydr. Chem. Biochem. 1981, 38, 323.
10. Anderson, L.; Unger, F. M. Bacterial Lipopolysaccharides: Structure, Synthesis and Biological Activities, American Chemical Society: Washington, DC 1983; ACS Symp. Ser. No. 231.
11. Bednarski, M.; Danishefsky, S. J. Am. Chem. Soc. 1986, 108, 7060.
12. Still, W. C.; Gennari, C. Tetrahedron Lett. 1983, 24, 4405.
13. Houk, K. N.; Moses, S. R.; Rondan, N. G.; Jager, V.; Schohe, R.; Fronczak, F. R. J. Am. Chem. Soc. 1984, 106, 3880.
14. For examples of traditional synthesis of KDO glycosides, see: Paulsen, H.; Hamauchi, Y.; Unger, F. M. Liebigs Ann. Chem. 1984, 1270-97.
15. Recent chemical syntheses of NANA glycosides include: Okamoto, K.; Kondo, T.; Goto, T. Tetrahedron Lett. 1986, 27, 5229; ibid., 5233-6.
16. For combined chemical and enzymatic synthesis, see: Sabesan, S.; Paulson, J. C. J. Am. Chem. Soc. 1986, 108, 2068.
17. Haverkamp, J.; van Halbeek, H.; Dorland, L.; Vliegenthart, J. F. G.; Pfeil, R.; Schauer, R. Eur. J. Biochem. 1982, 122, 305.
18. Schauer, R. Sialic Acids; Springer-Verlag: Wien, New York, 1982.
19. Ogawa, T.; Beppu, K.; Nakabayashi, S. Carbohydr. Res. 1981, 93, C6-C9.

RECEIVED October 14, 1988

Chapter 11

Total Synthesis of the Biologically Active Form of N-Acetylneuraminic Acid

A Stereospecific Route to the Construction of N-Acetylneuraminic Acid Glycosides

Samuel J. Danishefsky and Michael P. DeNinno

Department of Chemistry, Yale University, New Haven, CT 06511

The use of S-2-phenylselenopropanal in the synthesis of the naturally occurring (8R) antipode of the title compound is described. This chemistry was adapted to achieve a total synthesis of a sialic acid glycoside in a stereospecific fashion.

In the preceding chapter, (1-3) we described a total synthesis of racemic (±) N-acetylneuraminic acid (Neu5Ac). In that report we summarized the history of Neu5Ac and its involvement, via glycosides, in a host of important biological functions. We also described some preliminary studies which demonstrated the feasibility of exchanges of the methoxyl group of furyl glycosides (cf. **1** and **2**) with a variety of alcohols including primary OH groups of hexose derivatives (see formation of **3** and **4**). The furyl residues are readily oxidized to carboxylic acids (cf. **3** → **3a**). The sequence of exchange with a hexose based hydroxyl group, followed by oxidation, constituted a possible route to analogs of sialic acid glycosides. Adding to the attractiveness of this strategy was the finding that the exchange reaction was stereospecific, leading to equatorial glycosides of the type **4**.

Of course C-glycosylfurans bearing the functionality pattern suitable for construction of Neu5Ac were intermediates in our total synthesis (3). Recourse to such intermediates for the exchange reaction would provide products which are more realistic in terms of obtaining the full structural and stereochemical detail of Neu5Ac glycosides than was the case for model systems **1** and **2** described in Part I of this volume (1) Indeed, the absolute configurations of the pyranose rings of **1** and **2** are antipodal to the corresponding ring of Neu5Ac glycosides. However, the total synthesis was being conducted in the racemic series. Thus the proposition of using these advanced, fully synthetic intermediates as substrates for the exchange reaction would face the usual awkwardness in coupling a racemate with a single antipode.

The goal of a total synthesis of a biologically active natural product should never be regarded as fully realized until the relevant enantiomer is obtained in homogeneous form. In the case at hand, the attainment of the goal was a particularly urgent matter, since the synthesis of Neu5Ac glycosides was the primary thrust of the effort.

0097–6156/89/0386–0176$06.00/0

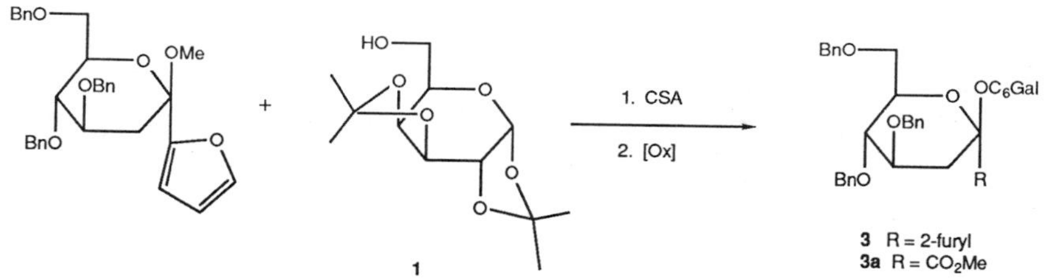

3 R = 2-furyl
3a R = CO₂Me

4 R = 2-furyl
4a R = CO₂Me

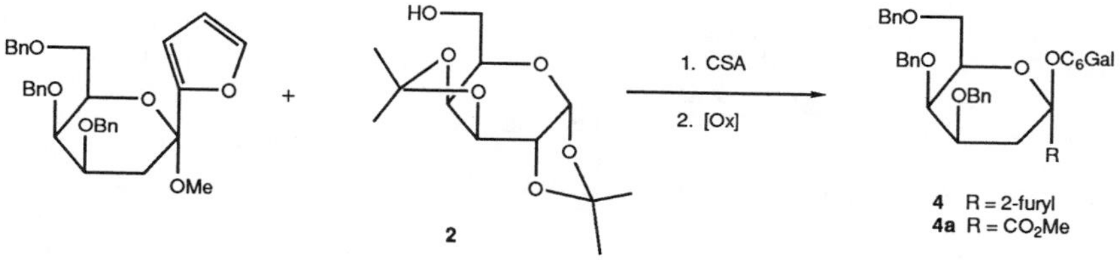

Neu5Ac

KDO

The convergence of two findings simplified the synthesis of naturally occurring Neu5Ac (with the C₈R configuration). The first discovery arose during our synthesis of racemic KDO, wherein the cyclocondensation of racemic **5** with diene **6** afforded, as the principal product, dihydropyrone **7**(4). Indeed compound **7** was accompanied by varying amounts of the corresponding *trans* disubstituted dihydropyrone. However, the important information (not obscured by the fact that the reaction was conducted on racemic material) was the total diastereofacial connectivity, in the Cram Felkin sense (5-6), between the selenium-bearing stereogenic center of **5** and the absolute configuration of the resultant pyranose. For the synthesis of racemic KDO this connectivity was helpful in that it simplified characterization of products, but clearly was not crucial. For the program at hand, it meant that access to the appropriate (S) antipode of the aldehyde (i.e., 5S) would allow for a route to the natural form of Neu5Ac.

The second finding was a disclosure from the laboratory of Paul Hopkins (7) which provided experimental protocols wherein either the S or R enantiomers of compound **5** could be obtained from ethyl lactate. *Thus the combination of the discoveries of Pearson (4) and Hopkins provided the basis for a synthesis of both the naturally occurring (8R) Neu5Ac and (7R) KDO.*

7 X = βOBz; αH
8 X = αOBz; βH

The Synthesis of Neu5Ac

Cyclocondensation of **5S** prepared according to Hopkins with diene **6** using $BF_3 \cdot OEt_2$ in methylene chloride at -78° C afforded a 5:1 ratio of **7** to its *trans* isomer, **8**. Compound **7** was substantially optically pure (enantiomeric homogeneity was eventually achieved after crystallization of enoate **15**). The next sequence of steps was closely patterned after those employed in the racemic KDO synthesis (4). Reduction with sodium borohydride-CeCl$_3$ (8) gave the equatorial alcohol **9** which upon methanolysis afforded **10**. It is seen that the success of this smooth methanolysis reaction was a favorable precedent for the feasibility of the exchange reaction (*vide infra*) in that it suggested stabilization by the furan of the oxonium ion character. Compound **10** was converted to its tert-butyldimethylsilyl ether **11**. Oxidative elimination of the selenoxide, derived from treating **11** with aqueous hydrogen peroxide, afforded overwhelmingly the vinyl compound **12**. Treatment of **12** with osmium tetroxide afforded **13** which, upon exposure to lead tetraacetate, gave aldehyde **14**. The racemic version of **14** was an intermediate in our preparation of racemic Neu5Ac (3). Accordingly, it was a simple matter to retrace these steps in the enantiomorphically homogeneous series to produce optically pure Neu5Ac.

It is of interest to consider in retrospect the logic of the chirality transfers involved in this synthesis. In essence stereochemical information was passed from the stereogenic center of **5S** to control the sense of emergence of the pyranose ring in compound **8**. Eventually, at the stage of Z-enoate **15**, stereochemical information is passed from the ring back to the double bond in the all critical osmylation reaction (9).

The Synthesis of KDO

This synthesis was accomplished using selenoaldehyde **5R**. Again, cyclocondensation with **6** gave ent **7**. Using the same steps as were used for the Neu5Ac synthesis, ent **7** was converted to ent **13**. From there the steps to the naturally occurring (**7R**) antipode of KDO merely involved retracing steps followed in the synthesis of the racemate.

The Synthesis of Sialic Acid Glycosides

After considerable trial and error, compound **16** was identified as the latest intermediate in the Neu5Ac synthesis on which one could achieve the exchange glycosylation. The exchange reaction was carried out using four substrate alcohols. The conditions and results are shown in Table I. In the three carbohydrate cases the furan ring was converted to the corresponding methyl ester by oxidation with ruthenium tetroxide followed by esterification with diazomethane. In each case the equatorial glycoside was obtained stereospecifically. This specificity might well be the result of thermodynamic equilibration at the stage of the furan.

Table I

R-OH	Yield	Yield
	70%	96%
	60%	91%
	73%	92% (compound **17**)
	53%	
	No Reaction	

Finally, in the case of compound **17**, the system was transformed to the sialic acid glycoside **20**. The key step in this regard was the benzoyl migration shown as **18 → 19**. This planned axial benzoate → equatorial benzoate rearrangement served to liberate the single axial alcohol at C5 for activation (triflylation) and displacement (tetra-N-butylammonium azide). An analogous sequence had been employed in the total synthesis of racemic Neu5Ac (3).

The successful application of compound **16** to the synthesis of sialic acid glycosides points to the need for a more direct and efficient route to such systems. Partial synthesis from carbohydrate sources would be a promising alternative to total synthesis. This goal and the goal of extending the scope of the exchange reaction to include secondary sugar alcohol substrates, are now important objectives of our laboratory.

18 R = H; R' = PHCO
19 R = PhCO; R' = H

Acknowledgments

This work was supported by PHS Grant AI 16943. A fellowship from the Corn Refiners Association, Inc. to M.P.D. is gratefully acknowledged. The KDO experiments were carried out by Mr. S-h. Chen of our laboratory. NMR spectra were obtained through the auspices of the Northeast Regional NSF/NMR Facility at Yale University, which is supported by NSF/Chemistry Division Grant CHE 7916210.

Literature Cited

1. Danishefsky, S. J.; DeNinno, M. P.; Audia J. E. Chapter 10 of this volume.
2. Danishefsky, S. J.; DeNinno, M. P.; Chen, S. J. Am. Chem. Soc. In Press.
3. Danishefsky, S. J.; DeNinno, M. P. J. Org. Chem. 1981, 51, 2615.
4. Danishefsky, S. J.; Pearson, W. H.; Segmuller, B. E. J. Am. Chem. Soc. 1985, 107, 1280.
5. Cf. Ahn, N. J. Top. Curr. Chem. 1980, 88 145.
6. Danishefsky, S. J. Aldrichimica Acta 1986, 19, 59.
7. Fitzner, J. N.; Shea, R. G.; Fankhausers, J. E.; Hopkins, P. B. J. Org. Chem. 1985, 50, 417.
8. Gemal, A. C.; Luche, J. L. J. Am. Chem. Soc. 1981, 103, 5454.
9. Danishefsky, S.J.; DeNinno, M.P.; Schulte, G.A. J. Am. Chem. Soc. In Press.

RECEIVED September 27, 1988

Chapter 12

Stereoselective Synthesis of Carbohydrates from Acyclic Precursors

R. R. Schmidt

Fakultät Chemie, Universität Konstanz, D–7750 Konstanz, Federal Republic of Germany

Inverse type hetero-Diels-Alder reactions between β-acyloxy-α-phenylthio substituted α,β-unsaturated cabonyl compounds as 1-oxa-1,3-dienes, enol ethers, α-alkoxy acrylates, and styrenes, respectively, as hetero-dienophiles result in an efficient one step synthesis of highly functionalized 3,4-dihydro-2H-pyrans (hex-4-enopyranosides). These compounds are diastereospecifically transformed into deoxy and amino-deoxy sugars such as the antibiotic ramulosin, in pyridines having a variety of electron donating substituents, in the important 3-deoxy-2-glyculosonates, in precursors for macrolide synthesis, and in $\underline{C}$-aryl-glucopyranosides.

Carbohydrate derivatives and related natural products are mainly synthesized by transformation of readily available sugars ($\underline{1}$-$\underline{3}$). However, the high denisty of functional groups of comparable reactivity requires regioselective protection and deprotection measures and stereospecific functional group exchange reactions which often result in multistep syntheses. Several methods have been developed over the last years which are useful in the diastereoselective and the enantioselective generation of new stereocenters ($\underline{4}$). Therefore "de novo syntheses" or "total syntheses" of carbohydrate derivatives and related natural products from achiral starting materials may be competitive or even superior ($\underline{5}$-$\underline{10}$).

NOTE: This chapter is part 31 in the series, "De Novo Synthesis of Carbohydrates and Related Natural Products".

0097–6156/89/0386–0182$06.00/0
© 1989 American Chemical Society

For the generation of several contiguous chiral centers, two alternative key step reactions can be applied: (i) stereoselective CC-bond formation; and (ii) stereoselective functionalization of compounds having already the required carbon skeleton. In the de novo-synthesis of carbohydrates and related natural products via inverse-type hetero-Diels-Alder reactions both of these principles are especially well documented as will be shown below (8-10).

Aiming at the pyranose form of sugars, normal type hetero-Diels-Alder reactions were extensively used for the synthesis of functionally substituted dihydropyran and tetrahydropyran systems (5-10) (see routes A - D in the general Scheme 1) which are also important targets in the "chiron approach" to natural product syntheses (3). Hetero-Diels-Alder reactions with inverse electron demand such as α,β-unsaturated carbonyl compounds (1-oxa-1,3-dienes) as heterodienes and enol ethers as hetero-dienophiles, are an attractive route for the synthesis of 3,4-dihydro-2$\underline{H}$-pyrans (11).

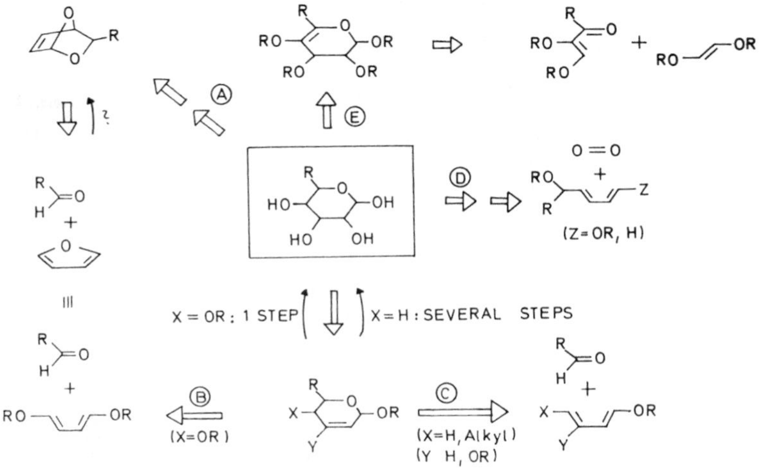

Scheme 1

Recent investigations have demonstrated that electron withdrawing substituents at the α-position increase the rate of this reaction strongly (12). This reaction would have great potential for natural product syntheses provided that additional electron donating functional substituents could be introduced in the β- and α-positions, that enol ethers, enediol ethers, ketene acetals could react as dienophiles (see route E in Scheme 1). In addition, high diastereoselectivities are required in the bond formations.

α-Acyl-β-methoxy Substituted 1-Oxa-1,3-dienes.

We first turned our attention to the reaction of α-methoxymethylene substituted carbonyl compounds such as 1-oxa-1,3-dienes (Scheme 2) with enediol ethers providing the desired 3,4-dihydro-2H-pyrans (hex-4-enopyranosides)in high yields (12-14). However, the *endo/exo*-selectivity was very often low. Correspondingly, low diastereoselectivities were already observed in more simple cases (11); they were explained in terms of secondary orbital overlap of the HOMO of the dienophile and the LUMO of the 1-oxa-1,3-diene (11). According to this principle the *endo*-selectivity could be increased by having better electron donating substituents at the dienophile and/or stronger electron withdrawing substituents at the heterodiene. This was exemplified in several reactions, for instance in the case of a α-nitro substituted α,β-unsaturated carbonyl compound reacting with methyl enediol ehter (10, 12) (Scheme 2).

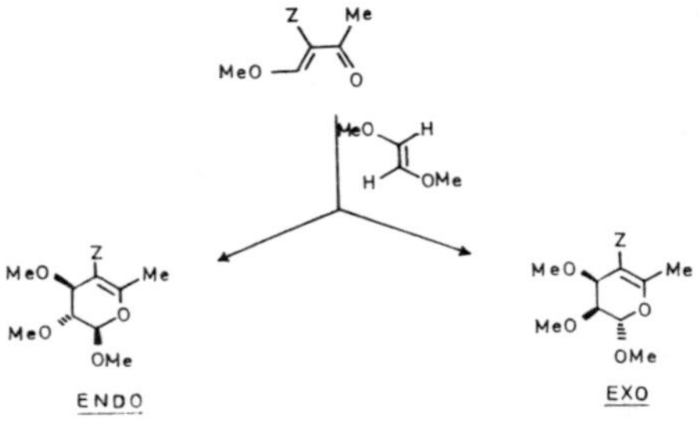

Z	YIELD	ENDO/EXO
-C-Me $\overset{\parallel}{O}$	71 %	1 : 1
- COOMe	69 %	2 : 1
-NO$_2$	46 %	ENDO

Scheme 2

The scope of this cycloaddition reaction was very promising. Subsequently, removal of the CC-double bond and stereoselective functionalizations at positions 4 and 5 (carbohydrate numbering) was investigated for the synthesis of carbohydrates and related natural products, to provide C-3 branched carbohydrate derivatives (or C-4 heteroatom substituted derivatives after carbon/heteroatom exchange reactions). However, the desired hydrogenation of such systems with various hydrogen donors has mainly resulted in low yields and/or side reactions due to the inherent stability of the formal CC-double bond (12, 15).

β-Acyloxy-α-phenylthio Substituted 1-Oxa-1,3-dienes

The aim of using the inverse type hetero-Diels-Alder strategy with 1-oxa-1,3-dienes for the synthesis of carbohydrates and related natural products was to make a carbon substituent at the 4-position redundant. Instead, a functional heterosubstituent would be more advantageous. Because β-alkoxy α,β-unsaturated carbonyl compounds having no electron withdrawing substituent at the α-position are very unreactive towards enol ethers (16), we undertook investigations to introduce a functional substituent at the α-position (appearing in 4-position of the cycloadduct) which (i) increases the rate and the diastereoselectivity of the cycloaddition reaction, and (ii) enables a straight forward introduction of hydroxy, amino, methyl, hydrogen, and perhaps other substituents in a diastereospecific manner. Results with β-acyloxy-α-phenylthio α,β-unsaturated carbonyl compounds as 1-oxa-1,3- dienes demonstrate that the α-phenylthio group in combination with a β-acyloxy group fulfills these requirements (Scheme 3) (12, 15, 17-19). 4-Functionalization and derivatization in general is predisposed by the phenyltioenol ether structure.

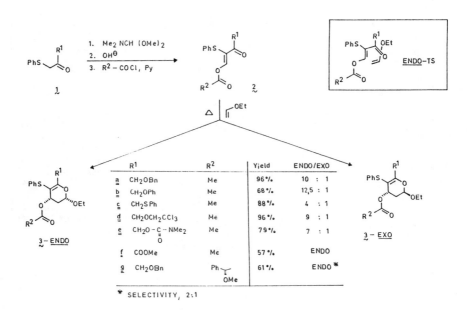

	R¹	R²	Yield	ENDO/EXO
a	CH₂OBn	Me	96%	10 : 1
b	CH₂OPh	Me	68%	12,5 : 1
c	CH₂SPh	Me	88%	4 : 1
d	CH₂OCH₂CCl₃	Me	96%	9 : 1
e	CH₂O-C-NMe₂ ‖ O	Me	79%	7 : 1
f	COOMe	Me	57%	ENDO
g	CH₂OBn	Ph⌣OMe	61%	ENDO *

* SELECTIVITY, 2:1

Scheme 3

The starting materials 2 were readily obtained from phenylthioacetyl derivatives 1. The 1-oxa-1,3-dienes 2a-e exhibited satisfactory reactivity at 50 to 70 °C providing the expected dihydropyrans 3a-e in high yields. Compound 2f furnished the correspondig dihydropyran 3f at room temperature demonstrating an exceptionally high reactivity for this starting material. The endo/exo-diastereoselectivity in these reactions, on the other hand, was found to vary; however, in the most important cases, such as a, d, and f, a high endo-preference was observed. From these results it can be concluded that it is not the extent of the electron withdrawing character of the carbonyl substituents, but the overall electron polarization in the 1-oxa-1,3-diene system, balanced by the attached substituents, and their stereochemical demand, that lead to a favorable HOMO-LUMO interaction in the endo-transition state (18, 19). With the O-methyl mandeloyl group as chiral auxiliary (compound 2g), preferential formation of one diastereomer could be achieved.

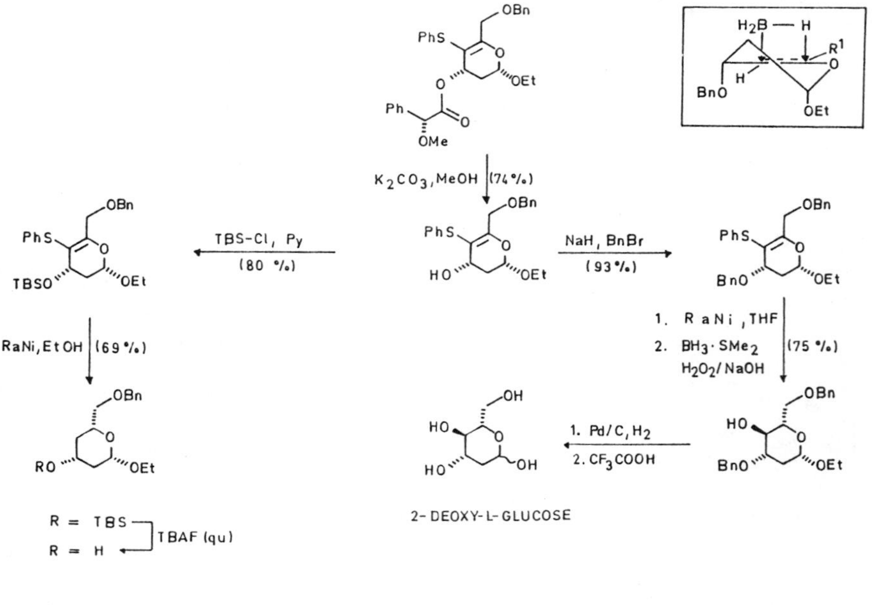

2-DEOXY-L-GLUCOSE

Scheme 4

As indicated in Scheme 4, removal of the 3-O-acyl group and protection of the 3-hydroxy group as tert.-butyldimenthylsilyl or benzyl ethers gave the two 3,4-

dihydropyran intermediates which were diastereospecifi-cally converted into 2,4-dideoxy-L-*threo*- and 2-deoxy-L-*arabino*-hexopyranosides (2-deoxy-L-glucopyranoside), re-spectively (17). As indicated in the box of Scheme 4, due to the anomeric and the allylic effect one conformer is preferred which is diastereospecifically attacked by the borane reagent and also by Raney nickel from the less hindered site. This reaction was recently extended to the use of 1-oxa-1,3-dienes having an acylamino instead of an acyloxy substituent which successfully provided 3-amino-3-deoxy sugars (20).

Thus, the inverse-type hetero-Diels-Alder reaction based hexopyranoside synthesis is high yielding and permits the diastereospecific generation of up to four chiral centers. Concomitant stereocontrolled generation of a fifth chiral center at the 2-position is also possible as is shown below. The direct access to partially O-protected derivatives with different O-protecting groups (see Scheme 4) is an additional advan-tage of this method because carbohydrates are usually required for regioselective glycoside bond formations in the form of glycosyl donors and acceptors.

Modification of the 1-Oxa-1,3-diene

The attachment of a chiral group in the 2-position of the 1-oxa-1,3-diene should be a simple means for diaste-reoselective syntheses of higher sugars. This is exempli-fied in Scheme 5 for the synthesis of a 2-deoxy-D-*galacto*-heptose derivative (De Gaudenzi, L; Apparao, S.; Schmidt, R.R.; unpublished results). From methyl glyce-rate the required optically active 1-oxa-1,3-diene with D-*glycero* configuration was obtained. With enol ether as a heterodienophile, the D-*erythro*-dihydropyran was formed preferentially. The diastereomer ratio could be even increased with the help of the O-methyl-L-mandeloyl group as chiral auxiliary by double diastereoselection. The usual transformations provided the 4-O-unprotected 2-deoxy-D-*galacto*-heptose derivative as an ethyl glycoside. The structure was assigned through transformation to the known ethyl 3,4,6,-tri-O-benzyl-2-deoxy-α-L-glucopyrano-side (15).

The general usefulness of the developed methodology can be verified also in different areas. Thus chemoselec-tive reactions could be also carried out with β-unsubsti-tuted α-phenylthio substituted α,β-unsaturated carbonyl compounds and enol ethers as outlined in Scheme 6 (21). Naturally occuring 2,3,6-trideoxy and 4-amino-2,3,4,6-tetradeoxy sugars are obtained quite readily from this methodology.

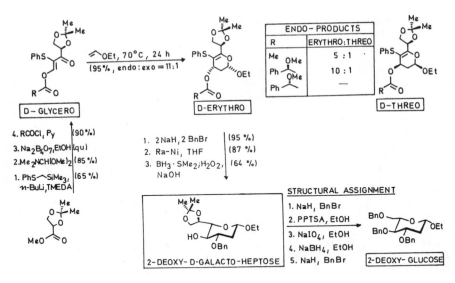

Scheme 5

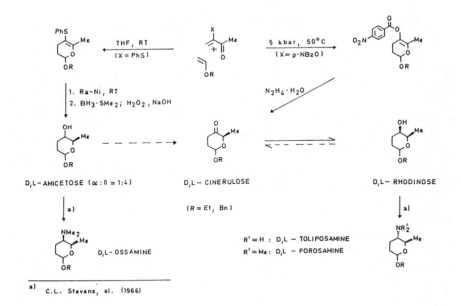

Scheme 6

The efficiency of this approach with carbon substituents at the β-position was demonstrated in the synthesis of the antibiotic ramulosin (Scheme 7) (<u>10</u>, <u>21</u>). This

approach should also be applicable to the synthesis of
the biogenetically related antibiotics hydroxyramulosin
and actinobolin which recently gained importance because
of interesting biological properties. Ramulosin inhibits
the germination of seeds and spores of microorganisms
(22). Actinobolin has broad spectrum antibiotic and moderate antitumor activity. It is a potential cariostatic
agent (23). This compound is structurally related to the
very promising antitumor agent bactobolin (24). As outlined in the retrosynthesis (Scheme 7) the required 1-
oxa-1,3-diene precursor B should be readily available
from a 4-formylbutanoate derivative and phenylthio-
acetone. Hetero-Diels-Alder reaction with enol ether or
ketene acetal derivatives should then provide the desired
dihydropyran intermediate A which after diastereoselective CC-double bond transformation and chemoselective
ring closure yields the target molecule.

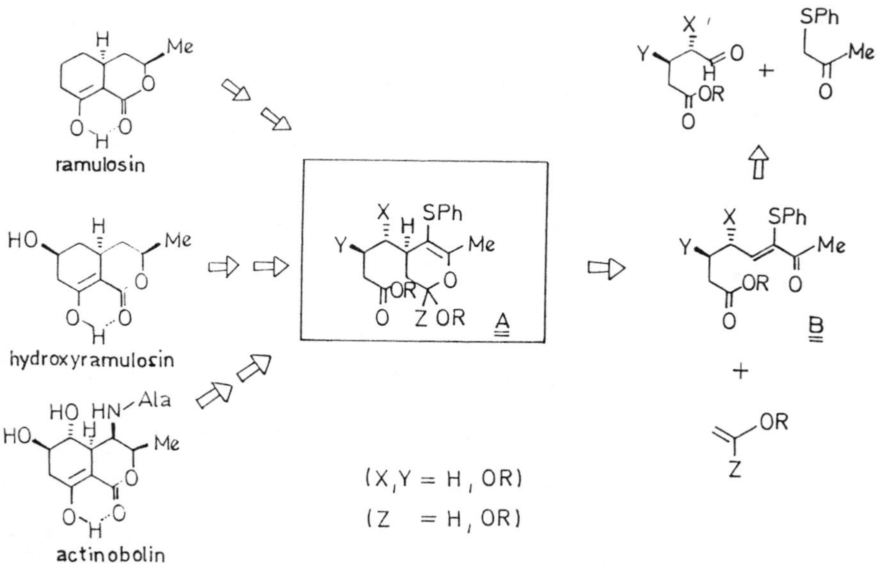

Scheme 7   (Reprinted with permission from ref. 21.   Copyright
1988 Pergamon.)

The required 1-oxa-1,3-diene precursor was synthesized according to the synthesis design (Scheme 8).
Cycloadditon with enol ether furnished exclusively the
endo-isomer. Raney nickel treatment in refluxing ethanol
yielded in one step the desired tetrahydropyran
derivative in a favorable 6:1 *cis/trans* ratio.
Transformation into the lactone and ring closure with
potassium tert.-butoxide afforded (±)-ramulosin.

Alternatively, in an even shorter route through ketene dithioacetal cycloaddition this molecule was obtained (Scheme 8) (10, 21).

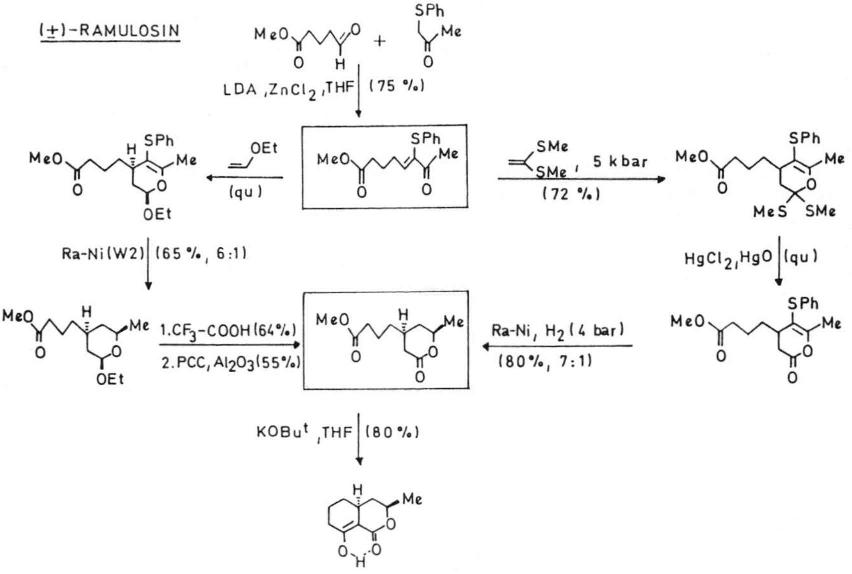

Scheme 8   (Reprinted with permission from ref. 21.   Copyright 1988 Pergamon.)

Modification of the Heterodienophile

The inverse type hetero-Diels-Alder reaction of functionally substituted α,β-unsaturated carbonyl compounds was also possible with C-alkyl substituted enol ethers (Scheme 9) (25). This was demonstrated in the synthesis of pyridines required for the formation of potential ATPase inhibitors. Deacylation of the cycloadducts, oxidation of the hydroxy group, alcohol elimination, and ammonia treatment provided γ-pyridones. They were transformed upon O-alkylation into pyridines possessing an interesting array of different electron donating substituents.

Also propiogeninic type macrolide moieties should be accessible via this inverse type hetero-Diels-Alder methodology. This was recently exhibited starting from cis-propenylether as heterodienophile (Schmidt, R.R.; Haag-Zeino, B.; Hoch, M.; Liebigs Ann. Chem., in press). In a highly endo-selective cycloaddition reaction and in subsequent diastereoselective transformations of the methyl substituted dihydropyran obtained a 2,4-dimethyl-

3,5,6-trihydroxy-hexanal structure could be generated with stereocontrol over five consecutive chiral centers. The relative stereochemistry of carbon atoms C-2 to C-5 matches a partial structure of rifamycin S.

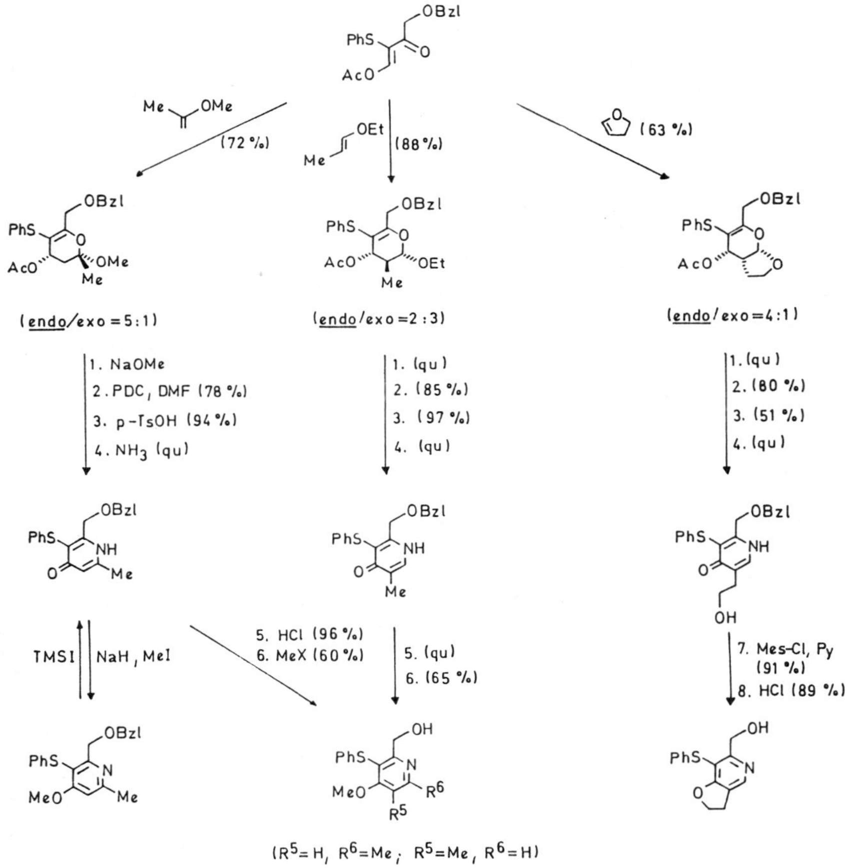

Scheme 9

3-Deoxy-2-glyculosonates are biochemically generated from phosphoenol pyruvate and phosphorylated sugars in an aldol type reaction (Scheme 10, routes A - C) (26). They are an important class of natural products: 3-deoxy-D-*arabino*-2-heptulosonate (pathway A) is a metabolic intermediate in the shikimate pathway, providing aromatic amino acids (Phe, Tyr, Trp) and finally proteins thereof; 3-deoxy-D-*manno*-2-octulosonate (KDO) (pathway B) is an integral constituent of the lipopoly-saccharides of Gram-negative bacteria (27); 5-acetylamino-3,5-dideoxy-D-

glycero-D-galcto-2-nonulosonate, commonly known by its
trivial name N-acetyl-neuraminic acid (NeuNAc) (pathway
C), is an integral part of many glycosphingolipids and
glycoproteins (26, 28).

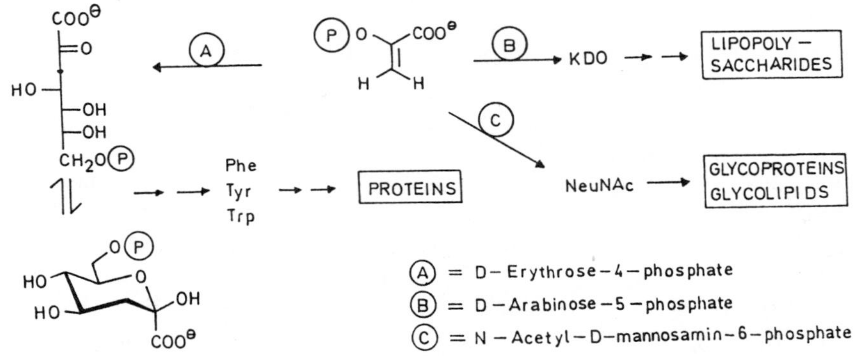

Scheme 10

The biological importance of these compounds prompted
development of several synthetic approaches, mainly for
KDO and NeuNAc and derivatives thereof (26, 27, 29-32).
We devised a de novo-synthesis approach to this class
of compounds based on the inverse-type hetero-Diels-Alder
methodology, which is exhibited in the retrosynthetic
Scheme 11.

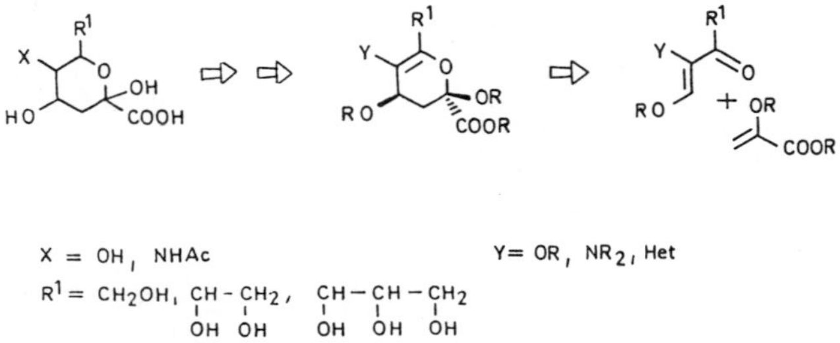

Scheme 11

The efficiency of this strategy for the synthesis of
3-deoxy-D-arabino-2-heptulosonates is outlined in Scheme
12 (53).

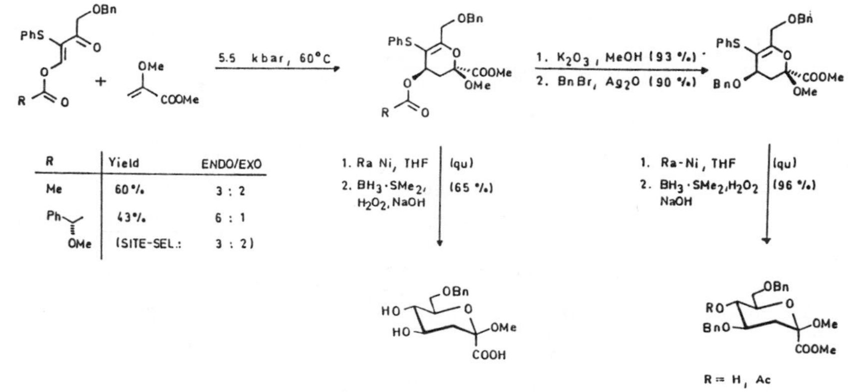

R	Yield	ENDO/EXO
Me	60%	3 : 2
Ph–CH(OMe)	43%	6 : 1
	(SITE-SEL.:	3 : 2)

Scheme 12

The basis of this strategy is a cycloaddition reaction between the β-acyloxy-α-phenylthio substituted 1-oxa-1,3-diene and α-alkoxy acrylates as heterodienophiles. Because of low reactivity, the reaction had to be carried out under higher pressure, thus providing regiospecifically and with *endo*-preference the desired dihydropyran. The usual transformations either via O-deacylation or directly in an overall three step route furnished the target molecule again in a valuable partly protected form. Preliminary investigations towards asymmetric induction of the cycloaddition reaction with the help of the chiral O-methyl mandeloyl group showed a good *endo*-preference; however, the site-selectivity in the *endo*-product was only 3:2 and has to be improved in further investigations (<u>33</u>). This general strategy is also exploited for the synthesis of KDO and of NeuNAc (Apparao, S.; De Gaudenzi, L.; Haag-Zeino, B; Schmidt, R.R. unpublished results).

<u>C</u>-Aryl-glycosides are wide-spread in Nature (<u>34</u>, <u>35</u>). The biological importance led to various synthetic approaches. Mainly glycosyl donors and activated benzene derivatives in Friedel-Crafts type reactions were applied (<u>36</u>, <u>37</u>), however, many difficulties were encountered in these reactions. Therefore we investigated the addition of styrenes as heterodienophile with the above described 1-oxa-1,3-dienes. Again under higher pressure an *endo*-specific cycloaddition reaction was observed, providing interesting intermediates for convenient <u>C</u>-aryl-glycoside syntheses. For instance, <u>C</u>-aryl glucopyranosides and 2-deoxyglucopyranosides were obtained via this route (<u>33</u>). This is demonstrated in Scheme 13 for a 2-deoxy-β-gluco-pyranosyl derivative of 1,2-dimethoxybenzene.

C—ARYL—2—DEOXY — GLUCOPYRANOSIDES

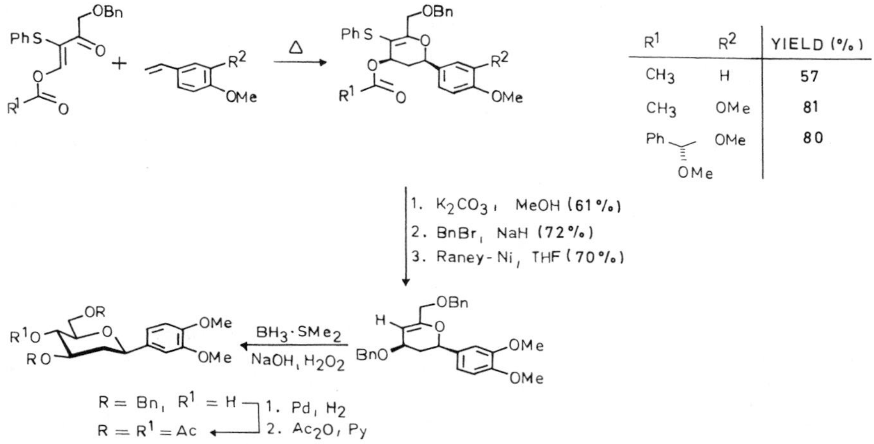

R^1	R^2	YIELD (%)
CH_3	H	57
CH_3	OMe	81
Ph⌣OMe	OMe	80

1. K_2CO_3, MeOH (61%)
2. BnBr, NaH (72%)
3. Raney-Ni, THF (70%)

$BH_3 \cdot SMe_2$
NaOH, H_2O_2

R = Bn, R^1 = H ┐ 1. Pd, H_2
R = R^1 = Ac ┘ 2. Ac_2O, Py

Scheme 13

Acknowledgments

The research described here was carried out by S. Apparao, L. De Gaudenzi, W. Frick, B. Haag-Zeino, M.E. Maier, and K. Vogt. I am particularly indebted to these coworkers. Financial support by the Deutsche Forschungsgemeinschaft and the Fonds der Chemischen Industrie is gratefully acknowledged.

Literature Cited

1. Hanessian, S. Acc. chem. Res. 1979, 12, 159-165.
2. Schmidt, R.R.; Maier, M. Synthesis 1982, 747-748.
3. Hanessian, S. "Total Synthesis of Natural Products: The Chiron Approach"; Pergamon Press: Oxford 1983.
4. Nogradi, M. "Stereoselective Synthesis"; VCH Verlagsgesellschaft mbH: D-6940 Weinheim 1987; and references cited therein.
5. Danishefsky, S.J. Acc. Chem. Res. 1980, 14, 400-407.
6. Zamojski, A.; Banaszek, A.; Grynkiewicz, G. Adv. Carbohydr. Chem. Biochem. 1982, 40, 1-129.
7. Hoppe, D. Nachr. Chem. Techn. Lab. 1982, 30, 403-406.

8. McGarvey, G.J.; Kimura, M.; Oh, T.; Williams, J.M. J. Carbohydr. Chem. 1983, 3, 125-188.

9. Schmidt, R.R. Acc. Chem. Res. 1986, 19, 250-259.

10. Schmidt, R.R. Pure Appl. Chem. 1987, 59, 415-424.

11. Desimoni, G.; Tacconi, G. Chem. Rev. 1975, 75, 651-692.

12. Maier, M.; Schmidt, R.R. Liebigs Ann. Chem. 1985, 2261-2284; and references cited therein.

13. Schmidt, R.R.; Maier, M. Tetrahedron Lett. 1982, 23, 1789-1792.

14. Tietze, L.-F.; Glüsenkamp, K.H. Angew. Chem. 1983, 95, 901-902; Angew. Chem., Int. Ed. Engl. 1983, 22, 887-888.

15. Maier, M. Ph.D. Thesis, Universität Konstanz, D-7750 Konstanz, 1985.

16. Berti, G.; Catelani, G; Colonna, F.; Monti, L. Tetrahedron 1982, 38, 3067-3072.

17. Schmidt, R.R.; Maier, M. Tetrahedron Lett. 1985, 26, 2065-2068.

18. Apparao, S.; Schmidt, R.R. Synthesis 1987, 896-899.

19. Apparao, S.; Maier, M.E.; Schmidt, R.R. Synthesis 1987, 900-904.

20. Tietze, L.-F.; Voß, E. Tetrahedron Lett. 1986, 27, 6181-6184.

21. Vogt, K.; Schmidt, R.R. Tetrahedron 1988, 44, 3271-3280.

22. Stodola, F.H.; Cabot, C.; Benjamin, C.R. Biochem. J. 1964, 93, 92-97.

23. Hunt, D.E.; Hunt, J.K. Arch. Oral Biol. 1980, 25, 431-434; and references cited therein.

24. Ishizaka, M.; Fukasawa, S; Masuda, T.; Sato, J.; Kanbayashi, N.; Takeuchi, T.; Umezawa, H. J. Antibiotics 1980, 33, 1054.

25. Haag-Zeino, B.; Maier, M.E.; Schmidt, R.R. Chem. Ber. 1987, 120, 1505-1509.

26. Schmidt, R.R. In "Organic Synthesis an Interdisciplinary Challenge, V th IUPAC Symposium; Streith, J.; Prinzbach, H.; Schill, G., Eds.; Blackwell Scientific Publications: Oxford 1985; pp. 281-294.

27. Unger, F.M. Adv. Carbohydr. Chem. Biochem. 1981, 38, 323-388.

28. Schmidt, R.R.; Angew. Chem. 1986, 98, 213-236; Angew. Chem.,Int. Ed. Angl. 1986, 25, 212-235; and references cited therein.

29. Schmidt, R.R.; Betz, R. Angew. Chem. 1984, 96, 420-421; Angew. Chem., Int. Ed. Engl. 1984, 23, 430-431.

30. Danishefsky, S.J.; Pearson, W.H.; Segmuller, B.E.

31. Schauer, R. Adv. Carbohydr. Chem. Biochem. 1982, 40, 131-234.

32. Danishefsky, S.J.; DeNinno, M.P. J. Org. Chem. 1986, 51, 2615-2617.

33. Schmidt, R.R.; Frick, W.; Haag-Zeino, B.; Apparao, S. Tetrahedron Lett. 1987, 28, 4045-4048.
34. Hagues, L.J. Adv. Carbohydr. Chem. Biochem. 1965, 20, 357-369.
35. Asakawa, J. Progr. Chem. Org. Nat. Prod. 1982, 42, 154-163.
36. Effenberger, G. Ph.D. Thesis, Universität Konstanz, D-7750 Konstanz, 1985.
37. Schmidt, R.R.; Effenberger, G. Carbohydr. Res. 1987, 171, 59-79.

RECEIVED October 14, 1988

Chapter 13

Synthesis of Carbohydrates and Derivatives from 7-Oxanorbornenes ("Naked Sugars")

Pierre Vogel, Yves Auberson, Mampuya Bimwala,
Etienne de Guchteneere, Eric Vieira, and Jürgen Wagner

Institut de Chimie Organique de l'Université, 2 Rue de la Barre,
CH 1005 Lausanne, Switzerland

The regioselectivity of electrophilic additions of the C=C double bond in 7-oxabicyclo[2.2.1]hept-5-en-2-yl (7-oxanorborn-5-en-2-yl) derivatives depends on the nature of the substituents at C(2). The adducts so-obtained can be transformed into the corresponding 5,6-disubstituted 7-oxanorbornan-2-ones, which can be mono-substituted at C(3) stereoselectively, giving products with the same stereochemical information as hexoses. Thus, optically pure 7-oxanorborn-5-en-2-yl derivatives can be viewed as "naked sugars" Applications to the total, asymmetric syntheses of L-daunosamine, 2-deoxy-L-fucose, D- and L-allose, D- and L-talose, D- and L-ribose, D- and L-methyl 2,5-anhydroallonate, cyclohexenepolyols, (4R,5R,6R)-2-crotonyloxymethyl-4,5,6-trihydroxycyclohex-2-enone, (+)- and (-)-methyl nonactate are presented.

An increasing number of rare, natural, carbohydrate derivatives are being discovered. Some of them show interesting biological properties. Others are part of important molecules such as antibiotics or enzyme inhibitors.[1] Because of their rarity, there is a need to develop synthetic technologies that allow one to prepare them in significant amounts. Furthermore, many unnatural carbohydrate derivatives also show interesting biological activities and thus justify the invention of new synthetic methods for their preparation. Many natural sugars (e.g., D-glactose, D-glucose) are obvious starting materials because they are inexpensive and optically pure. However, structural modification of their skeletons and/or changes in their substitution often requires many protection and deprotection steps that may lead to lengthy procedures of no practical value. Alternatively, more efficient procedures can be envisioned through total, asymmetric synthesis. The concept of "naked sugars" that is presented here is one approach towards that goal.

From MO Calculations to the "Naked Sugars"

Methyl propynoate adds to 5,6-dimethylidenebicyclo[2.2.1]heptan-2-one (1) with

0097–6156/89/0386–0197$12.25/0
© 1989 American Chemical Society

relatively good "para" regioselectivity, giving the major Diels-Alder adduct **2**.[2] The results were in agreement with predictions based on FMO theory which considers the HOMO(diene) - LUMO(dienophile) orbital interaction to control the regioselectivity of the cycloaddition.[3] The HOMO of **1** showed a significant difference in the size of the p atomic coefficients between those at $C(5)=CH_2$ and those at $C(6)=CH_2$.[2] Furthermore, the shape of this HOMO suggested also an important hyperconjugative interaction of $n(CO) \leftrightarrow \sigma C(1),C(2) \leftrightarrow \pi C(5),C(6)$ in **1** that makes the carbonyl group act as an electron-donating homoconjugated substituent. Photoelectron spectra of **1**, 2,3,5,6-tetramethylidenebicyclo[2.2.1]-heptan-7-one and related β,γ-unsaturated ketones and alkenes[4] confirmed the significant interactions between the lone-pair orbital n(CO) and the homoconjugated double-bond π orbitals[4,5] which are due mostly to a "through-bond" mechanism.[6] Semi-empirical calculations first,[7] and then *ab initio* STO 3G MO calculations,[8] confirmed the hypothesis that a homoconjugated carbonyl group can act as an electron-donating substituent, as it was found that the 6-oxobicyclo[2.2.1]hept-2-yl cation (**3**) is more stable than its 5-oxo isomer **4** (for completely optimized geometries).

	E		
1	E = COOCH$_3$	**2** (major)	

$$\Delta E = + \ 8 \text{ kcal/mol (MINDO/3)}^{7}$$
$$+ \ 2.8 \text{ kcal/mol (MNDO)}^{7}$$
$$+ \ 3.0 \text{ kcal/mol (AM 1)}$$
$$+ \ 7 \text{ kcal /mol (STO 3G)}^{8}$$

The relatively long C(1)-C(6) (2.247 Å) and short C(1)-C(2) (1.427 Å) bond calculated for **3** were in agreement with the hyperconjugative stabilizing interaction $n(CO) \leftrightarrow \sigma C(1),C(2) \leftrightarrow pC(6)$ which can be represented by the limiting structures **3** $\leftrightarrow$ **3'**. These features were not present in the case of 5-cyano- and 6-cyanobicyclo[2.2.1]hept-2-yl cations, the former ion being calculated to be more stable than the latter,[8] as expected for carbocations remotely perturbed by an electron-withdrawing group such as the CN-substituent (field effect).[9] Following these calculations, we predicted that additions of electrophiles E^+X^- to the olefinic moieties of bicyclo[2.2.n]alk-5-en-2-ones 5 - 7 should be regioselective and give preferentially adducts 11 under conditions of kinetic control. In contrast, their synthetic precursors such as the corresponding chlorocarbonitriles 12, 13 or cyano-acetates 14, 15 should give preferentially adducts 19.

5 Z = CH$_2$
6 Z = CH$_2$CH$_2$
7 Z = O

12 Y = Cl Z = CH$_2$ 16
13 Y = Cl Z = CH$_2$CH$_2$
14 Y = OAc Z = CH$_2$
15 Y = OAc Z = O

A sulfur or selenium electrophile, E^+, is expected to engender bridged cationic intermediates 9[10] when reacting with enones 5 - 7. The onium ions 9 are then attacked preferentially at C(6) by the nucleophile, or counter-ion, X^-, because the corresponding limiting structures 10 contribute more than 8, the former being stabilized by the electron-donating ability of the homoconjugated carbonyl group. In the case of the electrophilic additions of alkenes 12 - 15, the corresponding cationic intermediates 17 are expected to be attacked by the nucleophile X^- preferentially at C(5) because the corresponding limiting structures 18 should contribute more than 16 by virtue of the electron-withdrawing effect of the substituents at C(2). Alternatively, the regiochemistry of addition to 12 - 15 could be sterically controlled by repulsion between the incoming *endo* nucleophile and the *endo* C(2) substituent. Completely optimized geometries obtained by the *ab initio* STO 3G calculation technique for the model 2,3-episulfoniobicyclo[2.2.1]heptanes 20 - 24 (see Table 1) confirmed our hypotheses.[8]

The difference in bond length between $C(3)-S^+$ and $C(2)-S^+$ in 20 is calculated to be 0.004 Å. Although very small, it is indicative of the contribution from limiting structures of type 20 ↔ 25 to the stabilization of the ion. This interpretation was confirmed by the "extra" long C(4)-C(5) bond (1.573 Å) in 20, which should be compared with that calculated for the 5-dicyanomethylidene-2,3-episulfonionor-bornane (**21**, 1.54 Å). The data suggest that homoconjugation **20** ↔ **26** is of little importance in contrast with homoconjugation **22** ↔ **27** which was suggested for the 5-methylidene-2,3-episulfonionorbornane (**22**). In agreement with this latter hypothesis, the C(3)-S$^+$ bond in **22** was calculated to be 0.007 Å longer than the C(2)-S$^+$ bond. Furthermore, the C(3)-C(5) distance is significantly reduced in comparison with that in 5-*exo*-cyano-2,3-episulfonionorbornane (**23**) and the parent cation **24**.

Table 1. Ab initio STO 3G miminized geometries of episulfonium cations **20** - **24**. Bond lengths in Å [8]

Bond:	20	21	22	23	24
C(1)-C(2)	1.546	1.546	1.544	1.544	1.543
C(2)-S$^+$	1.869	1.868	1.870	1.870	1.871
C(3)-S$^+$	1.873	1.867	1.877	1.867	1.871
C(3)-C(4)	1.541	1.549	1.547	1.543	1.543
C(4)-C(5)	1.573	1.540	1.542	1.572	1.561
C(6)-C(1)	1.556	1.561	1.561	1.560	1.561
C(6)-C(2)	2.438	2.438	2.434	2.434	2.431
C(5)-C(3)	2.409	2.390	2.393	2.435	2.430

The episulfonium ions **20** - **23** are expected to be quenched by a nucleophile (X$^-$) onto the *endo* face at the most polarizable center, i.e. the center C(2) or C(3) attached to the sulfur atom with the longest bond. Thus **20** and **22** were predicted to add X$^-$ preferentially onto C(3) whereas **23** would add X$^-$ preferentially onto C(2). In the case of **21**, no regioselectivity was expected. These predictions were in agreement with the experimental regioselectivities observed for the additions of sulfur and selenium electrophilic agents to norborn-5-en-2-one (bicyclo[2.2.1]-hept-5-en-2-one (**5**)), bicyclo[2.2.2]oct-5-en-2-one (**6**),[11] 7-oxanorborn-5-en-2-one (**7**),[12] and to the corresponding synthetic precursors **12** - **15**[11,12] and to 5-dicyano-methylidenenorborn-2-ene.[13]

These results demonstrated that direct substitution of centers C(5) and C(6) in 7-oxanorborn-5-en-2-yl derivatives **7** and **15** can be carried out with high stereo- and regioselectivity.[12] As expected for bicyclo[2.2.1]hept-2-ene derivatives,[14] the electrophiles attack preferentially onto the *exo* face of the double bond. An exception to that rule has been observed for the PhSeCl addition to the *endo* alcohol 28 which gave adduct 29 as the major product.[15] Interestingly, addition of PhSCl to 28 preferred the *exo* face of the olefin and led to the tricyclic derivative 30.[15]

As we shall see, optically pure 7-oxanorborn-5-en-2-yl derivatives 31[16] and 32[17] can be obtained readily. Saponification of acetate 31 or camphanate 32 gives (1R)-7-oxanorborn-5-en-2-one ((+)-7). In the latter reaction (32 → (+)-7) the chiral auxiliary (R*OH = (1S)-camphanic acid) is recovered. It can be recycled into 32. The regioselectivity of the electrophilic additions are controlled by the nature of the substituents at C(2) (e.g. 32 → 33, (+)-7 → 34, Scheme 1). Saponification of adducts 33 gives the corresponding ketone 35 and the chiral auxiliary R*OH. 34 and 35 can be substituted at C(3) stereoselectively into derivatives 36 and 37, respectively, by direct procedures. The 7-oxanorborn-5-en-2-yl derivatives (+)-7, 31 and 33 possess the same stereochemical information as hexoses and, therefore, can be viewed as "naked sugars". Using (1R)-camphanic acid (R'OH) instead of (1S)-camphanic acid (R*OH) as chiral auxiliary, the "naked sugars" 38 and (-)-7 are also available.

Preparations of the "naked sugars" and their application to the synthesis of carbohydrate derivatives and other natural products are presented below.

Synthesis of Optically Pure 7-Oxanorbornane Derivatives

Derivatives of 7-oxanorbornane[18] have been used as starting materials in the synthesis of racemic carbohydrates,[19] nucleosides,[20] muscarine derivatives,[21] prostaglandins,[22] methyl shikimate,[23] anthracyclines,[24] cis-maneonenes-A and -B,[25] Citreoviral,[26] inositols[26b] and other products of biological interest.[27] Some 7-oxanorbornane derivatives have biological activities.[28] These bicyclic systems are available most simply through intermolecular *Diels-Alder* additions of furans to activated dienophiles,[29] and these often sluggish cycloadditions can be accelerated by applying very high pressure[30] or, more simply, by adding catalytical amounts of a *Lewis* acid.[16,17,31,32] Care must be taken when chosing an acidic catalyst as polymerization of the furans and/or rearrangement of the adducts can be competitive processes.[33] A priori, the ideal method for generating optically pure 7-oxanorborn-5-en-2-yl derivatives would be to add furan to a mono- or 1,1-disubstituted olefin in the presence of an optically pure, chiral catalyst capable of inducing an asymmetric cycloaddition.[34] Because the retro-*Diels-Alder* reaction of 7-oxanorbornenes can compete with the forward reaction (aromaticity of furan), causing equilibration of enantiomeric and diastereomeric adducts, the catalyst must be active at relatively low temperature (< 20 °C). To our knowledge no such catalyst has been found yet.

In the presence of a catalytical amount of ZnI_2, (-)-1-cyanovinyl camphanate, derived from the commercially available (-)-(1S)-camphanoyl chloride and 2-oxopropionitrile, added to furan (an inexpensive starting material derived from the biomass[35]) and afforded a 45:45:5:5 mixture of adducts 32/40/41/42 in 90 % yield from which 32 could be isolated (98 % d.e.) in 29 % yield after repetitive

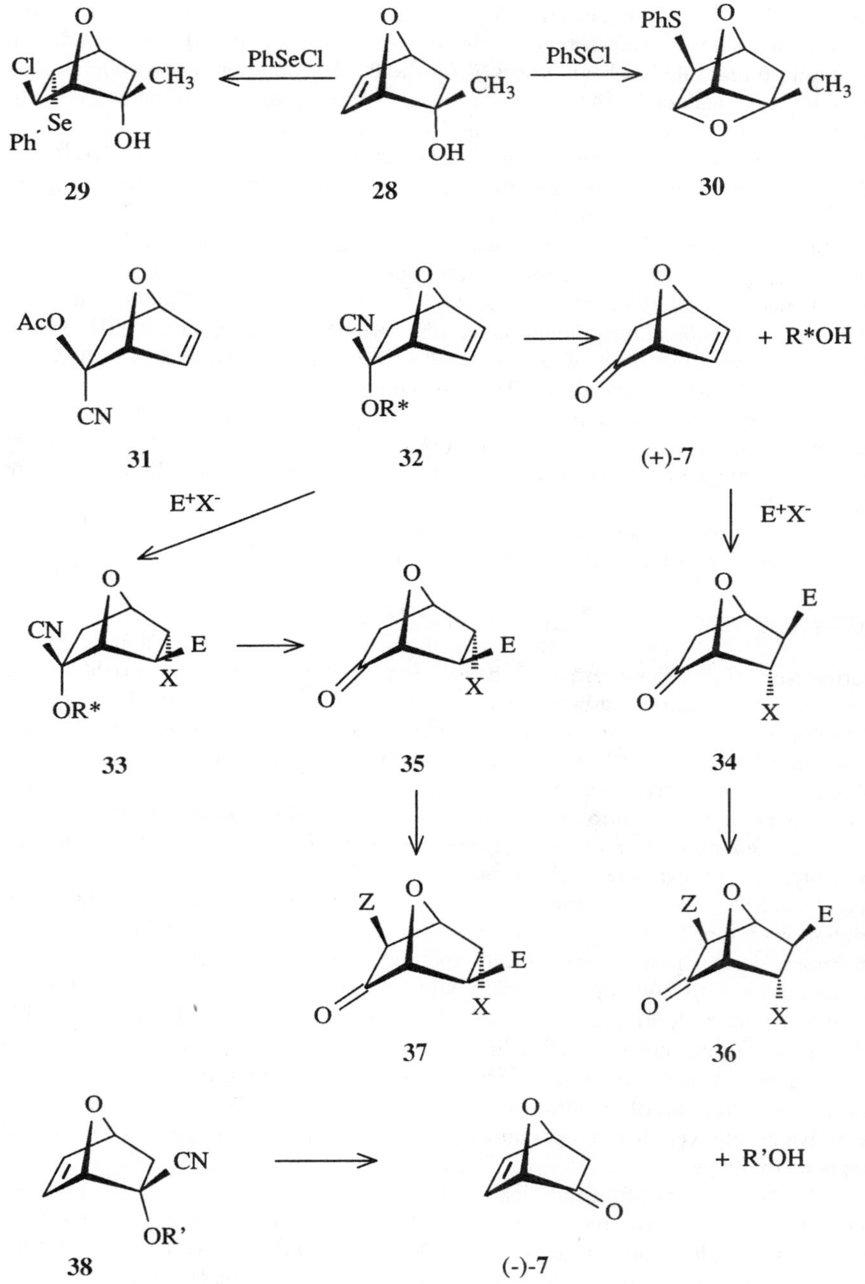

Scheme 1

recrystallization from hexane/ethyl acetate. The remaining adducts were recycled nearly quantitatively into furan and dienophile **39** on heating in toluene. Saponification of **32** furnished (+)-**7**(96 %) and (-)-camphanic acid (76 %).[17]

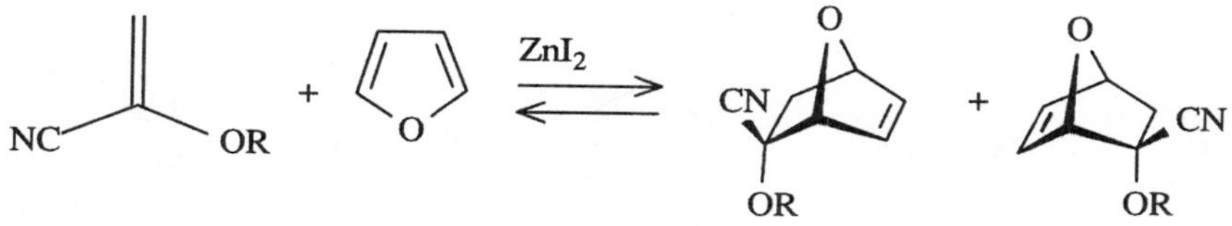

39   R = R*

32 R = R* **40** R = R*
43 R = H   **44** R = H
47 R = Ac **48** R = Ac

R* =

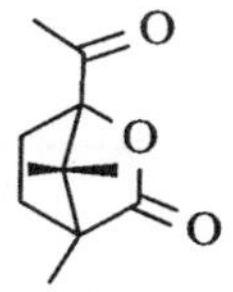

41 R = R* **42** R = R*
45 R = H   **46** R = H
31 R = Ac **49** R = Ac

Another practical method for the preparation of "naked sugars" (+)-**7** ((+)-(1R,4R)-7-oxabicyclo[2.2.1]hept-5-en-2-one) and **31** ((+)-(1R,2S,4R)-2-*endo*-cyano-7-oxabicyclo[2.2.1]hept-5-en-2-exo-yl acetate) involves the diastereo-selective formation of a brucine complex with the corresponding cyanohydrins **43** - **46**, obtained by saponification of the adducts **31**, **47** - **49** resulting from the ZnI$_2$-catalyzed *Diels-Alder* addition of furan to 1-cyanovinyl acetate.[16] When 0.5 equivalents of brucine was added to a basic, methanolic, solution of **43** - **46**, a white complex precipitated (36 %). Treatment with MeONa in MeOH (20 °C, 2 h) and then formalin (40 % aq. H$_2$CO) afforded (+)-**7** (93 %) in 86 % optical purity. When the above complex was treated with Ac$_2$O/pyridine in CHCl$_3$ (20 °C, 24 h) the acetate, **31** (>99 % e.e), was isolated in 20 % yield after recrystallization from ether/petroleum ether.[36] Brucine was also recovered. Mother-liquors were concentrated and treated with Ac$_2$O/pyridine, giving a mixture of **31**, **47** - **49** that could be thermolysed into furan and 1-cyanovinyl acetate. Optically pure (+)-**7** (>99 % e.e.) was also obtained by saponification of **31**.

High diastereoselectivity was reported recently by *Koizumi* and co-workers[37] for the uncatalyzed *Diels-Alder* addition (0 °C, 6 days) of 2-methoxyfuran (a diene more electron-rich and thus more reactive than furan[38]) to (S)-S-menthyl trifluoromethylpyridin-2-ylsulfinyl)acrylate (**50**) which afforded adduct **51** almost exclusively. The latter compound was transformed into Glyoxalase I inhibitor (52, COTC), a product isolated first from the culture broth of *Streptomyces griseosporens* by *Umezawa* et al.[39] and which has attracted considerable interest because of its cytotoxic and cancerostatic activity.[40]

R = menthyl

Enzymic discriminations of enantiopic groups of *meso*-compounds have been widely exploited in asymmetric synthesis.[41] *Ohno* and co-workers[42] reported a total synthesis of L- and D-ribose, showdomycin and cordycepin based on the optically active half-esters 56 and 57, which were obtained enzymatically (Pig Liver Esterase:PLE) from the corresponding *meso*-diesters 54 and 55, respectively, derived from the *Diels-Alder* adduct 53 of furan to dimethyl acetylenedicar-boxylate. The diesters 58 - 60 were also found to be good substrates for PLE leading to the corresponding half-esters 61 - 63 with high enantiomeric purities.[43]

E = COOCH$_3$ 55   57   (77 % e.e)

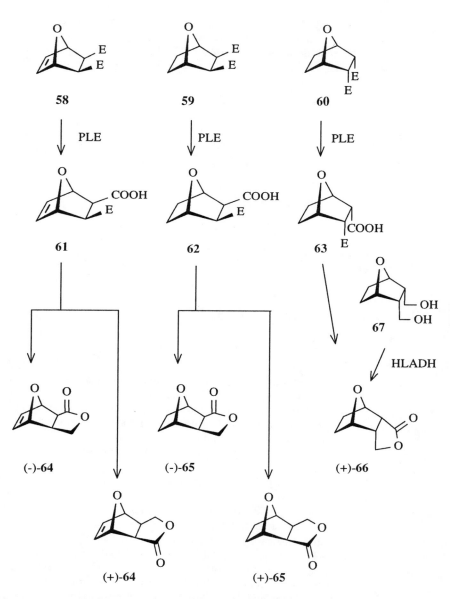

58   59   60

PLE   PLE   PLE

61   62   63

67

HLADH

(-)-64   (-)-65   (+)-66

(+)-64   (+)-65

The latter were transformed into (+)-**64**, (-)-**65** and (+)-**66**, respectively, in good yields. Lactone (+)-**66** was also prepared in 37 % yield[41a] by stereospecific Horse Liver Alcohol Dehydrogenase (HLADH) catalyzed oxidation[44] of the *meso*-diol **67**. Lactones (-)-**64** and (+)-**64** are potentially interesting starting materials for the asymmetric synthesis of natural products and branched carbohydrates as we have found that electrophilic additions to their double bonds are stereo- and regioselective, the major adducts being **68** (EX : Br_2, PhSCl, 2-NO_2-C_6H_4SCl, 2,4-$(NO_2)_2C_6H_3$SCl, PhSeCl). The best regioselectivity (**68/69** 8:1) was observed with the most electrophilic reagent EX = 2,4-(NO_2)-C_6H_3SCl (in CH_3CN, -40 °C), thus suggesting that electronic factors intervene in the long-range substituent effects.[45]

64   **68** (major)   **69**

70 R = Et   **71**   **72** (major)

73 R = H

74 R = menthyl

In contrast, electrophilic additions to the double bond of acetal **70** (derived from **64**[22]) gave adduct mixtures **71/72** with regioselectivities opposite to those of reactions **64** + EX → **68** + **69**, **72** being the major adducts. Tests were carried out to confirm that adducts **68** + **69** and **71** + **72** were formed under conditions of kinetic control. Acetal **70** was obtained optically pure via resolution of lactol **73** by medium pressure chromatographic (silica gel) separation of the diastereomeric acetals **74** derived from (-)-menthol.[45]

Optically pure methyl-(1R,4S)-7-oxanorborn-2-ene-carboxylate ((+)-**75**)[46] was obtained via resolution of the β-nitroacid **76** derived from the major Diels-Alder adduct **77** of furan to methyl β-nitroacrylate.[47] Coupling of racemic acid **76** with D-(-)-α-phenylglycinol gave amides **78** and **79** which were separated by preparative high pressure liquid chromatography. Hydrolysis of **78** gave ester **80** (48 % based on (±)-**76**). Elimination of nitrous acid from **80** provided (+)-**23** (75 %) which was used as the starting material in a total synthesis of the hypocholesterolemic agent (+)-Compactin.[46] Ester (-)-**75** was obtained in a similar way from amide **79**.

The method of *Johnson* and *Zeller*[48] for the optical resolution of ketones has been applied successfully to the preparations of (+) and (-)-7-oxanorbornan-2-one ((+)-**81**, (-)-**81**)[36] and (+) and (-)-5-*exo*,6-*exo*-isopropylidenedioxy-7-oxanorbornan-2-one ((+)-**82** and (-)-**82**).[49]

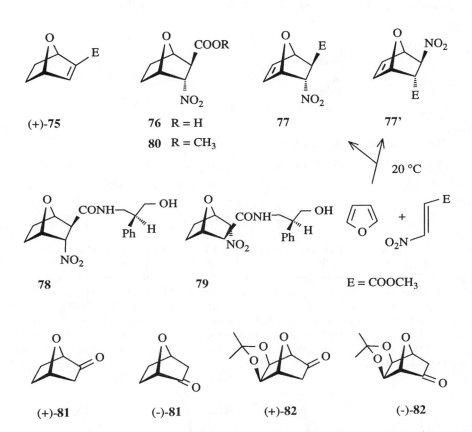

(+)-75   76 R = H   77   77'
          80 R = CH₃

20 °C

78   79   E = COOCH₃

(+)-81   (-)-81   (+)-82   (-)-82

The reaction of racemic 7-oxanorborn-5-en-2-one ((±)-7) with the conjugate base of (+)-(S)-N,S-dimethyl-S-phenylsulfoximide[50] gave a 1:1 mixture of corresponding diastereomeric sulfoximides 83 (43.4 %) and 84 (41.4 %) which were separated by low-pressure chromatography on silica gel (ΔR_f = 0.08). Thermolysis (120 °C/20 Torr) of 83 and 84 gave (+)-7 and (-)-7, respectively, in low yield (10 - 15 %) due to competitive decomposition. When the same technique was applied to (±)-81, the pure sulfoximides 85 (42.2 %) and 86 (41.5 %) were obtained (ΔR_f = 0.09). In this case, the thermolysis (180 °C/15 Torr) of 85 and 86 afforded ketones (+)-81 and (-)-81, respectively, in good yield (80 % based on (±)-81) and excellent optical purity (>99 % e.e.). (+)-81 and (-)-81 were also obtained by catalytical hydrogenation of the C-C double bond in (+)-7 and (-)-7, respectively. We shall see that (+)-81 and (-)-81 can be transformed in 5 synthetic steps to (+)- and (-)-methyl nonactate, respectively.[36] The sulfoximides 87 and 88 derived from (±)-82 were also readily separated by elution chromatography on silica gel (ΔR_f = 0.13) and afforded ketones (+)-82 and (-)-82, respectively, on thermolysis (230 °C/15 Torr). As we shall see, these compounds are starting material for the synthesis of a variety of natural compounds including D- and L-ribose, D- and L-allose, D- and L-talose, conduritols and COTC (52).

83	(1R,2R,4R)	Z = CH=CH	84	(1S,2S,4S)	Z = CH=CH
85	(1R,2R,4S)	Z = CH₂-CH₂	86	(1S,2S,4R)	Z = CH₂-CH₂
87	(1R,2R,4R,5R,6R)	Z = CH- CH	88	(1S,2S,4S,5S,6S)	Z = CH- CH

The optical resolution of (+)- and (-)-7-oxanorborn-5-en-2-*endo*-carboxylic acid ((+)-**89** and (-)-**89**) has been accomplished by the use of (+)-(R)- and (-)-(S)-α-methylbenzylamine, respectively. Treatment of (-)-**89** with 90 % aq. HCOOH and 35 % H_2O_2 gave lactone **90** (66 %). LiAlH₄ reduction followed by acetylation afforded the triacetate **91**. Acetolysis of **91** ($Ac_2O/AcOH/H_2SO_4$, 80 °C, 20 h) gave a mixture of the fully acetylated pseudo-sugars **92** (27 %) and **93** (34 %). Triacetate **91** was also transformed into the protected validamine **94**.[51] (-)-**89** was the starting material for total syntheses of (+)-pipoxide and (+)-β-senepoxide.[52]

(+)-**89**   (-)-**89**   **90**   **91**

92   **93**   **94**

The reaction of 7-oxabenzonorbornadiene **95** with (-)-diisopinocamphenyl-borane (**96**) gave the corresponding trialkylborane which, on treatment with acetaldehyde, followed by oxidation with $H_2O_2/NaOH$, afforded (+)-(1R,2S,4R)-7-oxabenzonorborn-5-en-2-*exo*-ol (**97**) in 80 % yield and 100 % enantiomeric purity.[53]

95   **96**   **97**

Total Synthesis of D- and L-Deoxy-Sugars

We present now some applications of the "naked sugars" to the total synthesis of important deoxycarbohydrates. L-Daunosamine (**98** 3-amino-2,3,6-trideoxy-L-*lyxo*-hexose)[54] is the carbohydrate component of antitumor anthracycline antibiotics such as Adriamycin and Daunomycin.[55] Several ingenious syntheses of **98** have been reported starting from carbohydrate[56] and noncarbohydrate substrates.[57] The technology described here (Scheme 2) is highly stereoselective and is amenable to large the scale preparation of **98** and of related derivatives such as 2-deoxy-L-fucose (**99**),[58] the carbohydrate component of a large number of antibiotics.[1b,59]

98   **99**

Addition of benzeneselenenyl chloride to camphanate **32** gave adduct **100** in 97 % yield. On treatment with a ten-fold excess of 30 % aqueous H_2O_2,[60] **100** was oxidized to **101** (92 %). Saponification followed by treatment with formaline yielded the β,γ-unsaturated chloro ketone **102** (99 %). The (-)-camphanic acid was recovered at this stage in 85 % yield. Quantitative and stereospecific hydrogenation of the chloroalkene **102** to the chloro ketone **103** was achieved with diimide.[59] Upon addition of a small excess of t-BuOK to a mixture of ketone **103** and methyl iodide in anhydrous THF, the monomethylated derivative **104** was obtained in 71 % yield. The 360 MHz ^{1}H-NMR spectrum confirmed the *exo* position of the methyl group at C(3) (no vicinal $^3J_{H,H}$ coupling constant observed between H-C(3) and H-C(4)). Pure **104** was isolated by crystallization from $CHCl_3$ at - 20 °C. HPLC of the mother liquor allowed one to isolate 4 - 8 % of a dimethylated derivative. No trace of the *endo*-monomethyl isomer of **104** could be detected in the reaction mixture.

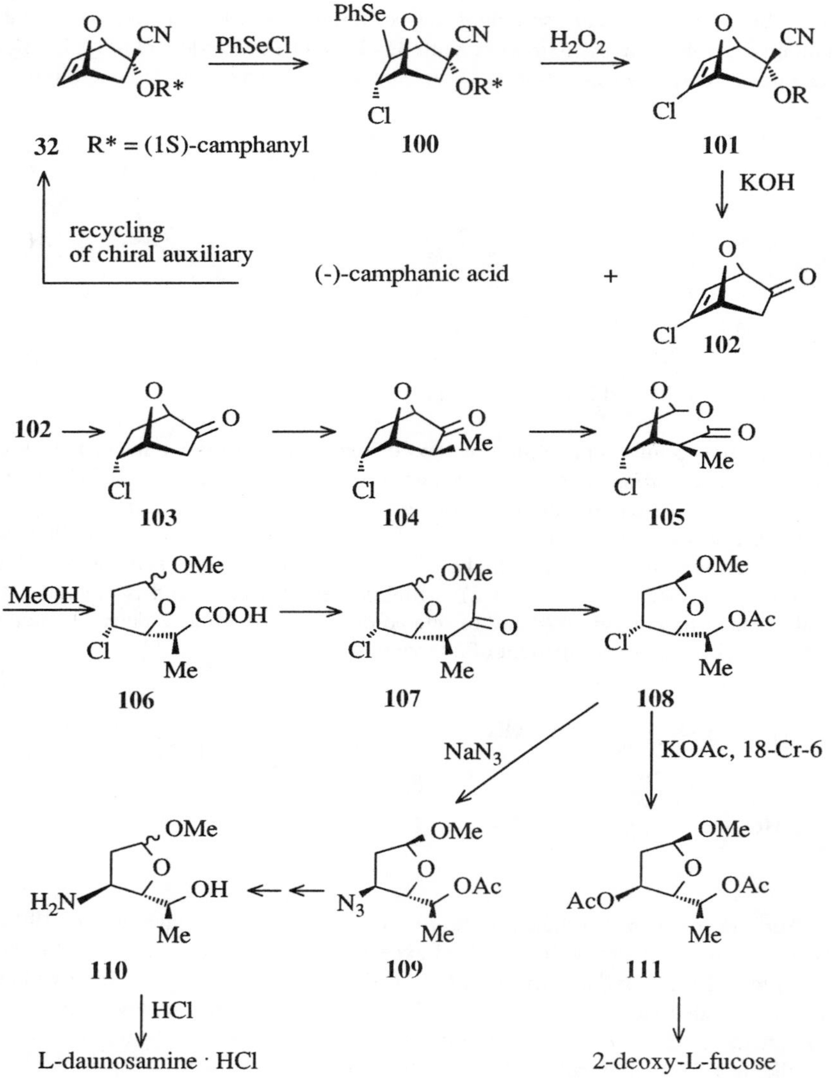

32   R* = (1S)-camphanyl   100   101

recycling
of chiral auxiliary

(-)-camphanic acid   +   102

102 →   103   104   105

MeOH→   106   107   108

NaN₃   KOAc, 18-Cr-6

110   109   111

HCl

L-daunosamine · HCl   2-deoxy-L-fucose

Scheme 2

Baeyer-Villiger oxidation of ketone **104** with metachloroperbenzoic (MCPBA) acid afforded lactone **105** (86 %). There was no detectable trace of the product resulting from oxygen insertion between centers C(2) and C(3).[60] On treatment with acidic methanol,[61] **105** was transformed into a 90:4 mixture of α- and β-acetal-acids **106** (94 %). Treatment with 2 equivalents of MeLi[62] afforded methyl ketone **107**

(63%). *Bayer-Villiger* oxidation of **107** yielded the desired acetate **108** (85 %) with complete retention of configuration at C(5).[63] Attempts to obtain **108** directly from the acids **106** through oxidative decarboxylation with Pb(OAc)$_4$[64] gave low yields of a mixture of acetates. The chloride **108** was displaced in a S$_N$2 fashion[56e,65] with NaN$_3$ (DMF, 120 °C, 12 h) to give the azide **109** (80 %). Catalytic hydrogenation[66] of **109** gave the corresponding free amine which, on ammonolysis,[57g] afforded **110** (94.5 %), the methyl furanoside form of L-daunosamine. **110** was transformed[56d] to the hydrochloride of L-daunosamine (67 %) on heating with HCl in aqueous tetrahydrofuran.

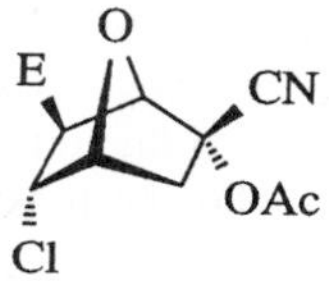

| 112 | E = PhSe |
| 113 | E = PhS |

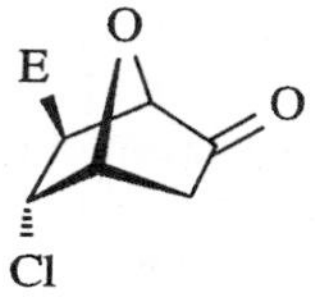

114

115

Hoping to improve or to shorten our total synthesis of L-daunosamine, we explored the following reactions with **112** and **113**, the adducts of the racemic 7-oxanorbornene derivative **47** + **48** to PhSeCl and PhSCl, respectively. Saponification of **112** and **113**, followed by treatment with formalin, afforded ketones **114** and **115**, respectively. Treatment of with tributyltin hydride in toluene/benzene (AIBN 1 - 2%, 80 °C)[67] gave the key intermediate (±)-**103** in 69 % yield. Under the same conditions, **115** was reduced to (±)-**103** in 40 - 45 % yield. Raney nickel reduction[68] of **115** afforded (±)-**103** in 50 % yield together with 40 % of (±)-7-oxabicyclo[2.2.1]heptan-2-one. However, we found the multistep procedure **32** → **100** → **101** → **102** → **103** easier to scale up. Several intermediates in our synthesis do not have to be isolated. For instance, transfomation of **32** into **102** can be carried out in the same pot in 94 % yield.

The "naked sugar" **32** has been transformed into methyl 3-amino-2,3,6-trideoxy-*lyxo*-L-hexofuranosides (**110**) in 21.8 % overall yield. The synthesis required the isolation of eight synthetic intermediates. Many of these are potential starting materials for the total synthesis of other natural products. For instance we found[69] that displacement of the chloride **108** with AcOK (0.6 equiv. 18-crown-6 ether, dimethylformamide, 110 °C) gave the diacetate **111** (35 - 40 %,[70] a precursor of 2-deoxy-L-fucose (**99**).[71] 2-Deoxy-L-fucose has been derived from other carbohydrates[72] or by using tartaric acid as starting material.[73] *Chmielewski*[74] reported a synthesis of 2-deoxy-DL-fucose starting with the *Diels-Alder* addition of 1-methoxybutadiene to tertiobutyl glyoxalate. The first total synthesis of 2-deoxy-D-fucose has been presented by *Roush* and *Brown*.[75] It is based on the enantioselective *Sharpless* epoxidation of allylic alcohols.[76] Using our " naked sugar" **38** (derived from furan and (+)-(1R)-camphanic acid) instead of **32**, the method outlined in Scheme 2 realizes a second total, asymmetric synthesis of 2-deoxy-D-fucose. Similarly, D-daunosamine can be derived from **38**.

The technology described in Scheme 2 is highly stereoselective, and the asymmetry is induced by commercially available chiral auxiliaries that are recovered at an earlier stage of the synthesis. Furthermore, it permits one to obtain the deoxysugars in both their furanose and pyranose form. It exploits the high stereoselectivity of the electrophilic additions of 7-oxanorborn-5-en-2-yl derivative **38** that give exclusively adducts where the electrophile (E) sits in the *exo* position of center C(6) and the nucleophile (X) at the *endo* position of center C(5) (**32** + EX → **33**, see Scheme 1). This high regioselectivity was attributed to the electron-withdrawing effect and/or to the bulk of the substituents at C(2). Ketone (+)-**7** (derived from **32**) adds electrophilic agents onto its homoconjugated double bond with opposite regioselectivity (giving exclusively adducts **34**, see Scheme 1). This was attributed to the electron-donating ability of the carbonyl group. We are exploring the application of that principle in a new approach to the total, asymmetric synthesis of D-lividosamine (**116** : 2-amino-2,3-dideoxy-D-*ribo*-hexopyranose) a component of antibiotics Lividomycins <u>A</u> and <u>B</u>.[1b] Our preliminary results are presented in Scheme 3.[79]

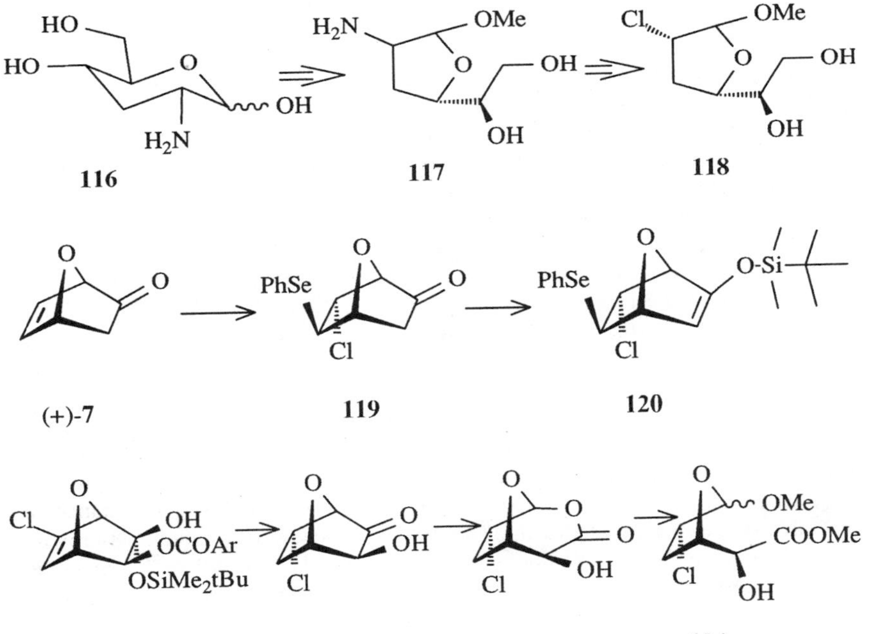

Scheme 3

Benzeneselenyl chloride added to (+)-**7** and gave adduct **119** nearly quantitatively. Treatment with <u>N</u>-methyl-<u>N</u>-<u>tert</u>butyldimethylsilyltrifluoroacetamide and triethylamine (dimethylformamide, molecular sieves, 40 °C, 15 h) afforded enol ether **120** in high yield (>90 %). Oxidation with an excess of MCPBA gave **121** resulting from oxidative elimination of the benzeneselenyl group and epoxidation of the enol ether. Catalytical hydrogenation of **121**, followed by treatment with K_2CO_3 in methanol furnished **122**. *Baeyer-Villiger* oxidation of **122** with MCPBA afforded **123**. Attempts to displace the chloride in **123** and in **124** (obtained by acidic methanolysis of **123**) led to epimerization of the alcohol. Studies are underway in our laboratory to explore the transformation of **124** into **118** which, by analogy with our synthesis of L-daunosamine (Scheme 2), should allow one to obtain the methyl furanoside of D-lividosamine.

Several syntheses of D-lividosamine and its derivatives have already been reported. They use carbohydrates[78,80] or optically pure but-3-ene-1,2-diol as starting materials.[81]

Total Synthesis of D- and L-Allose, D- and L-Talose and of D- and L-Ribose Derivatives

In their pioneering work, *Just* and co-workers[47] have described many interesting transformations of the *Diels-Alder* adducts of furan to methyl nitroacrylate (**77** + **77'**) and to dimethyl acetylenedicarboxylate (**53**). The mixture of racemic adducts **77** + **77'** was hydroxylated into the *exo-cis*-diols **125** + **125'**, separable by crystallization. Treatment of the isopropylidene acetal obtained from **125** with diazabicyclo[5.4.0]undec-5-ene (DBU) gave a high yield of alkene **126**. Ozonolysis followed by a reductive work-up with dimethylsulfide, then with NaBH₄, gave a mixture of epimeric triols **127**. Cleavage with sodium periodate afforded 2,5-anhydro-3,4-O-isopropylidene-DL-allose (**128**) in 15 % yield, based on methyl 2-nitroacrylate used. The same allose derivative was obtained from adduct **53**.[47,82]

125   125'   126   127

E = COOCH₃

128   129   130   131

Lactol **128** has been converted into a variety of racemic C-nucleosides.[83] The unstable aldehyde **130** was prepared[82b] from **128** by way of oxazolidine **129**. Lactone **131** was also derived from **128** and used as starting material in the synthesis of racemic C-nucleosides.[84] Adducts **77 + 77'** were transformed into epoxide **132**. Opening of the epoxide, followed by ozonolysis and reduction allowed one to

| **132** | **133** | **134** | **135** |

E = COOMe   R = tBuMe$_2$Si

prepare the keto-ester **133**. The latter was applied in the synthesis of DL-2-*epi*-showdomycin, and also of DL-2-*epi*-pyrazofurin A. Similarly, adduct **77** was transformed into keto-ester **135** via acetate **134**. **135** was applied in the synthesis of DL-2-deoxyshowdomycin.[20a]

Inspired by the work of *Just*[20a,b,82-84] and others[19,20c,d] we converted our "naked sugar" **32** into optically pure, protected methyl 2,5-anhydro-D-allonates **140**, **145**, the acetal **141** and lactones **131** and **146** (Scheme 4). These systems are potential precursors for the total asymmetric synthesis of C-nucleosides.[85,86] Stereospecific cis-hydroxylation of **32** with H$_2$O$_2$ and a catalytic amount of OsO$_4$[87] gave diol **136** which was isolated in the form of the corresponding acetonide **137** (65%) using acetone, 2,2-dimethoxypropane and para-toluenesulfonic acid in the work-up. Saponification of **137** (KOH/H$_2$O/THF), followed by treatment with formalin gave ketone **138** (92 %) and pure (-)-camphanic acid (83 %). Treatment of **138** with tBuMe$_2$SiN(Me)COCF$_3$ and NEt$_3$ in DMF (40 °C, 12 h) afforded the silyl enol ether **139** (85 %). Ozonolysis in MeOH/CH$_2$Cl$_2$ at -70 °C, followed by reduction with NaBH$_4$ (-70 °C, 1 h; then -10 °C, 30 min), acidification (pH = 3, HCl/MeOH), and treatment with diazomethane furnished **140** (65 %). Under more acidic work-up conditions and without addition of CH$_2$N$_2$, lactone **131** was isolated as the major product. Ozonolysis of **139**, followed by acidic work-up (HCl/MeOH) afforded the dimethylacetal derivative **141** (65 %). Protection of diol **136** by tBuMe$_2$Si groups gave **142** which was transformed, as above, into methyl 2,5-anhydroallonate **145** (73 %) via the synthetic intermediates **143** and **144**. The corresponding lactone **146** (65 %) was also derived from **144**.[88]

Silyl enol ether **139** has also been transformed into D-allose, as shown in Scheme 5. The same methods can be applied to the enantiomeric enol ether derived from camphanate **38**, and this allows one to prepare L-allose and its derivatives.[89] Oxidation of **139** with MCPBA in THF (20 °C) led to the product of epoxide acidolysis **147** (69 %) which yielded **148** on heating to 200 °C for 15 min.[90] Addition of 1.1 equiv. of MCPBA converted **148** into lactone **149** which in the presence of MeOH and K$_2$CO$_3$ (20 °C), gave selectively diester **150**. Reactions **147**

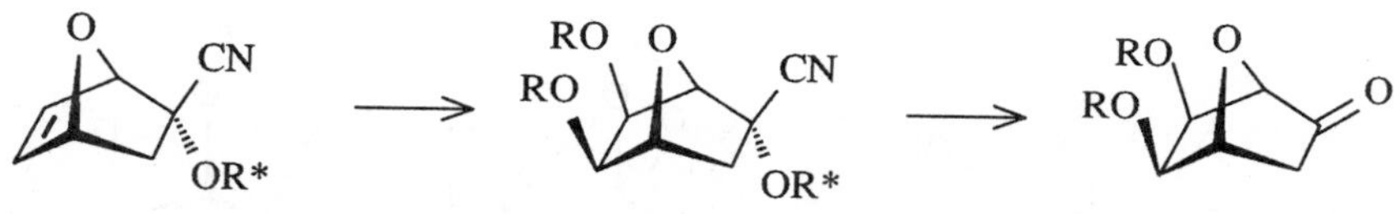

32

136 R = H
137 R,R = Me$_2$C
142 R = tBuMe$_2$Si

138 R,R = Me$_2$C

143 R = tBuMe$_2$Si

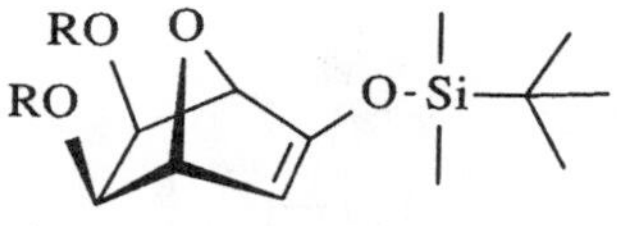

139 R,R = Me$_2$C
144 R = BuMe$_2$Si

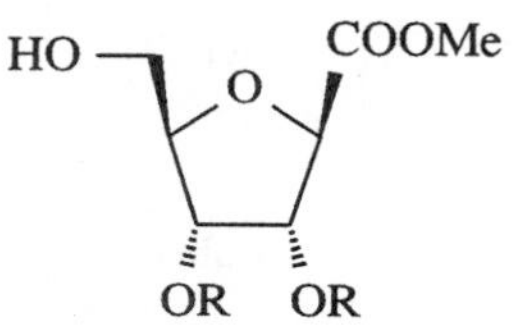

140 R,R = Me$_2$C
145 R = tBuMe$_2$Si

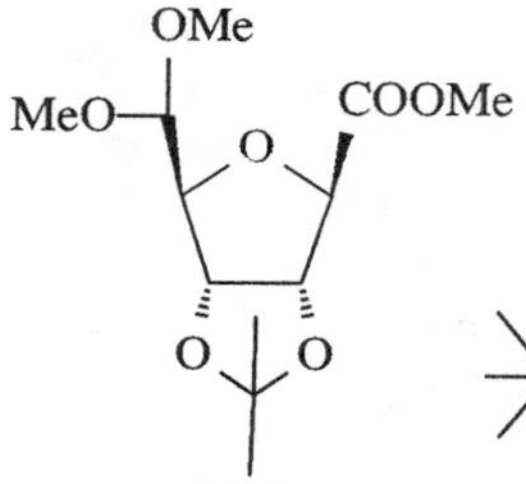

141

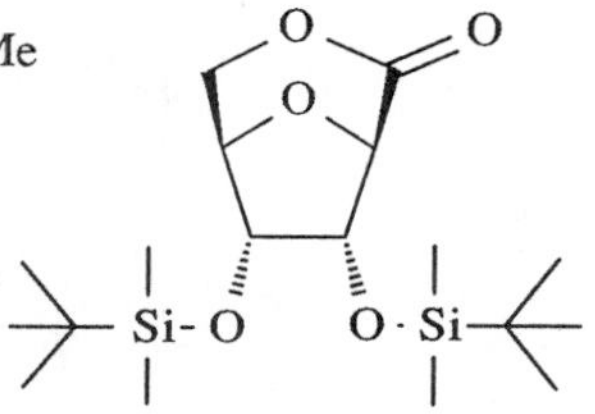

146

Scheme 4

$\rightarrow$ **148** $\rightarrow$ **149** $\rightarrow$ **150** were carried out in "one-pot" with an overall yield of 78 %. The methyl furanoside **151** (92 %) was obtained on acidic methanolysis of **150**. Reduction of **151** with 4.2 equiv. of LiAlH$_4$ (THF, 20 °C, 15 min) afforded **152** (71 %).[91] Acidic hydrolysis (2 % H$_2$SO$_4$ in H$_2$O, 100 °C, 2 h) of **152** gave D-allose.[92] L-allose[91] was prepared in the same manner starting with **38**.

In the presence of 1.1 equiv. of Br$_2$ in CH$_2$Cl$_2$ (-50 °C), **139** gave the α-bromoketone **153** (78 %). *Baeyer-Villiger* oxidation of **153** with CF$_3$CO$_3$H (CH$_2$Cl$_2$, Na$_2$HPO$_4$, 20 °C) afforded lactone **154** (85 %). As for reaction **148** $\rightarrow$ **149**, the oxidation was highly selective yielding exclusively the product of oxygen insertion between the bridgehead center C(1) and the carbonyl group. Methanolysis of **154** in MeOH saturated with K$_2$CO$_3$ (20 °C, 45 min) gave **155** (95 %). The reaction implies the intermediacy of hemiacetal **156** which, in the presence of a base (K$_2$CO$_3$) undergoes intramolecular S$_N$2 displacement of the bromide giving **155**. This hypothesis was confirmed by the isolation of **156** when **154** was treated with MeOH at 20 °C containing a small amount of NaHCO$_3$. **156** afforded **155** on treatment with MeOH and K$_2$CO$_3$. Reduction of **155** with LiAlH$_4$ (THF, 20 °C) gave 1,4-anhydro-2,3-O-isopropylidene-*talo*pyranose (**157**) in 82 % yield (for analogous 1,4-anhydropyranoses, see reference 93). Treatment with 1 N HCl (20 °C, 4 d) afforded L-talose.[94] D-talose and its derivatives can be obtained in the same

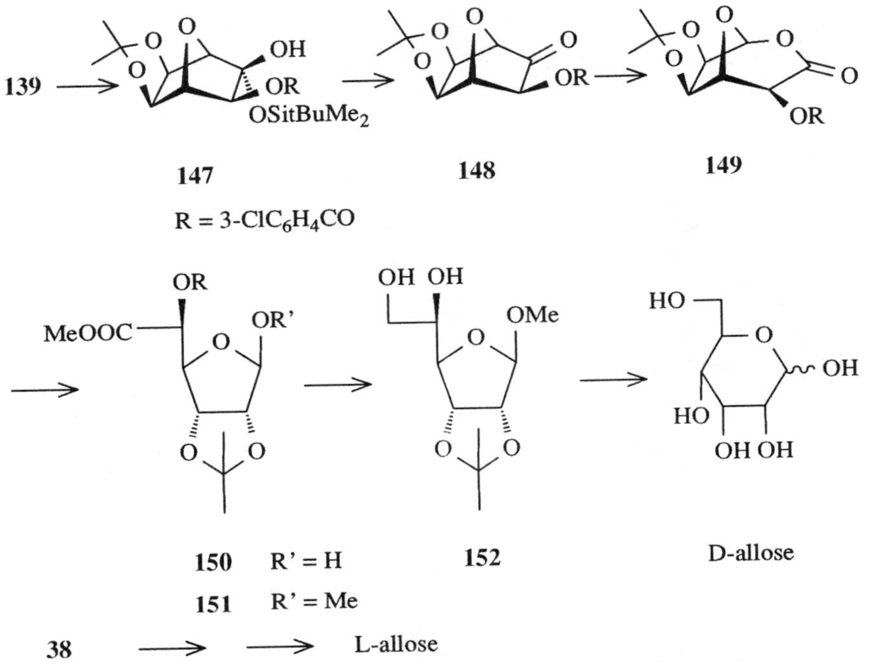

139 →

147　　　　**148**　　　　**149**

R = 3-ClC$_6$H$_4$CO

150 R' = H
151 R' = Me

152　　　　D-allose

38 ⟶ ⟶ L-allose

Scheme 5

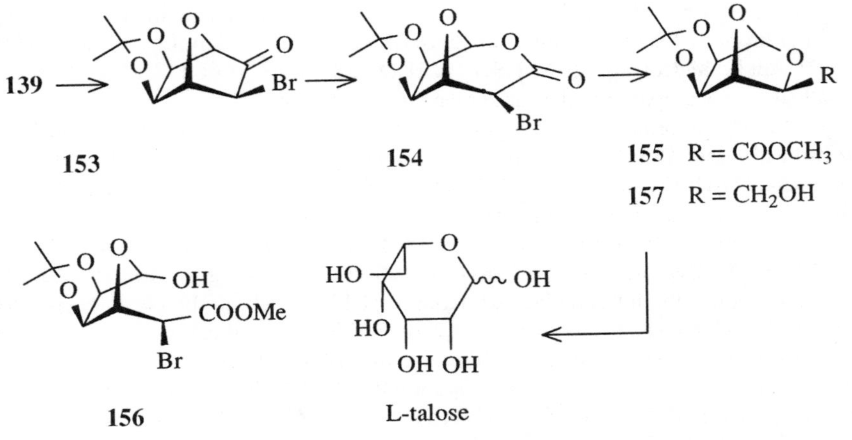

139 →

153　　　　**154**　　　　**155** R = COOCH$_3$
　　　　　　　　　　　　　　　　　157 R = CH$_2$OH

156　　　　L-talose

Scheme 6

manner starting with the "naked sugar" 38 derived from furan and (1R)-camphanic acid. (See Scheme 6.)

The best syntheses of D-ribose (an important component of nucleic acids, polysaccharides, nucleosides, vitamins, co-enzymes and many antibiotics, etc.) use natural, optically pure starting materials. *Stroh* and co-workers[95] on one hand, and *Kiss* and co-workers[96] on the other hand, transformed D-glucose into D-ribose in five synthetic steps in 24 % and 18.1 % global yield, respectively. *Yamada* and co-workers[97] transformed L-glutamic acid into a 2.7:1 mixture of methyl 5-*O*-benzyl-2,3-*O*-isopropylidene-β-D-*ribo*furanoside and methyl 5-*O*-benzyl-2,3-*O*-isopropylidene-α-D-*lyxo*furanoside, the former product being obtained in 13 % overall yield (7 steps). More recently, *Grignon-Dubois* and co-workers[98] have derived D-ribose from the inexpensive D-xylose in six synthetic steps and 21 % overall yield. Other syntheses[99-101] of D-ribose use 2,3-*O*-isopropylidene-D-glyceraldehyde[102] as starting material. Using a chemicoenzymatic approach, *Ohno* and co-workers[42] transformed dimethyl 7-oxanorborna-2,5-diene-2,3-dicarboxylate (53) into methyl 2,3-*O*-isopropylidene-β-D-*ribo*furanoside ((-)-162) in eight steps, 11.8 % overall yield and 95 % e.e. They also prepared methyl 2,3-*O*-isopropylidene-β-L-*ribo*furanoside ((+)-162), 7 steps, 15.1 %, 77 % e.e.). To our knowledge, this constitutes the first total, asymmetric synthesis of these ribose derivatives with good enantiomeric purity.

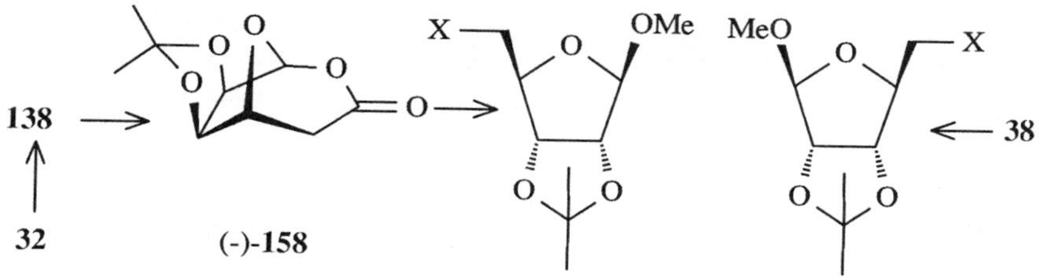

(-)-159	X = COOMe	(+)-159
(-)-160	X = COOH	(+)-160
(-)-161	X = Br	(+)-161
(-)-162	X = OH	(+)-162

Scheme 7

L-Ribose is quite rare and the only practical method for its preparation is the transformation of L-arabinose by the method of *Austin* and *Humoller*.[103] (4 steps, 9.5 % overall yield). L-Ribose has also been derived from, 2,3-*O*-isopropylidene-L-glyceraldehyde, (5 steps, 12 %) after separation from a mixture containing L-arabinose.[99] In Scheme 7 we summarize our total syntheses of D- and L-ribose derivatives using the " naked sugars" 32 and 38, respectively.[49] Ketone 138 (Scheme 4) was oxidized into the corresponding lactone (-)-158 with MCPBA[87] in 98 % yield. Treatment with anhydrous methanol, 2,2-dimethoxypropane and a small amount of methanesulfonic acid afforded the methyl 5-deoxy-D-allonate (-)-159

(82%). Saponification (KOH/H_2O/THF) gave (-)-**160** (98 %). The latter acid could also be obtained directly from (-)-**158** in an "one pot" procedure in 80 % yield. Oxidative decarboxylation of (-)-**160** with red HgO and Br_2 gave bromide (-)-**161** (60 %), a compound already prepared by *Kiss* and co-workers, that was shown to be an useful starting material for the preparation of 5-deoxy-D-ribose and its derivatives, including the anticancer agent 5'-deoxy-5-fluorouridine.[104] Buffered hydrolysis of bromide (-)-**161** gave the partially protected D-ribose derivative (-)-**162** (75 %, 21 % overall yield based on **32**). Starting with "naked sugar" **38**, the L-ribose derivatives (+)-**161** and (+)-**162** can be obtained in a similar way.[49]

Asymmetric Synthesis of Cyclohexenepolyols

In the presence of a strong base 7-oxanorbornane-2-carboxylates and 7-oxanorborn-2-yl alkyl ketones can be isomerized into the corresponding cyclohexenols.[23] In the presence of triethylamine the substituted 7-oxanorbornan-2-one **138** was not isomerized. The opening of the oxygen bridge O(7) occurred only if trimethylsilyl triflate (TMSOTf) was added to the reaction mixture. Under these conditions (0 °C, benzene) **138** was rearranged into a mixture of cyclohex-2-enone **163** and cyclohexa-1,3-diene **164**. After a prolonged reaction time and with a large excess of TMSOTf/Et_3N, the unstable diene **164** was the major product. The isomerization of the 7-oxanorbornane was much smoother and better yielding when the diol at C(5) and C(6) was protected with tBuMe$_2$Si groups. On treatment with two equivalents of TMSOTf and Et_3N) **143** gave cyclohex-2-enone **165**. Acid hydrolysis (MeOH/H_2O/HCl, 20 °C, 5 min) of **163** and **165** gave alcohols **166** (71 %) and **167** respectively.

138	R,R = Me$_2$C	**163**	R,R = Me$_2$C; R' = TMS	**164**
143	R = tBuMe$_2$Si	**165**	R = tBuMe$_2$Si; R' = TMS	
		166	R,R = Me$_2$C; R' = H	
		167	R = tBuMe$_2$Si; R' = H	
		168	R = R' = Ac	

Reduction of **166** with NaBH$_4$/CeCl$_3$ in aqueous MeOH[105] afforded the protected, chiral derivative **169** of Conduritol-D in 73 % yield. Conduritol-D (**170**) was formed nearly quantitatively upon acid hydrolysis of **169**. Treatment with Ac$_2$O/pyridine gave the crystalline tetracetate **171**.[106] Reduction of **165** with diisobutylaluminium hydride (DIBAL) in toluene (-80 °C, 6 h) gave **172**, another chiral derivative of Conduritol-D, in 97 % yield. Alternatively, **167** was reduced with LiAlH$_4$ (THF, -90 °C, 6 h) into **173** (95 %). Deprotection of **167** with tetrabutylammonium fluoride

(THF, 0 °C, 10 min), followed by acetylation gave all-*cis*-4,5,6-triacetoxycyclohex-2-enone (**168**, 95 %). Reduction of **168** with NaBH$_4$/CeCl$_3$ (-20 °C, 30 min), followed by acetylation gave **171** (51 %) and **171** (35 %). **174** is the tetracetate of Conduritol-C.[106a]

169 R,R = Me$_2$C; R' =H   **171**   **174**

170 R = R' = H

172 R = TBuMe$_2$Si; R' = TMS

173 R = tBuMe$_2$Si; R' = H

In 1985, *Vasella* and co-workers[107] provided an efficient synthesis of COTC ((4R,5R,6R)-2-crotonyloxymethyl-4,5,6-trihydroxycyclohex-2-enone : **52**), a glyoxalase I inhibitor.[39,40] Methyl α-D-mannopyranoside was transformed in three synthetic steps into the protected all *cis*-trihydroxycyclohex-2-enone **163** (34 %, overall) which was then transformed into COTC (**52**) in four steps. Thus, transformations of our "naked sugar" **32** to **137** → **138** → **163** realize a total, asymmetric synthesis of COTC. Another total synthesis of this natural compound has been reported recently by *Koizumi* and co-workers,[37] as mentioned earlier.

175   **176** R = tBuMe$_2$Si

In 1983, *Mori* and co-workers[108] isolated the cyanoglucoside **175** from a medicinal plant : *Ilex Warburgii Loesn*. In Scheme 8 we present our preliminary results on the synthesis of **176**, a partially protected form of the aglycone of **175**. *Wittig-Horner* reaction of ketone **143** with (EtO)$_2$POCH$_2$CN/NaH (25 °C, 5 min) afforded a mixture of (E)- and (Z)-olefins **177** (98 %). In the presence of 1.5 equivalents of (TMS)$_2$NLi (THF, 0 °C), **177** was isomerized exclusively into the (E)-cyanodiene **178** in 73 % yield. Irradiation of **178** (quartz vessel, low pressure Hg lamp, CH$_3$CN, 20 °C) gave a photoequilibrium **178** ⇌ **176**, the desired

(Z)-derivative **176** being separated from **178** by column chromatography on silica gel.[109] Glycosidation of **176** is under exploration in our laboratory.

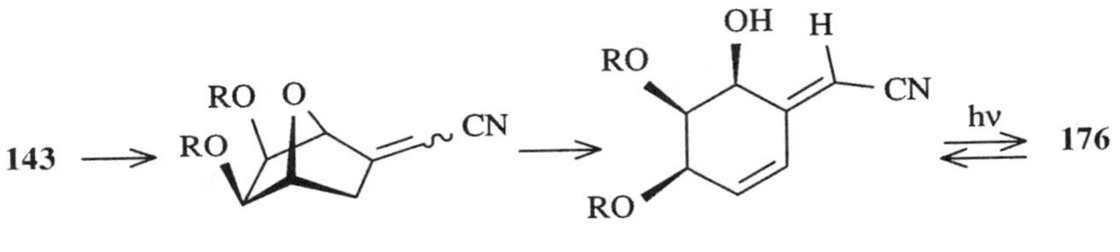

R = tBuMe$_2$Si   **177**   **178**

Scheme 8

Acid-catalyzed Rearrangements of 5,6-exo-Epoxy-7-oxanorbom-2-yl Derivatives

The 5,6-exo-epoxy-7-oxanorbornyl derivatives **179** - **184** (epoxides of "naked sugars") are less reactive toward protic acids and nucleophiles. The "super-acid" HSO$_3$F was required to induce ring opening of the oxiranes. Under these conditions all four reaction paths shown in Scheme 9 are possible for the protonated epoxides **185**. We have found that depending on the nature of the substituents at C(2), the 1,2-shifts (*Wagner-Meerwein* rearrangements) of alkyl groups (path I) or acyl groups (path II) compete with the 1,3-shifts of the *endo* substituents at C(2) (path III) and the σC(1), C(2) bond cleavage (path IV). Conditions have been found under which our "naked sugars" can be transformed with high stereoselectivity into polysubstituted cyclopentyl or 7-oxanorbom-2-yl derivatives that are potentially useful synthetic intermediates.[110]

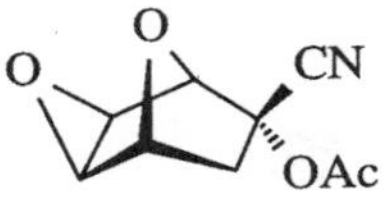

179 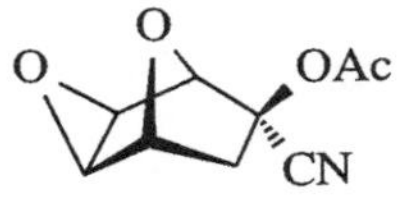   **180** 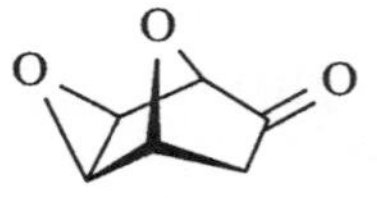   **181**

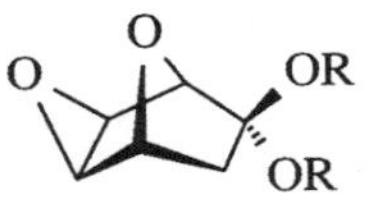

182 R = Me   **184**

183 R = CH$_2$Ph

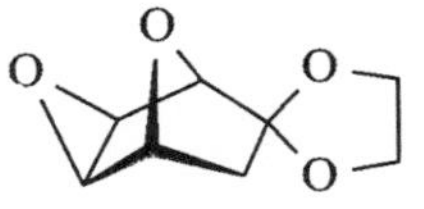

Racemic *Diels-Alder* adducts **47/48** and **31/49** reacted with MCPBA in chloroform and afforded epoxides **179** (82 %) and **180** (86 %), respectively. Saponification of **179** with K$_2$CO$_3$ in aqueous methanol in the presence of formalin gave epoxy-ketone **181** (79 %). Treatment of 7-oxanorbom-5-en-2-one (**7**) with

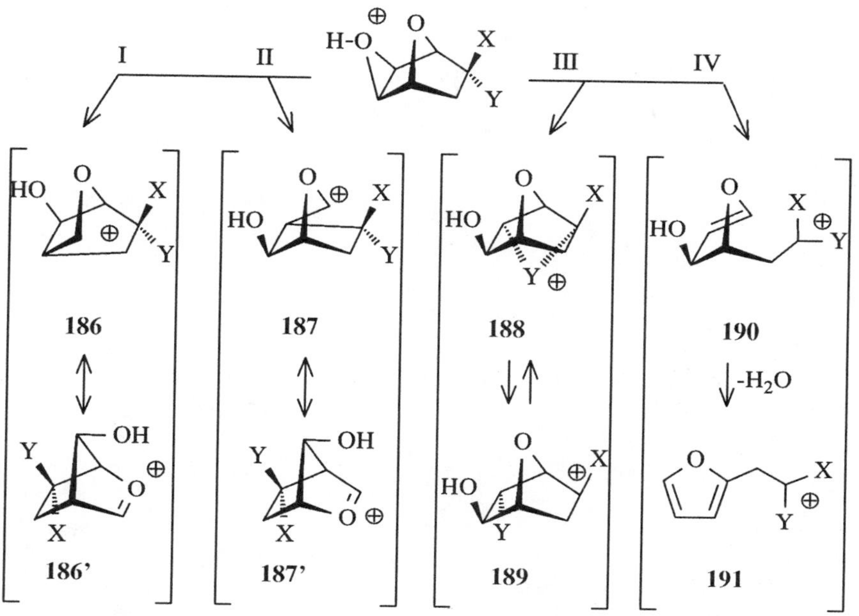

Scheme 9

MeOSiMe$_3$, PhCH$_2$OSiMe$_3$ and Me$_3$SiOCH$_2$CH$_2$OSiMe$_3$ in the presence of 8 % mol equiv. of Me$_3$SiOSO$_2$CF$_3$ in CH$_2$Cl$_2$ afforded the corresponding acetals which were oxidized into epoxides **182** (91 %), **183** (87 %), and **184** (88 %), respectively, with MCPBA in chloroform.

When **179** was treated with 0.2 equiv. of HSO$_3$F and 3 equiv. of Ac$_2$O in CH$_2$Cl$_2$ at -25 °C, products **192** - **194** of acetolysis and *Wagner-Meerwein* rearrangement involving migration of the alkyl group σC(3), C(4) (path I) were isolated in 43 %, 11 % and 7 % yield respectively, after aqueous (NaHCO$_3$) work-up and chromatography on silica gel. Under similar conditions (0.4 equiv. of HSO$_3$F, 12 equiv. of Ac$_2$O, CH$_2$Cl$_2$, 20 °C), epoxide **180** yielded acylal **195** in 45 % isolated yield. It also arises from a *Wagner-Meerwein* rearrangement involving σC(3), C(4) bond migration (path I). No other isomeric products could be isolated or detected in the crude reaction mixtures. Products **192** - **195** are variously protected forms of the carba-analogue of (±)-lyxose **196**.

The acid-catalyzed acetolysis of epoxy-ketone **181** (0.4 equiv. of HSO$_3$F, 6 equiv. of Ac$_2$O, CH$_2$Cl$_2$ -51 °C) gave major products **197** (51 %) and **198** (8 %) resulting from a *Wagner-Meerwein* rearrangement involving the migration of the acyl group (path II), and minor compounds **199** (1 %) and **200** (0.4 %) resulting from a *Wagner-Meerwein* rearrangement implying alkyl group migration (path I). The proportions of products **197/198/199/200** in the crude reaction mixture did not vary in a significant fashion by changing the amount of HSO$_3$F (0.1 to 0.4 equiv.), the temperature (-51° to -30 °C) and the reaction time (24 - 60 h).

192   **193**   **194**

195   **196**

197 (major)   **198**   **199** (minor)   **200**

Treatment of the dimethylacetal **182** with 0.2 equiv. of HSO$_3$F and 3 equiv. of Ac$_2$O in CH$_2$Cl$_2$ at -78 °C (12 h) afforded, after aqueous work-up (NaHCO$_3$), a mixture of **201** + **202** + **203** which were separated by column chromatography on silica gel and isolated in 30 %, 17 % and 5 % yield, respectively. The major products **201** and **202** arose from 1,3-migration of the *endo* MeO group (path III) and the minor product **203** from the *Wagner-Meerwein* rearrangement implying the migration of the unsubstituted alkyl group σC(3), C(4) (path I). When the reaction was run at -40 °C, the acylmethylacetal **201** was not observed and ketone **202** was isolated in 55 % yield. In the presence of one equivalent of CD$_3$OD there was no incorporation of deuterium in **201** - **203**, thus confirming the intramolecular nature of the methoxy group migration **182** → **185** → **188** → **189** → **201** + **202**. When **182** was treated with 0.3 equiv. of HSO$_3$F and 4 equiv. of MeOH in CH$_2$Cl$_2$ at 20 °C (36 h) the *trans* disubstituted 7-oxanorbornan-2-one **204** was obtained and isolated in good yield (83 %). Under the same conditions, the epoxy-ketone **181** gave first **182**, and then **204**. Similarly, when the dibenzylacetal **183** was treated with 0.6 equiv. of HSO$_3$F and 8 equiv. of PhCH$_2$OH in CH$_2$Cl$_2$ at 20 °C (14 h), the corresponding ketone **205** was formed and isolated in 88 % yield. Thus reactions 7 → **181** → **204**, 7 → **182** → **204** or 7 → **183** → **205** are efficient methods for introducing two different oxy functions at C(5) and C(6) of 7-oxanorbornan-2-one in a stereoselective, if not stereospecific, fashion.

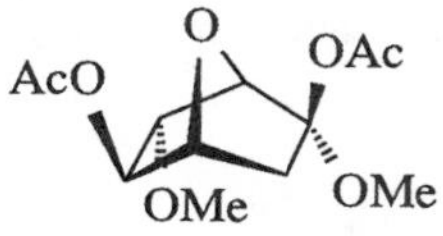

201

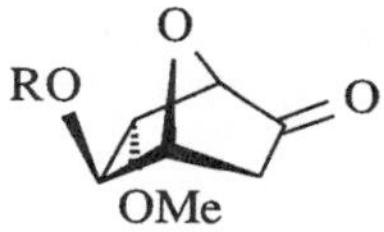

202 R = Ac
204 R = H

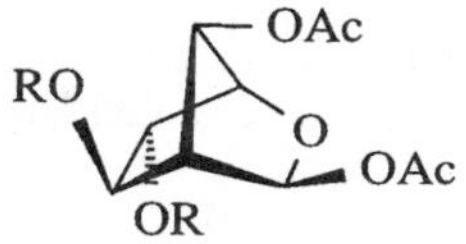

203 R = Me
207 R = PhCH$_2$

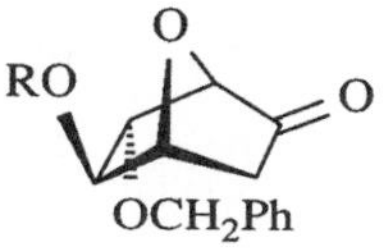

205 R = H
206 R = Ac

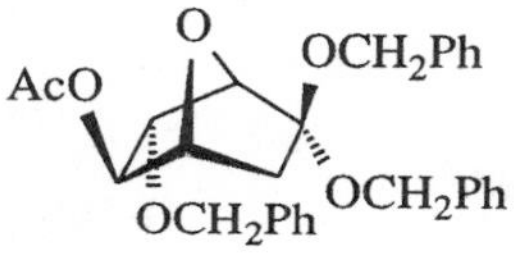

208

The treatment of the dibenzylacetal **183** with 0.5 equiv. of HSO$_3$F and 8 equiv. of Ac$_2$O in CH$_2$Cl$_2$ at -78 °C (90 min) gave a mixture from which ketone **206** and the product of acetolysis and rearrangement **207** were isolated in 57 % and 7 % yield, respectively. In some series, 1 - 8 % of the dibenzylacetal **208** was also isolated. Finally, treatment of the ethyleneacetal **184** with HSO$_3$F and Ac$_2$O in CH$_2$Cl$_2$ at -50 °C led to the formation of a mixture of products from which the furan derivative **209** (41 %) arising from C(1)-C(2) bond cleavage (path IV) and the product of acetolysis and rearrangement **210** + **211** (24 %) (path I) were isolated. **209** is an uncommon ethyleneacetal of the mixed anhydride derived from acetic acid and α-furyl acetic acid.[111]

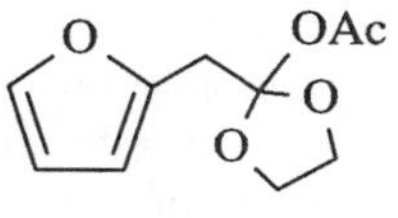

209

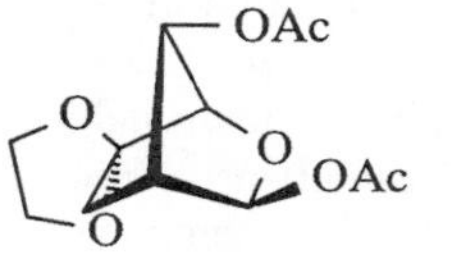

210

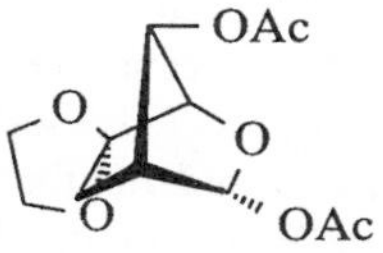

211

The necessity of "super acid" HSO$_3$F to induce ring opening of the oxiranes **179** - **184** demonstrates the lack of reactivity of these epoxides compared with that of the corresponding *exo*-epoxybicyclo[2.2.1]heptyl derivatives.[112] It can be attributed to the inductive effect of the oxygen bridge O(7) in **179** - **184**.[113] *Yur'ev* and *Zefirov*[114] were able to esterify diacid **212** into epoxydiester **213** in MeOH and oleum, without ring opening of the oxirane. They also found that HBr adds to **212** in AcOH to furnish bromohydrin **214** (62 %).[115] The relatively high energy barrier associated with the hypothetical rearrangement **215** → **216** (Scheme 10) must be attributed to the destabilizing effect of the electron-withdrawing COOH groups[116] in **215**. In media of lower nucleophilicity than HBr/AcOH, the *Wagner-Meerwein* rearrangement **215** → **216** was possible as product **217** was isolated in 65 % yield when **213** was treated with H_2SO_4 in Ac_2O at 70 °C.[117]

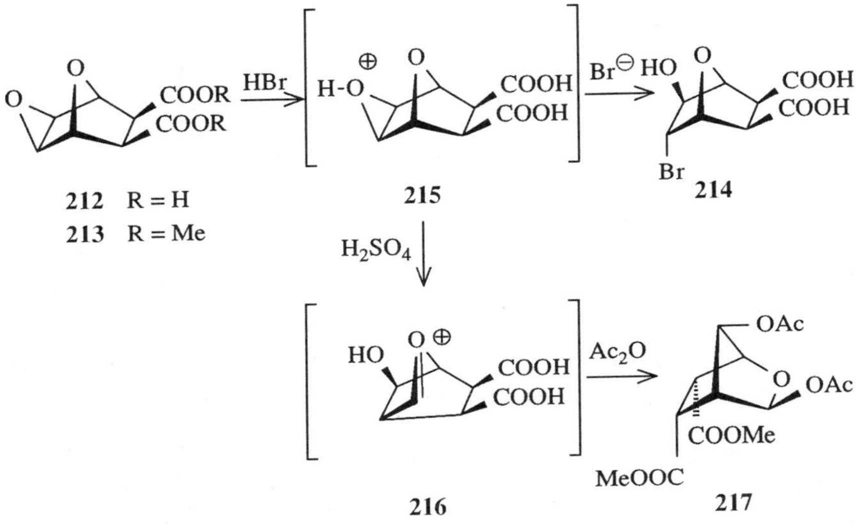

Scheme 10

Thus, in the light of these results, the preference for the cyanohydrin acetates **179** and **180** to undergo *Wagner-Meerwein* rearrangements with migration of the unsubstituted alkyl group (path I) rather than with migration of the substituted alkyl group (path II) is perfectly expected.

The preference for the 1,3-migration of an *endo*-alkoxy group (path III) in the case of the acetals **182** and **183** has been attributed to the relatively high stability of the 2-alkoxy-7-oxanorborn-2-yl cation intermediates **189**. Similar rearrangement had been reported for the buffered acetolysis of dimethyl acetal **218** which afforded a 11:20:39 mixture of **219:220:221**. The results were interpreted in terms of formation of the oxonium ion intermediate **222**,[117] analogous to **188** (Scheme 9). In the case of the ethyleneacetal **184**, the 1,3-migration of the *endo* alkoxy group is prohibited for steric reasons, thus making the bond breakage **185** → **190** the preferred process as **190**, a dialkoxy-substituted carbenium ion intermediate is expected to be a relatively stable species. The latter then eliminates one equiv. of water, giving **191** (Scheme 9) which is then quenched with acetic acid or acetic anhydride and furnishes the major product **209**. The minor products of the acid-promoted acetolyses of epoxy-ketals **182**, **183** and **184** arose from *Wagner-Meerwein* rearrangements (path I) implying migration of the non-substituted alkyl group, in agreement with the hypothesis that, because of the inductive effect of the acetal function, the substituted alkyl group

1,2-shift becomes a less favourable process. If the carbonyl group in epoxy-ketone **181** should be considered as an electron-withdrawing group similar to the ketal functions in **182** - **184** and acetoxy + cyano groups in **179** and **180** the acid-promoted rearrangement of **181** should have given preferentially the products of *Wagner-Meerwein* rearrangement **199** + **200** (path I) rather than products **197** and **198** (path II) the former arising from 1,2-shift of the alkyl group and the latter from 1,2-shift of the acyl group. Our results demonstrate, on the contrary, that the acyl group migration is definitely a faster process that the alkyl group migration.

1,2-Shifts of electron-withdrawing groups such as RCO[118,119] or COOR[120,121] groups to electron-deficient centers have been observed in several carbenium ion rearrangements. This was possible because alternative migrations of hydride, alkyl or aryl groups would have led to much less stable carbenium ion intermediates. Recently, *Gambacorta* and co-workers[122] have shown that acyl and alkyl group migrations were competitive processes in the acid-promoted rearrangements of 9-hydroxybicyclo[3.3.1]nonan-2-ones into *cis*-fused hexahydroinden-1-ones. Since the cationic intermediates **223** - **226** have similar stabilities (all methylcyclohexyl cations are perturbed at a remote position by the carbonyl group), it was possible that the proportion of products observed expressed the relative stabilities of **224**, **225** and **226** (equilibration of tertiary carbenium ions) rather than the relative rate of their formation (non-equilibrating ions). This ambiguity does not exist in the case of the acid-promoted acetolysis of epoxy-ketone **181** for the following reason. Both reaction **185** → **186** and **185** → **187** (Scheme 9) are expected to be highly exothermic because of the formation of the stable oxy-substituted carbenium ions **186** ↔ **186'** and **187** ↔ **187'**. Quantum mechanical calculations (*ab initio* STO 6-31 G*, completely optimized geometries[123]) suggested that **186** (X,Y = O) and **187** (X,Y = O) have similar stabilities, the oxo group having a smaller differential substituent effect than in 6-oxo- and 5-oxo-bicyclo[2.2.1]hept-2-yl cations (**3** and **4**) because of the delocalization of the positive charge in cations **186** ↔ **186'** and **187** ↔ **187'**. Therefore, the competition between *Wagner-Meerwein* rearrangements **185** → **186** (X,Y = O) (path I) and **185** → **187** (X,Y = O) (path II) depends on the intrinsic migratory aptitudes (kinetic effect) of acyl vs. alkyl groups. Consequently, the acid-promoted acetolysis of epoxyketone **181** is a unique situation which enables us to compare the "true" migratory aptitudes of acyl vs. alkyl groups in an "energetically unbiased" situation[124] because of the similar exothermicities of reaction **185** → **186** and **185** → **187** (X,Y = O). The facile acyl group 1,2-shift can be attributed to the electron-donating ability (polarizability) of the C=O function[8,11] which can be interpreted in terms of the canonical forms **3** ↔ **3'** in the case of the 6-oxonorborn-2-yl cation. It is not obvious though, why a similar hyperconjugative interaction does not dominate the *Wagner-Meerwein* tendencies (path I vs. path II) in the epoxyacetals **182 - 184**.

In summary, we have established the following order of intrinsic migratory aptitudes for 1,2-shifts toward electron-deficient centers : acyl > alkyl > alkyl α-substituted with inductive electron-withdrawing groups. This order is valid for competitive *Wagner-Meerwein* rearrangements involving equilibria between carbocation intermediates with similar exothermicities. The acid-promoted rearrangement of epoxy-acetals **182** and **183** is an efficient method to substitute C(5) and C(6) of our "naked sugars" in a *trans* fashion. Importantly, the two oxy functions introduced that way bear different protecting groups (Me, PhCH₂, Ac or none). Thus, optically pure 7-oxanorbornan-2-one derivatives **204** and **205** are potentially useful starting materials for the asymmetric synthesis of carbohydrates, nucleosides and other natural products.

Further Synthetic Application of the "Naked Sugars" : Total Synthesis of Nonactin

Nonactin (**227**) is the lowest homologue of the actin family of antibiotics[125] which has been isolated from a variety of *Steptomyces*.[126] It is a macrotetrolide composed of two subunits of (-)-nonactic acid ((-)-**228**) and two subunits of (+)-nonactic acid ((+)-**228**) arranged in an alternating order, as determined by *Prelog* and co-workers 25 years ago.[127] Their assignment (S₄ symmetry[128]) was confirmed by crystallographic studies.[129] Three syntheses of the natural ionophore **227** have been described. *Gerlach* and co-workers[17] assembled the linear tetramer from racemic nonactic acid monomers; macrocyclization of the 2-pyridinethiol ester then gave a mixture of stereoisomeric macrotetrolides, from which **227** could be isolated in 10 % yield. In 1975, the first synthesis of both enantiomers (-)- and (+)-**228** of nonactic acid was reported by *Schmidt* and coworkers.[131] They also realized the first "Reverse Coupe du Roi"[128] approach to the synthesis of **227** by coupling the tosylate of (+)-benzyl 8-epinonactate with the potassium salt of (-)-nonactic acid through a S_N2 displacement, to produce a dimer which was then condensed to the tetrameric product **227** in 20 % yield.[132] A similar approach to the synthesis of **227** has been

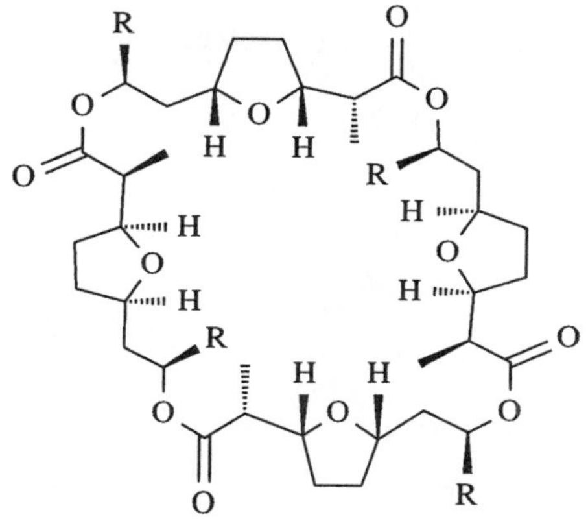

227 R=Me

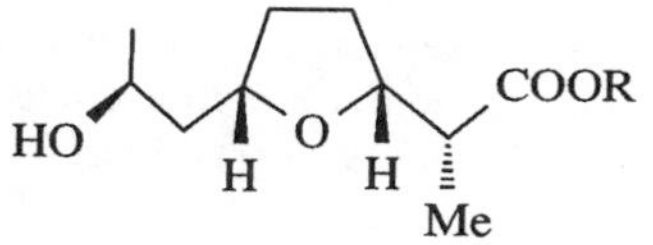

(-)-228 (2R,3R,6S,8S) R=H

(-)-229   R=Me

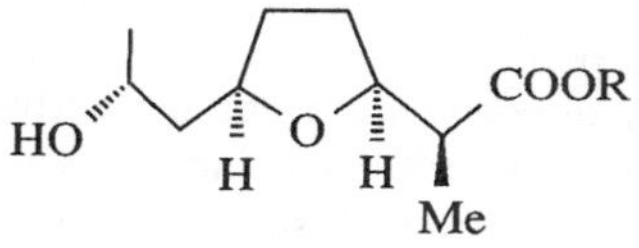

(+)-228 (2S,3S,6R,8R) R=H

(+)-229   R=Me

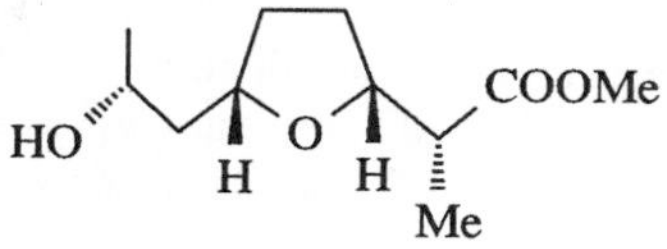

(-)-230

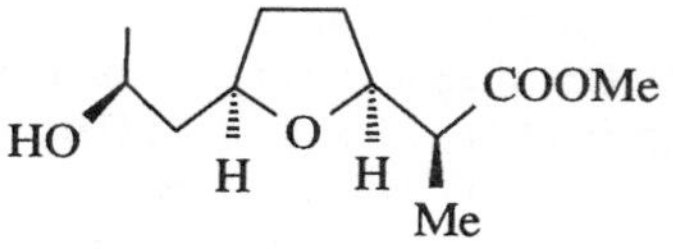

(+)-230

presented more recently by *Bartlett* and coworkers.[133] It requires the production of both (+)-nonactic acid ((+)-228) and the mesylate of (-)-methyl 8-epinonactate ((-)-230).[133]

Six syntheses of (-)-methyl nonactate ((-)-229) and (+)-methyl nonactate ((+)-229) or of other ester derivatives of 228 have already been described. *Schmidt* et al.[131] used (-)-(S)-1,2-epoxypropane (derived from ethyl lactate) as starting material for both (-)- and (+)-229 (separation of diastereoisomers required). *Sun* and *Fraser-Reid*[134] proposed an enantiodivergent approach[135] (5 steps in common) of (-)- and (+)-229 D-ribose. *Ireland* and *Vevert*[136] obtained a mixture of (-)-229 and (+)-230 from D-gulono-γ-lactone. *Bartlett* and coworkers[133] described an enantiodivergent synthesis (13 steps in common) of (-)- and (+)-229 starting with (-)-(S)-malic acid. *Page* et al.[137] derived (+)- and (-)-tert-butyl 8-O-[(tert-butyl)dimethylsilyl]nonactates from (E)-but-2-enal and but-3-enyl bromide through a technique involving a kinetic resolution of enantiomers. Finally, *Batmangerlich* and *Davidson*[138] prepared (+)- and (-)-tert-butyl nonactates and the corresponding 8-epi

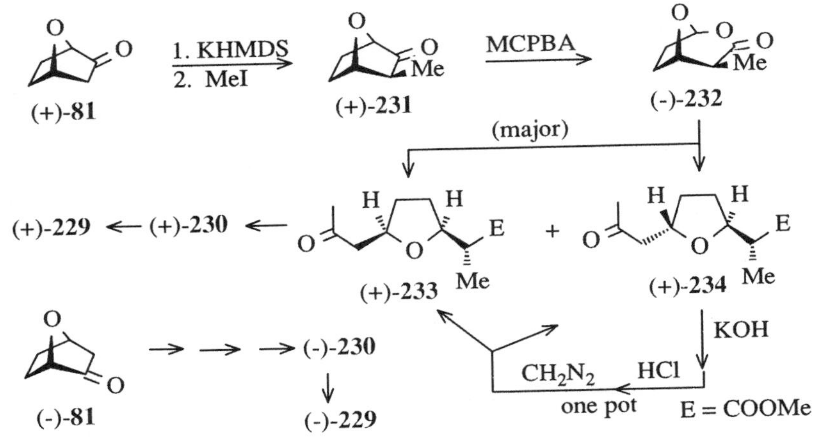

Scheme 11

isomers from L-glutamic acid in 8 steps. All but one of these syntheses required a chiral pool and, in some instances, difficult chromatographic separations. We report here a new and short stereoselective synthesis of (-)- and (+)-methyl 8-epinonactates ((-)- and (+)-230). The latter can be transformed to (-)- and (+)-229, respectively,[139] by applying the *Mitsunobu* displacement reaction.[140] Our approach[36] (Scheme 11) employs (-)- and (+)-7-oxabicyclo[2.2.1]heptan-2-ones ((-)-81 and (+)-81), derived from our naked sugars 38 and 32, respectively, or from optical resolution of (±)-81 using the *Johnson* and *Zeller* method[48] (for other syntheses of racemic nonacic acid derivatives, see reference 141).

Monomethylation of (+)-81 to (+)-231 was achieved in the following way. KHMDS (prepared from KH and hexamethyldisilazane) in THF was added to a 1:10 mixture of (+)-81 and MeI (THF, -50 °C). Workup with 2 N HCl and aq. Na$_2$S$_2$O$_3$ soln. gave (+)-231 in 63 % yield. In some runs, 10 - 15 % of 3,3-dimethyl-7-oxabicyclo[2.2.1]heptan-2-one was also formed. The latter compound was readily separated from 231 by medium-pressure chromatography (silica gel). *Baeyer-Villiger* oxidation with 1 equiv. of MCPBA and NaHCO$_3$ (CHCl$_3$, 12 °C) gave the unstable oxo acetal (-)-232 in 94 % yield. Addition of 1 equiv. of 2-(trimethyl-silyloxy)propene to a 1:1 mixture of (-)-232 and TiCl$_4$ (CH$_2$Cl$_2$, -78 °C, 3 h) furnished a 1:3 mixture of ketones (+)-233 and its *trans* isomer (+)-234. The latter was equilibrated into a 4:3 mixture (+)-233/(+)-234 on treatment with 2 N KOH (20 °C, 2 h), acidification, and esterification with CH$_2$N$_2$. The isomers (+)-233/(+)-234 were separated by medium-pressure chromatography (silica gel) and were isolated in 36 and 27 % yield, respectively. The minor product (+)-234 could be recycled as above into a 4:3 mixture (+)-233/(+)-234 in 85 % yield. The fact that a mixture of both possible stereoisomers is formed under conditions of kinetic control suggests the intervention of a S$_N$1 mechanism alone, or competing with a S$_N$2 mechanism. Steric effects are probably responsible for the favoured displacement reaction with inversion of configuration at the acetal C-centre (S$_N$2 mechanism and/or S$_N$1 mechanism with preferential quenching of the cationic intermediate onto the face *anti* with respect to the carboxylic group). Reduction of (+)-234 with L-Selectride[139]

(THF, -78 °C) gave a 10:1 mixture of (+)-**230** to (+)-**229**. Column chromatography (silica gel) afforded pure (+)-**230** in 82 % yield. Treatment of (+)-**230** with diethyl azodicarboxylate/triphenylphosphine/benzoic acid, followed by saponification (MeOH, MeONa), yielded (+)-**229** in 85 % yield[139]. The enantiomeric forms (-)-**229** and (-)-**230** can be derived in a similar way from (-)-**81**. Our approach[36] is, in principle, applicable to the synthesis of a variety of derivatives of methyl nonactates and methyl 8-epinonactate. This assumes that monoalkylation of 7-oxanorbornanone with groups larger than the methyl group can be achieved or/and that condensation (**232** → **233**) of trimethylsilyl ethers of enols derived from ketones homologous to acetone are possible. Furthermore, since centers C(5) and C(6) in 7-oxanorbornanones can be substituted stereoselectively by a variety of groups, our method also allows us to envision the total synthesis of nonactic acid derivatives substituted at C(4) and C(5) of the tetrahydrofuran ring.

Conclusion

Optically pure 7-oxanorborn-5-en-2-yl derivatives ("naked sugars") are readily available. Substitution of their centers C(3), C(5) and C(6) can be done with high stereo- and regioselectivity in a predictable fashion. The polysubstituted 7-oxanorbornan-2-ones so-obtained can be transformed into D- or L-carbohydrate derivatives, C-nucleoside precursors or polysubstituted cyclohex-2-enones and cyclohexenols (Scheme 12). Stereoselective rearrangements of the 7-oxanorborn-2-yl systems into polyhydroxylated cyclopentyl derivatives are also possible.

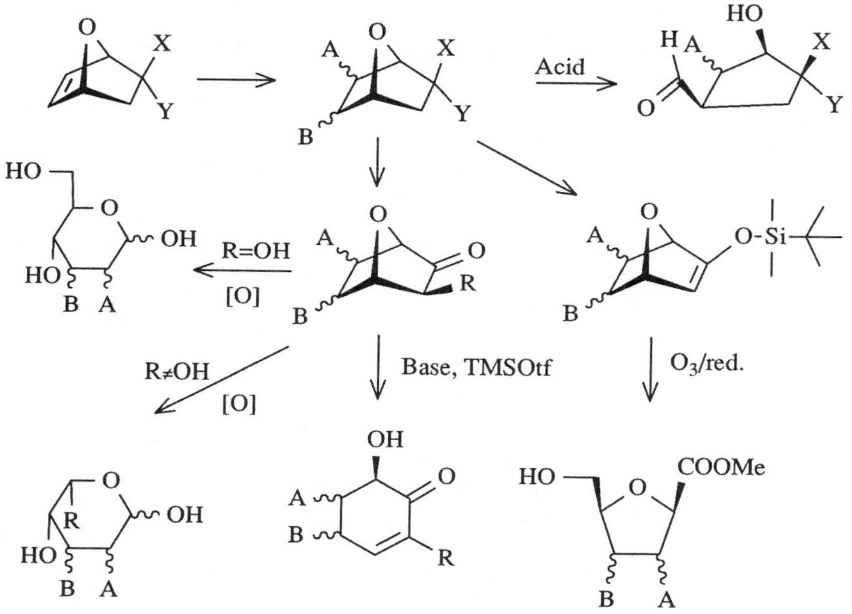

Scheme 12

Acknowledgments. We wish to thank the *Swiss National Science Foundation*, the *Fonds Herbette* (Lausanne), *Hoffmann-La Roche* & Co., AG (Basel), *Shell Agriculture Chemical* Co. (Modesto, Calif.) and *E. I. de Pont de Nemours* & Co. (Wilmington, Del.) for generous financial support. We are grateful also to the *Ecole Polytechnique Fédérale de Lausanne* and its Centre de Calcul (Mr. P. Santschi) for computing time.

Literature Cited

1. (a) Collins, P.M.; Munasinghe, V.R.N. in *"Carbohydrates"*, Ed. P.M. Collins, Chapman and Hall, London, 1987; (b) Glasby, J.S. in *"Encyclopaedia of Antibiotics"*, 2nd. Ed. J. Wiley & Sons, New York, 1979; Remers, W.A. in *"The Chemistry of Antibiotics"*, Vol. 1, J. Wiley & Sons, New York, 1979; (c) Fellows, L.E. *Chem. in Britain* 1987, 842.

2. Avenati, M.; Carrupt, P.-A.; Quarroz, D.; Vogel, P. *Helv. Chim. Acta* 1982, 65, 188.

3. (a) Klopman, G. *J. Am. Chem. Soc.* 1968, 90, 223; (b) Salem, L. *Ibid.* 1968, 90, 543, 553; (c) Fukui, K. *Acc. Chem. Res.* 1971, 4, 57; idem, *Bull. Chem. Soc. Jpn.* 1966, 39, 498; (d) Fleming, I. in *"Frontier Orbitals and Organic Chemical Reactions"*, Wiley, 1976; (e) Sustmann, R. *Pure Appl. Chem.* 1974, 40, 569; (f) Herndon, W.C. *Chem. Rev.* 1972, 72, 157; (g) Hudson, R.F. *Angew. Chem. Int. Ed. Engl.* 1973, 12, 36; (h) Epiotis, N.D. *J. Am. Chem. Soc.* 1973, 95, 5624; (i) Houk, K.N. *Acc. Chem. Res.* 1975, 8, 361; (j) Eisenstein, O.; Lefour, J.M.; Anh. N.T.; Hudson, R.F. *Tetrahedron* 1977, 33, 523; (k) Alston, P.V.; Ottenbrite, R.M.; Newby, J. *J. Org. Chem.* 1979, 44, 4939; (e) Dewar, M.J.S.; Dougherty, R.C. in *"The PMO Theory of Organic Chemistry"*, Plenum Press, New York, 1975.

4. Carrupt, P.-A.; Gabioud, R.; Rubello, A.; Vogel, P.; Honegger, E.; Heilbronner, E. *Helv. Chim. Acta,* 1987, 70, 1540.

5. (a) Chadwick, D.; Frost, D.C.; Weiler, L. *J. Am. Chem. Soc.*, 1971, 93, 4320, 4962; (b) Hentrich, G.; Gunkel, E.; Klessinger, J. *J. Mol. Struct.* 1974, 21, 231; (c) Schäfer, W.; Schmidt, H.; Schweig, A.; Hoffmann, R.W.; Kurz, H. *Tetrahedron Lett.* 1974, 1953; (d) Houk, K.N.; Lambert, L.S.; Engel, P.S.; Schexrayder, M.A.; Seeman, J.I.; Zifler, H. unpublished results.

6. (a) Hoffmann, R. *Acc. Chem. Res.* 1971, 4, 1; (b) Hoffmann, R.; Imamura, A.; Hehre, W.J. *J. Am. Chem. Soc.* 1968, 90, 1499; (c) Gleiter, R. *Angew. Chem.* 1974, 86, 770; (d) Heilbronner, E.; Schmelzer, A. *Helv. Chim. Acta* 1975, 58, 936.

7. Carrupt, P.-A. Ph.D. Dissertation, University of Lausanne, 1979.

8. Carrupt, P.-A.; Vogel, P. *Tetrahedron Lett.* 1984, 25, 2879.

9. (a) Apeloig, Y.; Arad, D.; Lenoir, D.; Schleyer, P.V.R. *Tetrahedron Lett.* 1981, 879; (b) Wilcox, C.F., Jr.; Tuszynski, W.J. *Ibid.* 1982, 23, 3119; (c) Lenoir, D. *Chem. Ber.* 1975, 108, 2055; (d) Gassman, P.G.; Tidwell, T.T. *Acc. Chem. Res.* 1983, 16, 279; (e) Tidwell, T.T. *Angew. Chem. Int. Ed. Engl.* 1984, 23, 20; (f) Charton, M. *Prog. in Phys. Org. Chem.*, R.W. Taft, Ed, Wiley, New York, 1981, 13, 119; (g) Grob, C.A.; Schaub, B.; Schlageter, M.G. *Helv. Chim. Acta,* 1980, 63, 57.

10. (a) Schmid, G.H.; Garratt, D.G. In *"The Chemistry of the Functional Groups*: suppl. A : *"The Chemistry of Double-Bonded Functional Groups"*, Patai, S. Ed. J. Wiley & Sons, London, 1977, pp. 828-854; (b) Fahey, R.C. *Top. Stereochem.* **1968**, *3*, 63; (c) Toshoshima, K.; Okuyama, T.; Fueno, T. *J. Org. Chem.* **1978**, *43*, 2789; (d) Raucher, S. *Ibid.* **1977**, *42*, 2950; (h) Schmid, G.H.; Modro, A.; Yates, K. *Ibid.* **1977**, *42*, 871; (f) Garratt, D.G.; Schmid, G.H. *Ibid.* **1977**, *42*, 1776; (f) Garratt, D.G.; Ryan, M.D.; Beaulieu, P.L. *Ibid.* **1980**, *45*, 839; (h) Schmid, G.H.; Yeroushalmi, S.; Garratt, D.G., *Ibid.* **1980**, *45*, 910; (i) Zefirov, N.S.; Sadovaja, N.K.; Novgorodtseva, L.A.; Achmedova, R. Sh.; Baranov, S.V.; Bodrikov, I.V. *Tetrahedron* **1979**, *35*, 2759; (j) Ruasse, M.F.; Argile, A.; Dubois, J.E. *J. Am. Chem. Soc.* **1978**, *100*, 7645; (k) Toshimitsu, A.; Aoai, T.; Uemura, S.; Okana, M. *J. Chem. Soc., Chem. Commun.* **1980**, 1041; (e) Jones, G.A.; Stirling, C.J.M.; Bromby, N.G.; *J. Chem. Soc. Perkin Trans. II*, **1983**, 385.

11. Carrupt, P.-A.; Vogel, P. *Tetrahedron Lett.* **1982**, *23*, 2563.

12. Black, K.A.; Vogel, P. *J. Org. Chem.* **1986**, *51*, 5341.

13. Vogel P. in *"Stereochemistry and Reactivity of Sytems Containing π-Electrons"*, Watson, W.H., Ed.; Verlag Chemie International : Deerfield Beach, FL, 1983; *"Methods in Stereochemical Analysis"*, pp 147-195.

14. (a) Diels, O.; Alder, K.; *Justus Liebigs Ann. Chem.* **1931**, *485*, 211; **1935**, *515*, 185; (b) Allen. A.D.; Tidwell, T.T. *J. Am. Chem. Soc.* **1982**, *104*, 3145, and references cited therein; (c) Garratt, P.J.; Hollowood, F. *J. Org. Chem.* **1982**, *47*, 68; (d) Schleyer, P.V.R. *J. Am. Chem. Soc.* **1967**, *89*, 3901-3903; (e) Burkert, U. *Angew. Chem., Int. Ed. Engl.* **1981**, *20*, 572; (f) Houk, K.N. *"In Reactive Intermediates"*; Jones, M.; Moss, R. A., Eds.; Wiley, New York, 1978; Vol. 1, p 326; (g) Inagaki, S.; Fujimoto, H.; Fukui, K. *J. Am. Chem. Soc.* **1976**, *98*, 4054; (h) Rondan, N.G., Paddon-Row, M.N.; Caramella, P.; Houk, K.N. *Ibid.* **1981**, *103*, 2436; (i) Huisgen, R. *Pure Appl. Chem.* **1981**, *53*, 171.

15. Arjona, O.; Fernández de la Pradilla, R.; Pérez, R.A.; Plumet, J.; Viso, A. *Tetrahedron Lett.* **1987**, *28*, 5546.

16. Black, K.A.; Vogel, P. *Helv. Chim. Acta* **1984**, *67*, 1612.

17. Vieira, E.; Vogel, P. *Helv. Chim. Acta* **1983**, *66*, 1865.

18. Lipshutz, B.H. *Chem. Rev.* **1986**, *86*, 795.

19. Zamojski, A.; Banaszek, A.; Grynkiewicz, G. *Adv. in Carbohydr. Chem. and Biochem.*, Acad. Press, New York, 1982, *40*, 1 and references cited therein.

20. (a) Just, G.; Lim, M.I. *Can J. Chem.* **1977**, *55*, 2993; (b) Just, G.; Liak, T.J.; Lim, M.I.; Potvin, P.; Tsantrizos, Y.S. *Ibid.* **1980**, *58*, 2024; (c) Kozikowski, A.P.; Floyd, W.C.; Kuniak, M.P. *J. Chem. Soc., Chem. Commun.* **1977**, 582; (d) Kozikowski, A.P.; Ames, A. *Ibid.* **1981**, *103*, 3923.

21. (a) Nelson, W.L.; Allen, D.R.; Vincenzi, F.F. *J. Med. Chem.* **1971**, *14*, 698; (b) Kotsuki, H.; Nishizana, H. *Heterocycles* **1981**, *16*, 1287; (c) Cowling, A.P.; Mann, J.; Usmani, A.A. *J. Chem. Soc., Perkin Trans. I* **1981**, 2116.

22. Eggelte, T.A.; de Koning, H.; Huisman, H.O. *J. Chem. Soc., Perkin Trans. I* **1978**, 980.

23. (a) Rajapaksa, D.; Keay, B.A.; Rodrigo, R. *Can. J. Chem.* **1984**, *62*, 826; (b) Campbell, M.M.; Kaye, A.D.; Sainsbury, M.; Yavarzadeh, R. *Tetrahedron* **1984**, *40*, 2461; (c) Campbell, M.M.; Sainsbury, M.; Yavarzadeh, R. *Ibid.* **1984**, *40*, 5063.

24. (a) Carrupt, P.-A.; Vogel, P. *Tetrahedron Lett.* **1979**, 4533; (b) Tamariz, J.; Vogel, P. *Tetrahedron* **1984**, *40*, 4549; (c) Tornare, J.-M.; Vogel, P. *Helv. Chim. Acta* **1985**, *68*, 1069; (d) Keay, B.A.; Rodrigo, R. *Can J. Chem.* **1983**, *61*, 637; (e) Wiersum, U.E. *Aldrichimica Acta* **1981**, *14*, 53.

25. Holmes, A.B.; Jennings-White, C.L.D.; Kendrick, D.A. *J. Chem. Soc., Chem. Commun.* **1983**, 415; **1984**, 1594.

26. (a) Shizuri, Y.; Nishiyama, S.; Shigemori, H.; Yamamura, S. *J. Chem. Soc., Chem. Commun.* **1985**, 292; (b) Kowarski, C.R.; Sarel, S. *J. Org. Chem.* **1973**, *38*, 117; (c) Yur'ev, Y.K.; Zefirov, N.S. *Zh. Obshchei Khim.* **1961**, *31*, 685; (d) Zefirov, N.S.; Yur'ev, V.K. *Ibid.* **1963**, *33*, 2153; (e) Criegee, R.; Becher, P. *Chem. Ber.* **1957**, *90*, 2516.

27. (a) Van Royen, L.A.; Mijngheer, R.; De Clercq, P.J. *Tetrahedron Lett.* **1983**, *24*, 3145; (b) Imagawa, T.; Sugita, S.; Akiyama, T.; Kawanisi, M. *Ibid.* **1981**, *22*, 2569; (c) Akiyama, T.; Fujii, T.; Ishiwari, H.; Imagawa, T.; Kawanisi, M. *Ibid.* **1978**, *19*, 2165; (d) Murai, A.; Takahashi, K.; Taketsuru, H.; Masamune, T. *J. Chem. Soc., Chem. Commun.* **1981**, 221; (e) Imagawa, T.; Nakagawa, T.; Matsuura, K.; Akiyama, T.; Kawanisi, M. *Chem. Lett.* **1981**, 903; (f) Mukaiyama, T.; Iwasawa, N.; Tsuji, T.; Narasaka, K. *Ibid.* **1979**, 1175; (g) Best, W.M.; Wege, D. *Aust. J. Chem.* **1986**, *39*, 647; (h) Hanessian, S.; Beaulieu, P.; Dubé, D. *Tetrahedron Lett.* **1986**, *27*, 5071; (i) Bloch, R.; Gilbert, L. *Ibid.* **1987**, *28*, 423.

28. (a) Chen, R.T.; Hua, Z.; Yang, J.-L.; Han, J.-X.; Zhang, S.-T.; Lu, F.-L.; Xu, B. *Chin. Med. J.* (Beijing, Engl. Ed.) **1980**, *93*, 183; (b) Wang, Z.; Leng, H.; Sha, K.; Liu, J. *Fenzi Kexue Yu Huaxue Yanjin* **1983**, *3*, 25; *Chem. Abstr.* **1984**, *100*, 17225h; (c) Anderson, W.K.; Dewey, R.H.; Mulumba, B. *J. Med. Chem.* **1979**, *22*, 1270; (d) Anderson, W.K.; Dewey, R.H. *Ibid.* **1977**, *20*, 306; (e) Fryer, R.I.; Boris, A.; Earley, J.V.; Reeder, E. *Ibid.* **1977**, *20*, 1268; (f) Hall, S.E.; Han, W.C.; Haslarger, M.F.; Harris, D.N.; Ogletree, M.L. *Ibid.* **1986**, *29*, 2335; (g) Smith, E.C.R.; Riley, T.N.; Borne, R.F.; Waters, I.W. *Ibid.* **1987**, *30*, 1105; (h) Sprague, P.W.; Heikes, J.E.; Gougoutas, J.Z.; Macley, M.F; Harris, D.N.; Greenberg, R. *Ibid.* **1985**, *28*, 1580.

29. (a) Acheson, R.M. *"An Introduction to the Chemistry of Heterocyclic Compounds"*, Wiley-Interscience, New York 1960; (b) Bosshard, P.; Eugster, C.H. *Adv. Heterocycl. Chem.* **1966**, *7*, 377, Academic Press New York; (c) Ouellette, R.J.; Rosenblum A.; Booth, G. *J. Org. Chem.* **1968**, *33*, 4302; (d) Kienzle, F. *Helv. Chim. Acta* **1975**, *58*, 1180; (e) Vogel, P.; Willhalm, B.; Prinzbach, H. *Ibid.* **1969**, *52*, 584; (f) McCullogh, A.W.; Smith, D.G.; McInnes, A.G. *Can. J. Chem.* **1974**, *52*, 1013; (g) Scharf, H.D.; Friedrich, P.; Linckens, A. *Synthesis* **1976**, 256; (h) Maier, G.; Jung, W.A. *Chem. Ber.* **1982**, *115*, 804; (i) Schuda, P.F.; Bennett, J.M. *Tetrahedron Lett.* **1982**, *23*, 5525; (j) Klein, L.L.; Deeb, T.M. *Ibid.* **1985**, *26*, 3935; (k) Mirsaedhi, S.; Rickborn, B. *J. Org. Chem.* **1985**, *50*, 4340.

30. (a) Dauben, W.C.; Krabbenhoft, H.O. *J. Am. Chem. Soc.* 1976, *98*, 1992; (b) Rimmelin, J.; Jenner, G.; Rimmelin, P. *Bull. Soc. Chim. Fr.* 1978, II-461; (c) Kotsuki, H.; Kondo, A.; Nishizawa, H.; Ochi, M.; Matsuoka, K. *J. Org. Chem.* 1981, *46*, 5454; (d) Jurczak, J.; Kozluk, T.; Filipek, St.; Eugster, C.H. *Helv. Chim. Acta* 1982, *65*, 1021; (e) Kotsuki, H.; Nishizawa, H. *Heterocycles* 1981, *16*, 1287; (f) Kotzuki, H.; Nishizawa, H.; Ochi, M.; Matsuoka, K. *Bull. Chem. Soc. Jpn.* 1982, *55*, 496; (g) Isaacs, N.S.; Van der Beeke, P. *Tetrahedron Lett.* 1982, *23*, 2147; (h) Dauben, W.G.; Gerdes, J.M.; Smith, D.B. *J. Org. Chem.* 1985, *50*, 2576; (i) Okamoto, Y.; Giandinoto, S.; Bochnik, M.C. *J. Org. Chem.* 1983, *48*, 3830: (j) Smith, A.B.; Liverton, N.J.; Hrib, N.J.; Sivaramakrishnan, H.; Winzenberg, K. *J. Am. Chem. Soc.* 1986, *108*, 3040; (k) Matsumoto, K.; Sera, A. *Synthesis* 1985, 999; (l) Jurczak, J.; Kawczynski, A.L.; Kozluk, T. *J. Org. Chem.* 1985, *50*, 1106.

31. (a) McCulloch, A.W.; Smith, D.G.; McInnes, A.G. *Can. J. Chem.* 1973, *51*, 4125; (b) McCulloch, A.W.; Smith, D.G.; McInnes, A.G. *Ibid.* 1974, *52*, 1013; (c) Hall, R.H.; Harkema, S.; den Hertog, H.J.; van Hummel, G.J.; Reinhoudt, D.N. *Rec. Trav. Chim. Pays-Bas* 1981, *100*, 312; (d) Kotsuki, H.; Asao, K.; Ohnishi, H. *Bull. Chem. Soc. Jpn.* 1984, *57*, 3339; (e) Kotsuki, H.; Kondo, A.; Nishizawa, H.; Ochi, M.; Matsuoka, K. *J. Org. Chem.* 1981, *46*, 5454; (f) Jurczak, J.; Belniak, S.; Kozluk, T.; Pikul, S.; Salamski, P. *Bull. Pol. Acad. Sci. Chem.* 1984, *32*, 135; (g) Brion, F. *Tetrahedron Lett.* 1982, *23*, 5299; (h) Campbell, M.M.; Kaye, A.D.; Sainsbury, M. *Ibid.* 1983, *24*, 4745; (i) Moore, J.A.; Partain, E.M. *J. Org. Chem.* 1983, *48*, 1105; (j) Nugent, W.A.; McKinney, R.J.; Harlow, R.L. *Organometalics*, 1984, *3*, 1315; (k) Laszlo, P.; Lucchetti, J. *Tetrahedron Lett.* 1984, *25*, 4387.

32. Vieira, E.; Vogel, P. *Helv. Chim. Acta* 1982, *65*, 17C0.

33. (a) McCulloch, A.W.; McInnes, A.G. *Can. J. Chem.* 1974, *52*, 143; (b) Gomes, L.M.; Aicart, M. *Compt. Rend. Hebd. Acad. Sc.* Ser. C 1977, *285*, 571; (c) Abbott, P.J.; Acheson, R.M.;. Flowerday, R.F. *J. Chem. Soc., Perkin Trans. I*, 1974, 1177; (d) Huang, N.Z.; Xing, Y.D.; Yung Ye, D. *Synthesis* 1982, 1041; (e) Kibayashi, T.; Ishii, Y.; Ogawa, M. *Bull. Chem. Soc. Jpn.* 1985, *58*, 3627.

34. (a) Oppolzer, W. *Angew. Chem. Int. Ed. Engl.* 1984, *23*, 876; (b) ApSimon, J.W.; Collier, T.L. *Tetrahedron* 1986, *42*, 5157; (c) Oppolzer, W. *Tetrahedron* 1987, *43*, 1669; (d) Narasaka, K.; Inoue, M.; Yamada, T. *Chem. Lett.* 1986, 1967; (e) Reetz, M.T.; Kyung, S.H.; Bolm, C.; Zierke, T. *Chem. Ind.* (London) 1986, *23*, 824; (f) Chapuis, C.; Jurczak, J. *Helv. Chim. Acta* 1987, *70*, 436; (g) Seebach, D.; Beck, A.K.; Imwinkelried, R.; Roggo, S.; Wonnacott, A. *Ibid.* 1987, *70*, 954.

35. (a) King. H.M. in *"Resources of Organic Matter for the Future"*, Multi-Science Publication Ltd., Montréal, 1978, pp. 55; (b) Goheen, D.W. *Chem. Educ.* 1981, *58*, 465, 544.

36. Warm, A.; Vogel, P. *Helv, Chim. Acta* 1987, *70*, 690.

37. (a) Takayama, H.; Hayashi, K.; Koizumi, T. *Tetrahedron Lett.* 1986, *27*, 5509; see also : (b) Takayama, H.; Lyobe, A.; Koizumi, T. *J. Chem. Soc., Chem. Commun.* 1986, 771; (c) Arai, Y.; Hayashi, Y.; Yamamoto, M.; Takayama, H.; Koizumi, T. V. *Chem. Lett.* 1987, 185.

38. (a) Iten, P.X.; Hofmann, A.A.; Eugster, C.H. *Helv. Chim. Acta* **1978**, *61*, 430; (b) Eugster, C.H.; Balmer, M.; Prewo, R.; Bieri, J.H. *Ibid.* **1981**, *64*, 2636; (c) Brownbridge, P.; Chan, T.-H. *Tetrahedron Lett.* **1980**, 3423, 3427 and 3431; (d) Murai, A.; Takahashi, K.; Taketsuru, H.; Masamune, T. *J. Chem. Soc. Commun.* **1981**, 221; (e) Gravel, D.; Deziel, R.; Brisse, F.; Hechler, L. *Can. J. Chem.* **1981**, *59*, 2997; (f) McDonald, E.; Suksamrarn, A.; Wylie, R.D. *J. Chem. Soc. Perkin I*, **1979**, 1893; (g) Pelter, A.; Al-Bayalti, R.; Lewis, W. *Tetrahedron Lett.* **1982**, *23*, 353; (h) Kraus, G.A.; Roth, B. *J. Org. Chem.* **1978**, *43*, 2072; (i) Matsumoto, K.; Ikemi, Y.; Hashimoto, S.; Lee, H.S.; Okamoto, Y. *J. Org. Chem.* **1986**, *51*, 3729; (j) Jung, M.E.; Street, L.J.; Usui, Y. *J. Am. Chem. Soc.* **1986**, *108*, 6810; (k) Camici, L.; Ricci, A.; Taddei, M. *Tetrahedron Lett.* **1986**, *27*, 5155.

39. (a) Takeuchi, T.; Chimura, H.; Hamada, M.; Umezawa, H.; Yoshioka, O.; Oguchi, N.; Takahashi, Y.; Matsuda, A. *J. Antibiot.* **1975**, *28*, 737; (b) Nakamura, H.; Takita, T.; Takeuchi, T.; Umezawa, H.; Kato, K.; Saito, S.; Tomizawa, T.; Litaka, Y. *Ibid.* **1975**, *28*, 743.

40. (a) Sugimoto, Y.; Suzuki, H.; Yamaki, H.; Nishimura, T.; Tanaka, N. *J. Antibiot.* **1982**, *35*, 1222; (b) Douglas, K.T.; Shinkai, S. *Angew. Chem. Int. Ed. Engl.* **1985**, *24*, 31.

41. (a) Jones, J.B. *Tetrahedron* **1986**, *42*, 3351; (b) Guanti, G.; Banfi, L.; Narisano, E.; Riva, R.; Thea, S. *Tetrahedron Lett.* **1986**, *27*, 4639; (c) Laumen, K.; Schneider, M. *Ibid.* **1985**, *26*, 2073; (d) Schneider, M.; Engel, N.; Hönicke, P.; Heinemann, G.; Görisch, H. *Angew. Chem. Int. Ed. Engl.* **1984**, *23*, 67; (e) Wang, Y.-F.; Sin, C. J. *Tetrahedron Lett.* **1984**, *25*, 4999; (f) Kasel, W.; Hultin, P.G.; Jones, J.B. *J. Chem. Soc., Chem. Commun.* **1985**, 1563; (g) Kamiyama, K.; Kobayashi, S.; Ohno, M. *Chem. Lett.* **1987**, 29; (h) Seebach, D.; Roggo, S.; Maetzke, T.; Braunschweiger, H.; Cercus, J.; Krieger, M. *Helv. Chim. Acta* **1987**, *70*, 1605.

42. (a) Ito, Y.; Shibata, T.; Arita, M.; Sawai, H.; Ohno, M. *J. Am. Chem. Soc.* **1981**, *103*, 6739; (b) Ohno, M.; Ito, Y.; Arita, M.; Shibata, T.; Adachi, K.; Sawai, H. *Tetrahedron* **1984**, *40*, 145.

43. (a) Bloch, R.; Guibe-Jampel, E.; Girard, C. *Tetrahedron Lett.* **1985**, *26*, 4087; (b) Jones, J.B.; Francis, C.J. *Can. J. Chem.* **1984**, *62*, 2578.

44. (a) Jakovac, I.J.; Goodbrand, H.B.; Lok, K.P.; Jones, J.B. *J. Am. Chem. Soc.* **1982**, *104*, 4659; (b) Lok, K.P.; Jakovac, J.; Jones, J.B. *Ibid.* **1985**, *107*, 2521.

45. Ferrari, T.; Vogel, P. *Tetrahedron Lett.* **1986**, *27*, 5507.

46. Grieco, P.A.; Lis, R.; Zelle, R.E.; Finn, J. *J. Am. Chem. Soc.* **1986**, *108*, 5908.

47. Just, G.; Martel, A. *Tetrahedron Lett.* **1973**, 1517.

48. Johnson, C.R.; Zeller, J.R. *J. Am. Chem. Soc.* **1982**, *104*, 4021.

49. Wagner, J.; Vieira, E.; Vogel, P., *Helv. Chim. Acta* **1988**, *71*, 624.

50. (a) Johnson, C.R.; Haake, M.; Schroeck, C.W. *J. Am. Chem. Soc.* **1970**, *92*, 6594; (b) Johnson, C.R.; Schroeck, C.W. *Ibid.* **1973**, *95*, 7418; (c) Johnson, C.R.; Schroeck, C.W.; Shanklin, J.R. *Ibid.* **1973**, *95*, 7424.

51. Ogawa, S.; Iwasawa, Y.; Nose, T.; Suami, T.; Ohba, S.; Ito, M.; Saito, Y. *J. Chem. Soc., Perkin Trans.* **1985**, 903.
52. Ogawa, S.; Takagaki, T. *J. Org. Chem.* **1985**, *50*, 2356.
53. Brown, H. C.; Prasad, J.V.N.V. *J. Am. Chem. Soc.* **1986**, *108*, 2049.
54. Arcamone, F.; Franceschi, G.; Orezzi, P.; Cassinelli, G.; Barbieri, W.; Mondelli, R. *J. Am. Chem. Soc.* **1964**, *86*, 5335.
55. (a) Carter, S.K. *J. Natl. Cancer Inst.* **1975**, *55*, 1265; (b) Skovsgaard, T.; Nissen, N.I. *Dan. Med. Bull.* **1975**, *22*, 62; (c) Arcamone, F. *Med. Res. Rev.* **1984**, *4*, 153; (d) DiMarco, A.; Gaetani, M.; Scarpino, B. *Cancer Chemother. Rep.* **1969**, *53*, 33.
56. (a) Richardson, A.C. *J. Chem. Soc., Chem. Commun.* **1965**, 627; (b) Richardson, A.C. *Carbohydr. Res.* **1967**, *4*, 422; (c) Marsh, J. P.; Mosher, C.W.; Acton, E.M.; Goodman, L. *J. Chem. Soc., Chem. Commun.* **1967**, 973; (d) Horton, D.; Weckerle, W. *Carbohydr. Res.* **1975**, *44*, 227; (e) Yamaguchi, T.; Kojima, M. *Ibid.* **1977**, *59*, 343; (f) Medgyes, G.; Kuszmann, J. *Ibid.* **1981**, *92*, 225; (g) Grethe, G.; Mitt, T.; Williams, T.H.; Uskokovic, M. R. *J. Org. Chem.* **1983**, *48*, 5309; (h) Baer, H.H.; Capek, K.; Cook, M.C. *Can. J. Chem.* **1969**, *47*, 89; (i) Crugnola, A.; Lombardi, P.; Gandolfi, C.; Arcamone, F. *Gazz. Chim. Ital.* **1981**, *111*, 395; (j) Pauls, H.W.; Fraser-Reid, B. *J. Chem. Soc., Chem. Commun.* **1983**, 1031; (k) Hirama, M.; Nishizaki, I.; Shigemoto, T.; Itô, S. *Ibid.* **1986**, 393; (l) Picq, D.; Carret, G.; Anker, D.; Abou-Assali, M. *Tetrahedron Lett.* **1985**, *26*, 1863; (m) Pelyvas, I.; Hasegawa, A.; Whistler, R. L. *Carbohydr. Res.* **1986**, *146*, 193; (n) Servi, S. *J. Org. Chem.* **1985**, *50*, 5865; (o) Gurjar, M.K.; Pawar, S.M. *Tetrahedron Lett.* **1987**, *28*, 1327; (p) Kita, Y.; Itoh, F.; Tamura, O.; Ke, Y.Y.; Tamura, Y. *Ibid.* **1987**, *28*, 1431.
57. (a) Iwataki, I.; Nakamura, Y.; Takahashi, K.; Matsumoto, T. *Bull. Chem. Soc. Jpn.* **1979**, *52*, 2731; (b) Fronza, G.; Fuganti, C.; Grasselli, P. *J. Chem. Soc., Chem. Commun.* **1980**, 442; (c) Dyong, I.; Wiemann, R. *Chem. Ber.* **1980**, *113*, 2666; (d) Wovkulich, P. M.; Uskokovic, M. R. *J. Am. Chem. Soc.* **1981**, *103*, 3956; (e) Fuganti, C.; Grasselli, P.; Pedrocchi-Fantoni, G. *J. Org. Chem.* **1983**, *48*, 909; (f) Grethe, G.; Sereno, J.; Williams, T.H.; Uskokovic, M. R. *Ibid.* **1983**, *48*, 5315; (g) Hauser, F.M.; Rhee, R.P.; Ellenberger, S.R. *Ibid.* **1984**, *49*, 2236; (h) Hauser, F.M.; Ellenberger, S.R.; Glusker, J.P.; Smart, C.J.; Carrell, H. L. *Ibid.* **1986**, *51*, 50; (i) DeShong, P.; Leginus, J.M. *J. Am. Chem. Soc.* **1983**, *105*, 1686; (j) Mukaiyama, T.; Goto, Y.; Shoda, S. *Chem. Lett.* **1983**, 671; (k) Hiyama, T.; Nishide, L.; Kobayashi, K. *Ibid.* **1984**, 361; (l) Hamada, Y.; Kawai, A.; Shioiri, T. *Ibid.* **1984**, *25*, 5409; (m) Hanessian, S.; Kloss, J. *Ibid.* **1985**, *26*, 1261; (n) Sammes, P.G.; Thetford, D. *J. Chem. Soc., Chem. Commun.* **1985**, 352; (o) DeShong, P.; Dicken, C.M.; Leginus, J.M.; Whittle, R.R. *J. Am. Chem. Soc.* **1984**, *1061*, 5598; (p) Danishefsky, S.J.; Maring, C. J. *Ibid.* **1985**, *107*, 1269; (q) Wovkulich, P.M.; Uskokovic, M.R. *Tetrahedron* **1985**, *41*, 3455; (r) Monneret, C.; Coureur, C.; Khuong, H.Q. *Carbohydr. Res.* **1978**, *68*, 35; (s) Hauser, F.M.; Ellenberger, S.R.; Glusker, J.P.; Smart, C.J.; Carrell, H.L. *J. Org. Chem.* **1986**, *51*, 50; (t) Hirama, M.; Nishizaki, I.; Shigemoto, T.; Itô, S. *J. Chem. Soc., Chem.*

Commun. **1986**, 393; (u) Banfi, L.; Cardani, S.; Potenza, D.; Scolastico, C.; *Tetrahedron* **1987**, *43*, 2317.

58. (a) Thiele, J. *Liebigs Ann. Chem.* **1892**, *271*, 127; (b) Hamersma, J.W.; Snyder, E.I. *J. Org. Chem.* **1965**, *30*, 3985; see also : (c) Adam, W.; Eggelte, H.J. *Angew. Chem., Int. Ed. Engl.* **1978**, *17*, 765; (d) Baird, W.C., Jr.; Franzus, B.; Surridge, J.H. *J. Am. Chem. Soc.* **1967**, *89*, 410.

59. Meinwald, J.; Frauenglass, E. *J. Am. Chem. Soc.* **1960**, *82*, 5235.

60. Sharpless, K.B.; Young, M.W.; Lauer, R.F. *Tetrahedron Lett.* **1973**, 1979.

61. Loev, B. *Chem. Ind.* (London) **1964**, 193.

62. (a) Jorgenson, M. *J. Org. React.* (N.Y.) **1970**, *18*, 1; (b) DePuy, C.H.; Dappen, G.M.; Eilers, K.L.; Klein, R.A. *J. Org. Chem.* **1964**, *29*, 2813; (c) Bare, T.M.; House, H.O. *Org. Synth.* **1969**, *49*, 81.

63. Emmons, W.D.; Lucas, G.B. *J. Am. Chem. Soc.* **1955**, *77*, 2287.

64. Sheldon, R.A.; Kochi, J.K. *Org. React.* (N.Y.) **1972**, *19*, 279.

65. (a) Malik, A.; Afza, N.; Voelter, W. *J. Chem. Soc., Perkin Trans. I* **1983**, 2103; (b) Horton, D.; Wander, J.D. "In Amino Sugars, Carbohydr.: Chem., Biochem.", 2nd ed.; Pigman, W.W., Horton, D., Eds.; Academic Press : New York, 1980; see also : (c) Nayak, U.G.; Whistler, R.L. *J. Org. Chem.* **1969**, *34*, 3819.

66. Moore, A.T.; Rydon, H.N. *Org. Synth.* (N.Y.) **1965**, *45*, 47.

67. (a) Kuivila, H.G. *Acc. Chem. Res.* **1968**, *1*, 299; (b) Menapace, L.W.; Kuivila, H.G. *J. Am. Chem. Soc.* **1964**, *86*, 3047; (c) Gutierrez, C.G.; Summerhays, L.R. *J. Org. Chem.* **1984**, *49*, 5206; (d) Pang, M.; Becker, E.I. *Ibid.* **1964**, *29*, 1948; (e) Haskell, T.H.; Woo, P.W.K.; Watson, D.R. *Ibid.* **1977**, *42*, 1302.

68. (a) Pettit, G.R.; van Tamelen, E.E. *Org. React.* (N.Y.) **1962**, *12*, 356; (b) Berson, J.A.; Jones, W.M. *J. Am. Chem. Soc.* **1956**, *78*, 6045; (c) Stork, G.; van Tamelen, E.E.; Friedman, L.J.; Burgstahler, A.W. *Ibid.* **1953**, *75*, 384.

69. Durgnat, J.-M.; Warm, A.; Vogel, P., in preparation; Durgnat, J.-M., Travail de diplôme, University of Lausanne, 1986.

70. Kaiser, H.; Keller-Schierlein, W. *Helv. Chim. Acta* **1981**, *64*, 407.

71. (a) Brockmann, H.; Waehneldt, T. *Naturwissenschaften*, **1961**, *48*, 717; (b) Brockmann, H.; Waehneldt, Niemeyer, J. *Tetrahedron Lett.* **1969**, 415; (c) Keller-Schierlein, W.; Richle, W. *Antimicrob. Agenta Chemother.* **1970**, 68; (d) Richle, W.; Winkler, E.K.; Hawley, D.M.; Dobler, M.; Keller-Schierlein, W. *Helv. Chim. Acta* **1972**, *55*, 467; (e) Brockmann, H.; Scheffer, B.; Stein, C. *Tetrahedron Lett.* **1973**, 3699; (f) Oki, T.; Matzuzawa, Y.; Yoshinuoto, A.; Humesawa, H.; Ishizuka, M.; Sada, H.; Takeuki, T. *J. Antibiot.* **1975**, *28*, 830; (g) De Bruyn, A.; Anteunis, M.; Garegg, P.J.; Norberg, T. *Acta Chem. Scand. Ser. B.* **1976**, *B30*, 820; (h) Nettleton, D.E.; Bracker, W.J.; Bash, J.A.; Coon, A.B.; Moseley, J.E.; Myllymaki, R.W.; O'Heiron, F.A.; Schreiber, R.H.; Vulcano, A.L. *J. Antibiot.* **1977**, *30*, 525; (i) Fercks, F.; Horton, D.; Weckerle, W.; Winter-Mihaly, E. *J. Med. Chem.* **1979**, *22*, 406.

72. (a) Iselin, B.; Reichstein, T. *Helv. Chim. Acta* **1944**, *27*, 1900; (b) Brockmann, H.; Waehneldt, T. *Naturwissenschaften* **1963**, *50*, 92; (c) David, S.; De Sennyey, G. *J. Chem. Soc., Perkin Trans. I* **1982**, 385; (d) Nelson, V.; El Khadem, H.S. *Carbohydr. Res.* **1983**, *124*, 161.

73. (a) Fronza, G.; Fuganti, C.; Grasselli, P.; Pedrocchi-Fantoni, G. *Tetrahedron Lett.* **1982**, *23*, 4143; (b) Fronza, G.; Fuganti, C.; Grasselli, P.; Marinoni, G. *Ibid.* **1979**, *20*, 3883.

74. Chmielewski, M. *Tetrahedron* **1979**, *35*, 2067.

75. Roush, W.R.; Brown, R.J. *J. Org. Chem.* **1983**, *48*, 5093.

76. (a) Katsuki, T.; Sharpless, K.B. *J. Am. Chem. Soc.* **1980**, *102*, 5974; (b) Katsuki, T.; Lee, A.W.M.; Ma, P.; Martin, V.S.; Masamune, S.; Sharpless, K.B.; Tuddenham, D.; Walker, F.J. *J. Org. Chem.* **1982**, *47*, 1373; (c) Ma, P.; Martin. V.S.; Masamune, S.; Sharpless, K.B.; Viti, S. *Ibid.* **1982**, *47*, 1378; (d) Lee, A.W.M.; Martin, V.S.; Masamune, S.; Sharpless, K.B.; Walker, F.J. *J. Am. Chem. Soc.* **1982**, *104*, 3515; see also : (e) Babine, R.E. *Tetrahedron Lett.* **1986**, *27*, 5791 and ref. cited; (f) Roush, W.R.; Straub, J.A. *Ibid.* **1986**, *27*, 3349.

77. Oda, T.; Mori, T.; Kyotani, J. *J. Antibiot.* **1971**, *24*, 503.

78. Kuhn, R.; Weiser, D.; Fischer, H. *Liebigs Ann. Chem.* **1959**, *628*, 207.

79. de Guchteneere, E.; Vogel, P. in preparation.

80. (a) Meyer zu Reckendorf, W.; Bonner, W.A. *Tetrahedron* **1963**, *19*, 1711; (b) Arita, H.; Fukukawa, K.; Matsushima, Y. *Bull. Chem. Soc. Jpn.* **1972**, *45*, 3614; (c) Jegon, E.; Brewer, C.L.; Guthrie, R.D. *J. Chem. Soc., Perkin Trans. I* **1974**, 657; (d) Cléophax, J.; Lebuol, J.; Gero, S.D. *Carbohydr. Res.* **1975**, *45*, 323; (e) Yamasaki, T.; Kubota, Y.; Tsuchiya, I.; Umezawa, S. *Bull. Chem. Soc. Jpn.* **1976**, *49*, 3190; (f) Oida, S.; Saeki, H.; Okashi, Y.; Ohki, E. *Chem. Pharm. Bull.* **1975**, *23*, 1547; (g) Saeki, H.; Takeda, N.; Shimada, Y.; Ohki, E. *Ibid.* **1976**, *24*, 724; (h) Haskell, T.H.; Woo, P.W.K.; Watson, D.R. *J. Org. Chem.* **1977**, *42*, 1302; (i) Tsuchiya, T.; Watanabe, J.; Yoshida, M.; Nakamura, F.; Usui, T.; Kitamura, M.; Umezawa, S. *Tetrahedron Lett.* **1978**, 3365; (j) Lemieux, R.U.; Georges, F.F.Z.; Smiatacz, Z. *Heterocycles* **1979**, *13*, 169; (k) Hanessian, S.; Vatele, J.-M. *Ibid.* **1981**, 3579; (l) Miyake, T.; Tsuchiva, T.; Takahashi, Y.; Umezawa, S. *Carbohydr. Res.* **1981**, *89*, 255; (m) Miyashita, M.; Chida, N.; Yoshikohsi, A. *J. Chem. Soc., Chem. Commun.* **1982**, 1354.

81. Jäger, V.; Schohe, R. *Tetrahedron* **1984**, *40*, 2199.

82. (a) Just. G.; Grozinger, K. *Tetrahedron Lett.* **1974**, 4165; *Can. J. Chem.* **1975**, *53*, 2701; (b) Just, G.; Martel, A.; Grozinger, K.; Ramjeesingh, M. *Ibid.* **1975**, *53*, 131.

83. Just, G.; Ramjeesingh, M.; Liak, T. J. *Can. J. Chem.* **1976**, *54*, 2940.

84. Just, G., Ramjeesingh, M. *Tetrahedron Lett.* **1975**, 985.

85. (a) Trumlitz, G.; Moffat, J.G. *J. Org. Chem.* **1973**, *38*, 1841; (b) Pommian, M.S.; Nowoswiat, E.F. *Ibid.* **1977**, *42*, 1109; **1980**, *45*, 203; (c) Fleet, G.W.J.; Seymour, L.C. *Tetrahedron Lett.* **1987**, *28*, 3015; (d) Riley, T.A.; Hennen, W.J.; Dalley, N.K.; Wilson, B.E.; Robins, R.K.; Larson, S.B. *J. Heterocycl. Chem.* **1987**, *24*, 955 and ref. cited.

86. Hanessian, S.; Pernet, A.G., *Adv. in Carbohydr. Chem. and Biochem.*, Ed. Tipson, R.S.; Horton, D., Academic Press, New York, **1976**, *33*, 111.

87. Schmidt, R.R.; Beitzke, C.; Forrest, A.K. *J. Chem. Soc., Chem. Commun.* **1982**, 909.

88.   Bimwala, R.; Vogel, P., unpublished results.

89.   Auberson, Y.; Vogel, P. in preparation.

90.   Colvin, E.W. *Chem. Soc. Rev.* **1978**, *7*, 15.

91.   Brimacombe, J.S.; Hunedy, F.; Husain, A. *J. Chem. Soc.* (C) **1970**, 1273.

92.   (a) Beylis, P.; Howard, A.S.; Perold, G.W. *J. Chem. Soc., Chem. Commun.* **1971**, 597; (b) Wei-Shin, C.; Shi-de, L.; Breitmaier, E. *Liebigs Ann. Chem.* **1981**, 1893.

93.   (a) Brimacombe, J. S.; Minshall, J.; Tucker, L. C. N. *J. Chem. Soc. Perkin. Trans. I*, **1973**, 2691; (b) Köll, P.; John, H.-G.; Kopf, J. *Liebigs Ann. Chem.* **1982**, 639; (c) Dessinges, A.; Castillon, S.; Olesker, O.; Ton, T.T.; Lukacz, G. *J. Am. Chem. Soc.* **1984**, *106*, 450; (d) Higuchi, R.; Tokimitsu, Y.; Hamada, N.; Komori, T.; Kawasaki, T. *Liebigs. Ann. Chem.* **1985**, 1192; (e) Higuchi, R.; Tokimitsu, Y.; Komori, T. *Ibid.* **1988**, 249.

94.   Tipson, R.S.; Isbell, H.S. *Meth. Carbohydr. Chem.* **1962**, *1*, 157.

95.   Stroh, H.H.; Dargel, D.; Haüssler, R. *J. prakt. Chem.* 4 Reihe, **1964**, *23*, 309.

96.   Kiss, J.; D'Souza, R.; Taschner, P. *Helv. Chim. Acta* **1975**, *58*, 311.

97.   Taniguchi, M.; Koga, K.; Yamada, S. *Tetrahedron* **1974**, *30*, 3547.

98.   Feniou, C.; Pontagnier, H.; Grignon-Dubois, M.; Lacourt, B.; Rezzonico, B. Brevet français No. 8514098 (1985); extended to Eur. Community, Appl. No. 86450016 (1985); USA appl. No 909911.

99.   Yamaguchi, M.; Mukaijanea, T. *Chem. Lett.* **1981**, 1005.

100.   Dondoni, A.; Fantin, G.; Fogagnolo, M.; Medici, A. *Angew. Chem., Int. Ed. Engl.* **1986**, *25*, 835.

101.   Masamune, S.; Choy, W.; Petersen, J.S.; Sita, L.R. *Angew. Chem., Int. Ed. Engl.* **1985**, *24*. 1.

102.   Jurczak, J.; Pikul, S.; Bauer, T. *Tetrahedron* **1986**, *42*, 447.

103.   Austin, W.C.; Humoller, F.L. *J. Am. Chem. Soc.* **1934**, *56*, 1152.

104.   Kiss, J; D'Souza, R.; van Koeveringe, J.A.; Arnold, W. *Helv. Chim. Acta* **1982**, *65*, 1522.

105.   (a) Luche, J.L. *J. Am. Chem. Soc.* **1978**, *100*, 2226; (b) Luche, J.L.; Gemal, A.L. *Ibid.* **1979**, *101*, 5848.

106.   (a) Posternak, T. in *"Les Cyclitols"*, Hermann, Paris, 1962, and ref. cited therein; (b) see also : Kindl, H.; Hoffmann-Ostenhof, O. *Monatsch. Chem.* **1970**, *101*, 1704; (c) Nagabhushan, T.L. *Can. J. Chem.* **1970**, *48*, 383; (d) Legler, G.; Herrchen, M. *Fed. Eur. Biochem. Soc. Lett.* **1981**, *135*, 139; (e) Knapp, S.; Ornaf, R.M.; Rodriques, K.E. *J. Am. Chem. Soc.* **1983**, *105*, 5494; (f) Lee, K.J.; Boyd, S.A.; Radin, N.S. *Carbohydr. Res.* **1985**, *144*, 148; (g) Nakajima, M.; Tomida, I.; Takei, S. *Chem. Ber.* **1957**, *90*, 246.

107.   Mirza, S.; Molleyres, L.-P.; Vasella, A. *Helv. Chim. Acta* **1985**, *68*, 988.

108.   Ueda, K.; Yasutomi, K.; Mori, I. *Chem. Lett.* **1983**, 149.

109.   Vieira, E. Ph. D. Dissertation, University of Lausanne, Dec. 1986.

110.   Le Drian, C.; Vogel, P. *Tetrahedron Lett.* **1987**, *28*, 1523; *Helv. Chim. Acta* **1987**, *70*, 1703.

111.   Van Melick, J.E.W.; Scheeren, J.W.; Nivard, R.J.F. *Rec. trav. Chim. Pays-Bas* **1973**, *92*, 775.

112. (a) Kwart, H.; Vosburgh, W.G. *J. Am. Chem. Soc.* **1954**, *76*, 5400; (b) Walborsky, H.M.; Loncrini, D.F. *Ibid.* **1954**, *76*, 5396; (c) Bartlett, P.D.; Giddings, W.P. *Ibid.* **1960**, *82*, 1240; (d) Winstein, S.; Stafford, E.T. *Ibid.* **1957**, *79*, 505; (e) Walborsky, H.M.; Loncrini, D.F. *J. Org. Chem.* **1957**, *22*, 1117; (f) McDonald, R.N.; Tabor, T.E. *Ibid.* **1968**, *33*, 2934; (g) Gray, A.P.; Heitmeier, D.E. *Ibid.* **1969**, *34*, 3253; (h) Baker, R.; Halliday, D.E.; Mason, T.J. *Tetrahedron Lett.* **1970**, 591; (i) Berti, G.; Bottari, F.; Macchia, B. *Gazz. Chim. Ital.* **1960**, *90*, 1763; (j) Johnson, B.L.; Langley, J.W.; Raston, C.L.; White, A.H. *Aust. J. Chem.* **1977**, *30*, 2729; (k) Kas'jan, L.A.; Gnedenko, L. Yu.; Galafeeva, M.F.; Kornilov, M.Yu.; Krasutsky, P.A.; Averina, N.V.; Zefirov, N.S. *Tetrahedron Lett.* **1986**, *27*, 2921; (l) Kas'yan, L.I.; Samitov, Yu.Yu. *Epoksidnye Monomery Epoksidnye Smoly*, **1975**, 87; *Chem. Abstracts* **1976**, *85*, 32718m.

113. (a) Martin, J.C.; Bartlett, P.D. *J. Am. Chem. Soc.* **1957**, *79*, 2533; (b) Lambert, J.B.; Larson, E.G. *Ibid.* **1985**, *107*, 7546; (c) Spurlock, L.A., Fayter, R.G.Jr. *Ibid.* **1972**, *94*, 2707; (d) Paquette, L.A.; Dunkin, I.R. *Ibid.* **1973**, *95*, 3067; (e) Gergely, V.; Akhavin, Z.; Vogel, P. *Helv. Chim. Acta* **1975**, *58*, 871.

114. Yur'ev, Yu. K.; Zefirov, N.S. *Zh. Obshch. Khim.* **1961**, *31*, 840.

115. (a) Yu'rev, Yu.K.; Zefirov, N.S. *Zh. Obshch. Khim.* **1961**, *31*, 1125; see also: (b) Malinovskii, M.S.; Kas'yan, L.I.; Ovsyanik, V.D.; Nepokrytaya, T.D.; Terent'ev, P.B.; Zamurenko, V.A. *Vopr. Stereokhim.* **1974**, *4*, 71; *Chem. Abstrats* **1975**, *83*, 58225g.

116. (a) Berson, J.A., in *"Molecular Rearrangements"*, Ed. P. de Mayo, Wiley-Interscience, New York, **1963**, p. 168; (b) Christol, H.; Coste, J.; Plénat, F. *Bull. Soc. Chim. Fr.* **1970**, 2005 and ref. cited therein; see also : (c) Quarroz, D.; Vogel, P. *Helv. Chim. Acta* **1979**, *62*, 335; (d) Kirmse, W.; Mrotzeck, U.; Siegfried, R. *Angew. Chem., Int. Ed.* **1985**, *24*, 55 and ref. cited therein; (e) Christol, H.; Coste, J.; Plénat, F.; Renard, G. *Bull. Soc. Chim. Fr.* **1979**, II -421.

117. (a) Yur'ev, Yu.K.; Zefirov, N.S. *Zh. Obshoh. Khim.* **1962**, *32*, 773; see also : (b) Zefirov, N.S.; Ivanova, R.A.; Kecher, K.M.; Yur'ev. Yu.K. *Ibid.* **1965**, *35*, 61; (c) Gassman, P.G.; Marshall, J.L. *Tetrahedron Lett.* **1968**, 2429.

118. (a) Eschenmoser, A.; Schinz, H.; Fischer, R.; Colonge, J. *Helv. Chim. Acta* **1951**, *34*, 2329; (b) House, H.O. *J. Am. Chem. Soc.* **1954**, *76*, 1235; (c) Domagala, J.M.; Bach. R.D. *Ibid.* **1979**, *101*, 3118; (d) idem, *J. Org. Chem.* **1979**, *44*, 2429, and ref. cited therein; see also : (e) Bach, R.D.; Domagala, J.M. *J. Chem. Soc., Chem. Commun.* **1984**, 1472.

119. (a) Peters, J.A.; van der Toorn, J.M.; van Bekkum, H. *Recl. Trav. Chim. Pays-Bas*, **1975**, *94*, 122; (b) Hart, H.; Shih, E.M. *J. Org. Chem.* **1975**, *40*, 1128; (c) Hart, H.; Huang, I.; Lavrik, P. *Ibid.* **1974**, *39*, 999; (d) Zaidi, J.H.; Waring, A.J. *J. Chem. Soc., Chem. Commun.* **1980**, 618; For a review on 1,2-shifts in carbocations, see e.g. : (e) Shubin, V.G. *Topics Curr. Chem.* **1984**, *117*, 267.

120. (a) Blaise, E.E.; Courtot, A. *Bull. Soc. Chim. Fr.* **1906**, *35*, 360; (b) Acheson, R.M. *Acc. Chem. Res.* **1971**, *4*, 177; (c) Berner, D.; Dahn, H.; Vogel, P. *Helv.*

Chim. Acta **1980**, *63*, 2538; (d) Berner, D.; Cox, D.Ph.; Dahn, H. *J. Am. Chem. Soc.* **1982**, *104*, 2631 und ref. cited therein.

121. (a) Marx, J.N.; Argyle, J.C.; Norman, L.R. *J. Am. Chem. Soc.* **1974**, *96*, 2121; (b) Abbott, P.J.; Acheson, R.M.; Flowerday, R.F.; Brown, G.W. *J. Chem. Soc., Perkin Trans. I* **1974**, 1177; (c) Suehiro, Yamazaki, S. *Bull. Chem. Soc. Jpn.* **1975**, *48*, 3655; (d) Dagli, D.J.; Gorski, R.A.; Wemple, J. *J. Org. Chem.* **1975**, *40*, 1741 and ref. cited therein.

122. (a) Alessandri, P.; De Angelis, F.; Gambacorta, A. *Tetrahedron* **1985**, *41*, 2831; see also : (b) Gambacorta, A.; Turchetta, S. XVI *Convegno Nazionale Della Divisione Di Chimica Organica Società chimica italiana*, Urbino, Sept, 7 - 12, 1986, Commun. No. 037, p. 242.

123. (a) Carrupt, P.-A.; Le Drian, C.; Vogel, P. *Satellite Symposium of the 5th International Congress on Quantum Chemistry*, Aug. 25 - 19, 1985, University of Toronto, Ontario, Canada; (b) Carrupt P.-A.; Vogel, P. *J. Phys. Org. Chem.* **1988**, *1*, to appear.

124. (a) Howells, D.; Warren, S. *J. Chem. Soc., Perkin Trans. II*, **1973**, 1645; (b) Brownbridge, P.; Hodgson, P.K.G.; Shepherd, R.; Warren, S. *J. Chem. Soc., Perkin Trans. I*, **1976**, 2024 and ref. cited therein.

125. (a) Keller-Schierlein, W.; Gerlach, H. *Fortschr. Chem. Org. Naturst.* **1968**, *26*, 161; (b) Keller-Schierlein, W. *Ibid.* **1973**, *30*, 313; (c) Dobler, M. *"Ionophores and their Structure"*, Wiley, New York, 1981.

126. Corbaz, R.; Ettlinger, L.; Gäumann, E.; Keller-Schierlein, W.; Kradolfer, F.; Neipp, L.; Prelog, V.; Zähner, H. *Helv. Chim. Acta* **1955**, *38*, 1445.

127. (a) Dominguez, J.; Dunitz, J.D.; Gerlach, H.; Prelog, V. *Helv. Chim. Acta* **1962**, *45*, 129; (b) Gerlach, H.; Prelog, V. *Liebigs Ann. Chem.* **1963**, *669*, 121.

128. (a) Mislow, K. *Croat. Chem. Acta* **1985**, *58*, 353; (b) Anet, F.A.L.; Miura, S.S.; Siegel, J.; Mislow, K. *J. Am. Chem. Soc.* **1983**, *105*, 1419.

129. Kilbourn, B.T.; Dunitz, J.D.; Pioda, L.A.R.; Simon, W. *J. Mol. Biol.* **1967**, *30*, 559.

130. (a) Gerlach, H.; Oertle, K.; Thalmann, A.; Servi, S. *Helv. Chim. Acta* **1975**, *58*, 2036; (b) Gerlach, H.; Wetter, H., *Ibid.* **1974**, *57*, 2306.

131. Zak, H.; Schmidt, U. *Angew. Chem.* **1975**, *87*, 454.

132. (a) Schmidt, U.; Gombos, J.; Haslinger, E.; Zak, H. *Chem. Ber.* **1976**, *109*, 2628; (b) Gombos, J.; Haslinger, E.; Zak, H.; Schmidt, U. *Tetrahedron Lett.* **1975**, 3391; see also : (c) Schmidt, U.; Werner, J. *J. Chem. Soc., Chem. Commun.* **1986**, 996; *Synthesis* **1986**, 986.

133. Bartlett, P.A.; Meadows, J.D.; Ottow, E. *J. Am. Chem. Soc.* **1984**, *106*, 5304.

134. Sun, K.M.; Fraser-Reid, B. *Can. J. Chem.* **1980**, *58*, 2732.

135. Trost, B.M.; Timko, J.M.; Stanton, J.L. *J. Chem. Soc., Chem. Commun.* **1978**, 436.

136. Ireland, R.E.; Vevert, J.-P. *Can. J. Chem.* **1981**, *59*, 572.

137. Page, P.C.B.; Carefull, J.F.; Powell, L.H.; Sutherland, I.O. *J. Chem. Soc., Chem. Commun.* **1985**, 822.

138. Bartmangherlich, S.; Davidson, A.H. *J. Chem. Soc., Chem. Commun.* **1985**, 1399.

139. Arco, M.J.; Trammell, M.H.; White, J.D. *J. Org. Chem.* **1976**, *41*, 2075.
140. Mitsunobu, O. *Synthesis* **1981**, 1.
141. (a) Beck, G.; Henseleit, E. *Chem. Ber.* **1971**, *104*, 21; (b) Bartlett, P.A.; Jernstedt, K.K. *Tetrahedron Lett.* **1980**, *21*, 1607; (c) Barrett, A.G.M.; Sheth, H.G. *J. Org. Chem.* **1983**, *48*, 5017; (d) *J. Chem. Soc., Chem. Commun.* **1982**, 170; (e) Baldwin, S.W.; McIver, J.M. *J. Org. Chem.* **1987**, *52*, 322; (f) Lygo, B.; O'Connor, N. *Tetrahedron Lett.* **1987**, *28*, 3597; see also : (g) Still, W.C.; MacPherson, L.J.; Harada, T.; Callagan, J.F.; Rheingold, A.L. *Tetrahedron* **1984**, *40*, 2275.

RECEIVED August 30, 1988

Chapter 14

Applications of Allylboronates in the Synthesis of Carbohydrates and Polyhydroxylated Natural Products

William R. Roush

Department of Chemistry, Indiana University, Bloomington, IN 47405

Contributions from our laboratory concerning the use of allylboronates in the synthesis of carbohydrates and other polyhydroxylated compounds are reviewed. In work directed towards the synthesis of D-fucose derivative **3**, the reaction of γ-methoxyallylboronate **14** and α,β-dialkoxyaldehyde **13** proceeded with excellent diastereoselectivity (>20:1) leading to **15**. This compound served as a key intermediate in a recently completed total synthesis of olivin. A similar reaction (**18** + **19** → **20**) figured prominently in the highly diastereoselective synthesis of compound **22** corresponding to the B ring of sesbanimide. Next, the stereochemistry of the reactions of substituted allylboronates and α-chiral aldehydes is discussed. Since many of these reactions are not sufficiently diastereoselective to be synthetically useful, the tartrate allylboronates **36-38** were developed and found to be exceptionally useful in controlling diastereofacial selectivity via the principle of a double asymmetric synthesis. The application of these reagents towards the synthesis of the C(19)- C(29) segment of the rifamycin S ansa chain is discussed. Tartrate allylboronate **36** also has been applied in the synthesis of the AB disaccharide unit of olivomycin A and in a completely general synthetic approach to 2-deoxyhexoses via the reactions with chiral, nonracemic 2,3-epoxyaldehydes. Finally, the origin of asymmetry of the tartrate allylboronates is discussed and illustrated by the design of a new auxiliary, N,N'-dibenzyl-N,N'-ethylenetartramide (**88**), that is substantially more enantioselective than the parent tartrate esters.

0097–6156/89/0386–0242$10.00/0
© 1989 American Chemical Society

The stereoselective synthesis of carbohydrates from acyclic precursors is a research topic that has attracted considerable attention over the past decade.[1] Efforts in this area are easily justified and have maximum impact particularly when directed toward rare sugars or other polyhydroxylated molecules that are not conveniently accessed via classical "chiron" approaches.[2] An underlying theme of such efforts, of course, is the development of practical synthetic methodology that will find broad application in the enantio- and diastereoselective synthesis of natural products, their analogues, and other compounds of biological interest.

We have been particularly interested for several years in the use of allylboron compounds as reagents for acyclic diastereoselective synthesis,[3] and are pleased to have this opportunity to summarize our efforts in the carbohydrate arena. It is appropriate that a focal point of this discussion will be olivomycin A (Figure 1), a member of the aureolic acid family of antitumor antibiotics,[4] since it was during our initial studies on the synthesis of the aglycone, olivin, that our interest in allylboron chemistry began to emerge. As will be shown subsequently, the allylboration reaction now also plays a central role in our work on the synthesis of the di- and trisaccharide units of this complex antibiotic. As an outgrowth of these studies, we have developed an exceedingly brief, highly selective and completely general synthesis of 2-deoxyhexoses and are currently exploring extensions of this chemistry to the parent hexoses themselves. Thus, we believe that the allylboration reaction will have as great an impact on the chemistry of polyglycolates as it has had on the polypropionates.[3h,i]

Synthesis of a Functionalized D-Fucose Derivative: Initiation of a Program in Allylboronate Chemistry

Several years ago we embarked on a total synthesis of olivin following the strategy summarized in Figure 2.[3a,5] The first problem we faced was how to synthesize D-fucose derivative 3 in an efficient manner. It is ironic that while the overall focus of this presentation is on the use of allylboron compounds in the diastereoselective synthesis of monosaccharides, 3 was first synthesized in our laboratory via a classical "chiron" approach using D-galactose as the starting material. At the time this chemistry was initiated, it was not at all obvious to us that any diastereoselective synthesis could be more efficient than one originating from this readily available hexose. In fact, we initially regarded D-galactose to be the ideal starting material since each of the six carbon atoms and the four chiral centers mapped directly into 3; only the hydroxyl group at C(6) would need to be removed. That is, this seemed to be a situation where a 'chiron' approach was clearly called for.

The synthesis of 3 started from galactopyranoside 5, which was prepared from commercially available methyl β-D-galactopyransonide (4) by using slight modifications of a literature procedure (Figure 3).[6] The free hydroxyl group was then methylated and the C(6) bromomethyl group reduced with LiAlH$_4$ to give 6. After hydrolysis of the acetonide unit, the axial C(3)-hydroxyl group of 7 was selectively benzylated via the intermediacy of a 3,4-dibutylstannylene derivative.[7] At this stage, we had hoped to perform Wittig reactions on the free sugars prepared from either 7 or 8 (e.g., 10) as a means of generating unsaturated esters (e.g., 11) or enones desired for subsequent C-C bond forming reactions. Unfortunately, attempts to condense 10a with Ph$_3$P=CHCO$_2$Me under a

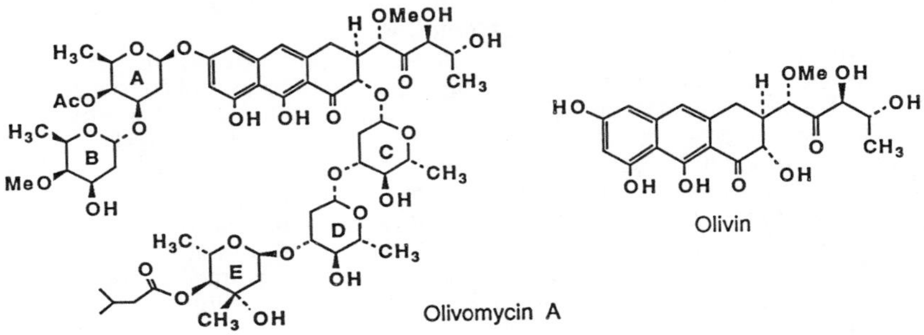

Figure 1. Structures of olivomycin A and olivin.

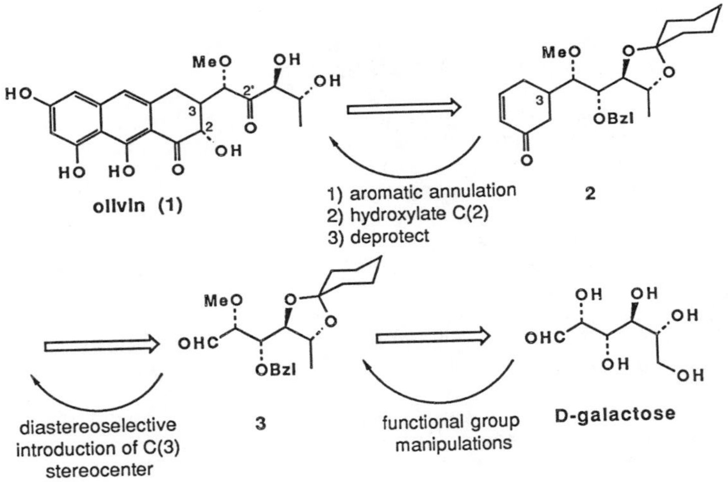

Figure 2. Original strategy for synthesis of olivin.

variety of conditions led to a mixture of pyran and furan derivatives, **12p** and **12f**, resulting from competitive internal Michael reactions,[8] while **10b** failed to react to any significant extent even when benzoic acid was added to catalyze the reaction (Figure 4).[8a]

These problems were circumvented by protecting the C(4),C(5) diol prior to Wittig olefination step (Figure 3). Thus, treatment of **10b** (a mixture of pyranose and furanose anomers prepared by hydrolysis of **8** with aqueous trifluoroacetic acid) with excess EtSH and concentrated HCl (as solvent) at 0°C[9] provided dithioacetal **9** in 50% yield, along with 25% of a mixture of thiopyranosides and thiofuranosides that was recycled to **10b** in high yield by treatment with $HgCl_2$ and $CaCO_3$ in aqueous CH_3CN. Finally, the diol unit was protected as a cyclohexylidene ketal, and then the thioacetal was hydrolyzed under oxidative conditions to arrive at the key aldehyde intermediate **3**.

Although we had achieved a synthesis of the targeted D-fucose derivative, we were not satisfied with what had been accomplished. First, this synthesis required 11 steps from D-galactose and was not nearly so efficient as we would have liked (14% yield overall from commercially available β-methyl galactopyranoside). Second, no new chemistry had been developed. And, third, the brutally harsh conditions[9] required for the conversion of **8** to **9** prevented introduction of more desirable protecting groups for C(3)-OH. Intermediates containing a TBDMS ether at this position ultimately were used in completing the olivin synthesis.[5]

We thus turned to alternative strategies for synthesizing aldehyde **3**. Particularly attractive was the proposal that sugar-like materials could be constructed via the reaction of an allyl ether anion and an α-alkoxyaldehyde (Figure 5).[10,11] For this approach to be successful, it would be necessary to control (i) the regioselectivity of the reaction of the allyl ether anion,[10] (ii) the syn (threo) or anti (erythro) relationship generated in concert with the new C-C bond, and (iii) this new C(2)-C(3) relationship with respect to the chiral center (C(4)) already present in the aldehyde reaction partner.

Solutions to problems (i) and (ii) were already available as a result of studies by Hoffmann and Wuts on the reactions of γ-alkoxyallyl-boronates with achiral aldehydes (Figure 6).[11-13] Relatively little information was available, however, regarding the stereochemistry of such reactions with chiral aldehydes. Hoffman had published several examples of reactions of (E)- and (Z)-crotylboronates (methyl replacing OMe in Figure 6) with chiral aldehydes such as 2-methylbutanal, but the best diastereofacial selectivity that had been reported was only 83:17.[14] Thus, it was by no means certain that the chemistry summarized in Figure 7 would be successful.[3a]

Aldehyde **13**, readily prepared by a four step synthesis from L-threonine,[3a,15] was treated with the known (Z)-γ-methoxyallylboronate **14**[12a,c]. This reaction, as with other reactions of pinacol allylboronates, was relatively slow and required 24-48 h at room temperature to reach completion. It was, however, extremely selective and provided homoallyl alcohol **15** in 70% yield with greater than 95% diastereoselectivity. The stereochemistry of this compound was quickly verified by conversion to **3** as shown in Figure 7.[3a] We now believe that this reaction proceeds by way of the Conforth-like transition state depicted in Figure 7, and not by way of a Felkin transition state as suggested in our original publication, since a serious nonbonded interaction exists between the (Z)-methoxyl group and the C(3) substituents of **13** in the Felkin transition state. A

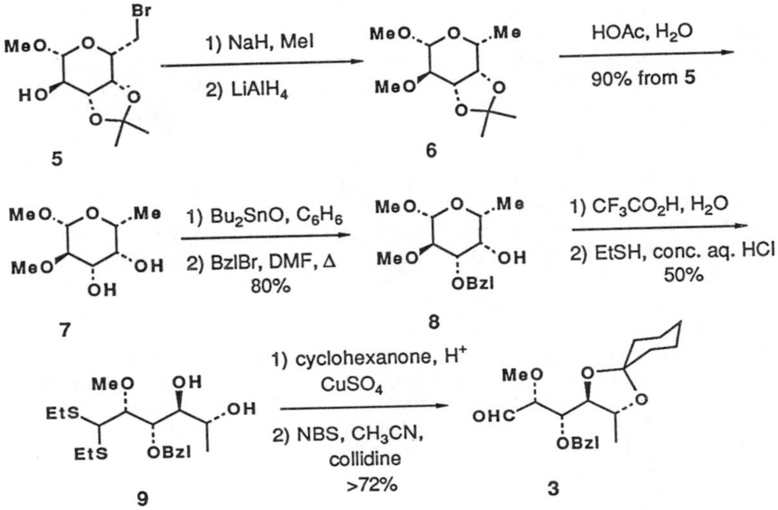

Figure 3. Synthesis of D-fucose derivative **3**.

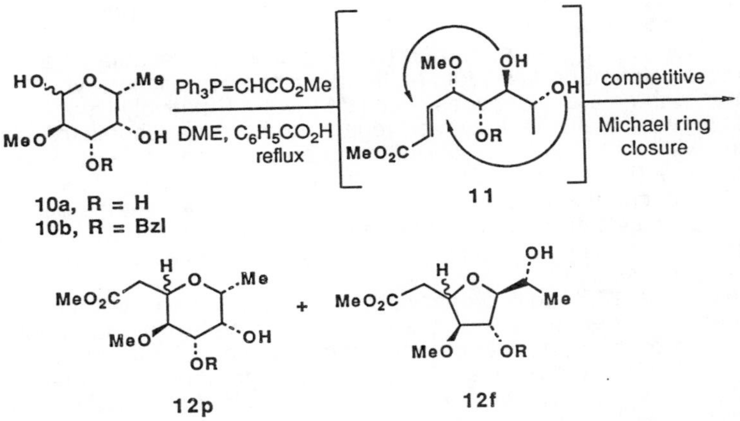

Figure 4. Condensation of **10a** with
Ph₃P=CHCO₂Me.

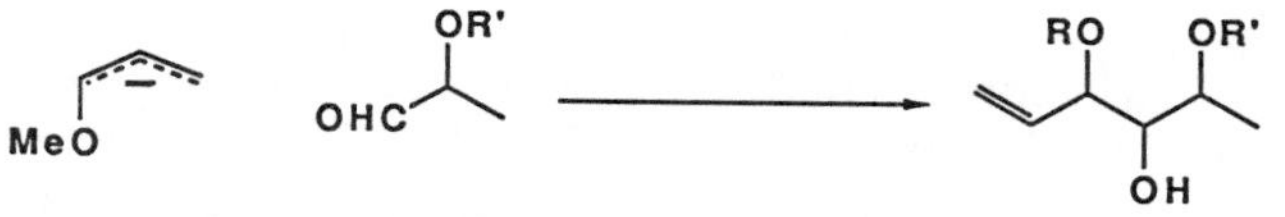

Figure 5. A new strategy for carbohydrate synthesis.

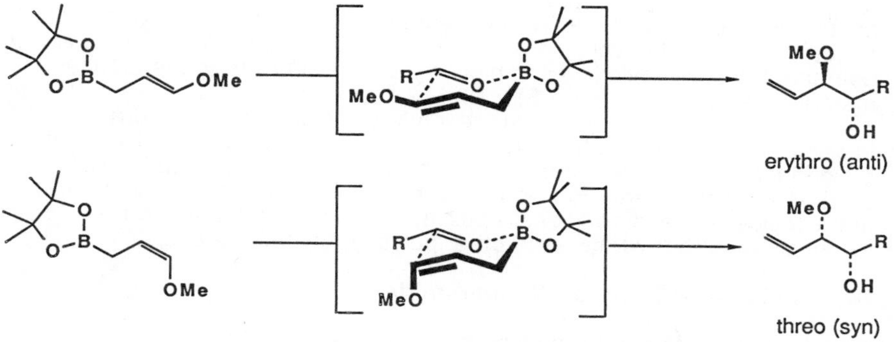

Figure 6. Reactions of γ-alkoxyallylboronates with achiral aldehydes.

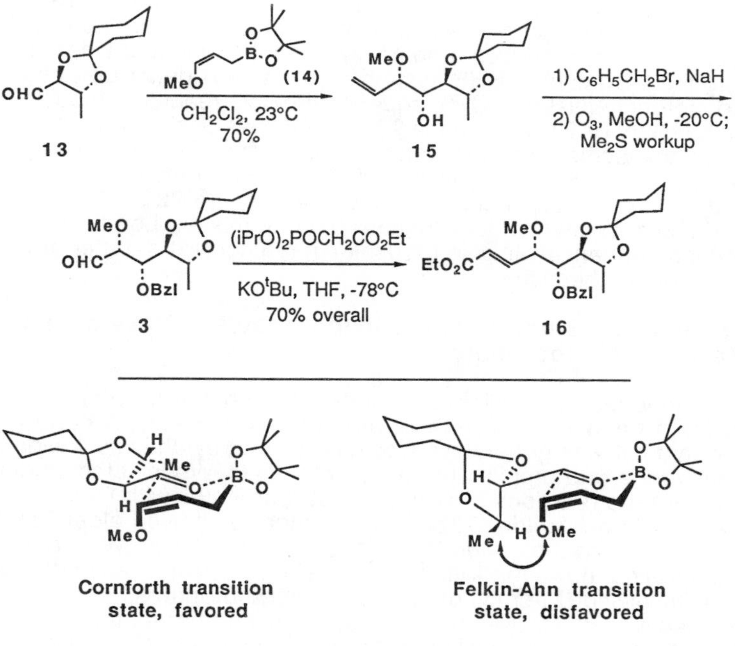

Figure 7. Preparation of homoallyl alcohol **15** via a Conforth-like transition state.

detailed analysis of our thoughts concerning 1,2-diastereoselection in the reactions of allylboronates and chiral aldehydes appears in reference 3c, and the reader is referred to this source for additional discussion of this point. This diastereoselective synthesis of **3** is relatively brief (seven steps from L-threonine) and considerably more efficient (25% overall) in comparison to the D-galactose based synthesis described at the outset. One problem with this new sequence, however, was the synthesis of reagent **14** which proved to be low yielding, tedious, and not readily amenable to scale up.[12a,c] We subsequently found that in situ generated dimethyl (Z)-γ-methoxyallylboronate (**17**)[12b] is extremely convenient to use and actually provides **15** in higher yield (75-83%) than the original method involving **14** (Figure 8).

The synthesis of olivin was recently completed in our laboratories at Indiana University using homoallyl alcohol **15** as a key intermediate.[5] It is beyond the scope of this presentation, however, for us to discuss this synthesis in detail here. For now, therefore, we leave the topic of olivin and consider instead additional applications of allylboronates in the synthesis of carbohydrates and other polyoxygenated materials.

Synthesis of B-Ring of Sesbanimide

A second application of a reaction of a γ-alkoxyallylboronate and an α,β-dialkoxyaldehyde was developed in connection with our work on the total synthesis of sesbanimide (Figure 9).[3g] In this case, the reaction of in situ generated allylboronate **18** and glyceraldehyde cyclohexyl ketal (**19**) provided **20** as the only observed stereoisomer. This reaction establishes the erythro relationship between C(8) and C(9) of the natural product target and, further, produces a homoallylic alcohol unit that when oxidized by using the VO(acac)$_2$-TBHP system[16] yields epoxide **21** possessing the desired stereochemistry at C(7), again as the sole reaction product. The stereochemistry of **21**, and hence **22** as well, was assigned by conversion to glucitol hexaacetate as indicated at the bottom of Figure 9. Thus, the combination of these two highly diastereoselective transformations enabled us to gain very rapid access to intermediate **22** containing the B ring of the sesbanimides. It is conceivable that this allylboration-epoxidation sequence will also be useful in the context of other problems in carbohydrate chemistry.

Stereochemical Studies with Allylboronates.   Development of the Tartrate Allylboronates

We were intrigued by the high level of selectivity of the reactions of aldehydes **13** and **19** with γ-alkoxyallylboronates **14** and **18** and realized that if it was general we would be able to synthesize a wide range of carbohydrate and propionate derived materials. We initiated studies, therefore, on the reactions of allyl and crotylboronates with D-glyceraldehyde acetonide (**23**) and the threonine derived aldehyde **13** as a means of probing the generality of these earlier results. We were surprised to find, however, that high diastereoselectivity was unique to reactions involving (Z)-crotyl or (Z)-γ-alkoxyallylboronates. Stereoselectivity diminished or disappeared altogether as the C(3) substituent was removed (allyl reagent **25**) or inverted ((E)-crotylboronate **26**; see Figure 10).[3e]

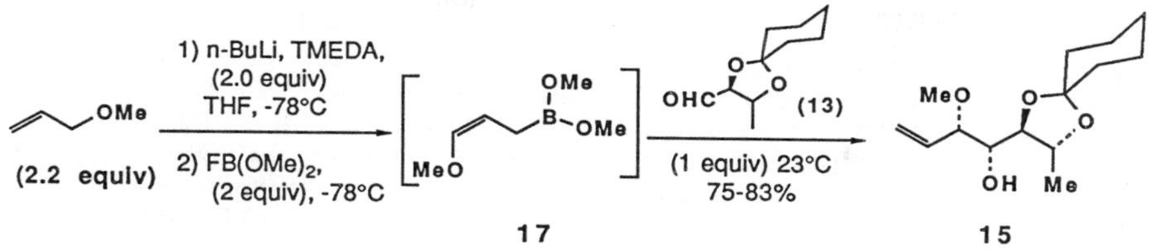

Figure 8. Preparation of homoallyl alcohol **15** via in situ generated allylboronate **17**.

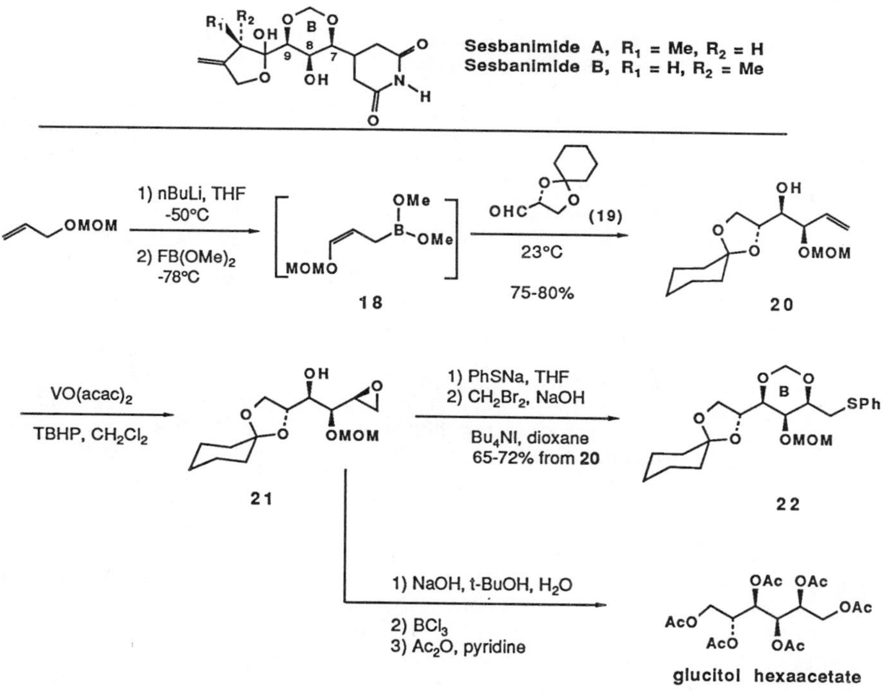

Sesbanimide A, R₁ = Me, R₂ = H
Sesbanimide B, R₁ = H, R₂ = Me

Figure 9. Preparation of intermediate **22** containing the B ring of the sesbanimides.

That aldehyde diastereofacial selectivity is dependent on the substitution pattern and geometry of the allylboron reagent appears to be general. Table I summarizes additional results published by Hoffmann and Wuts that support this thesis.[17] These data show further that diastereofacial selectivity also depends on the electronic makeup of the aldehyde reaction partner.[3e,17a] Notice that the percent of anti diastereoface selectivity decreases as one moves to the right or down any column in the Table. Both we and Hoffman have come to the same conclusion that diastereofacial selectivity is governed, in part, by minimization of nonbonded interactions between olefinic substituents on the allylboronates and substituents α to the aldehydic carbonyl.[3e,17] In the reaction of α-oxygenated aldehydes like 23 and 34, the transition states that appear to be lowest in energy correspond to the Cornforth model (for one example, refer to Figure 7). The increased anti selectivity with 34 and 23 presumably reflects an electronic activation of the favored Cornforth transition state.[3e]

In spite of the poor diastereoselectivity realized in reactions with most chiral aldehydes, allylboronates are highly attractive reagents for organic synthesis.[3,11,12,17] Most are easily prepared in large quantities, and are convenient to use.[18] They are nonbasic, relatively non-nucleophilic, and hence are highly chemoselective in their reactions. From all perspectives they are well behaved chemical entities.

The poor diastereoselectivity of the reactions of chiral aldehydes and achiral allylboronates appeared to be a problem that could be solved by recourse to the strategy of double asymmetric synthesis.[19] Our studies thus moved into this new arena of asymmetric synthesis, our objective being the development of a chiral allylboron reagent capable of controlling the stereochemical outcome of reactions with chiral aldehydes independent of any diastereofacial preference on the part of the carbonyl reaction partner.

Here, too, our work was preceded by that of Hoffmann, who had examined a number of terpene derived chiral diols[20] and had shown that chiral allylboronates incorporating endo-3-phenyl-exo-2,3-bornandiol were moderately successful in increasing the diastereofacial selectivity of several aldehyde addition reactions (matched cases).[14,17b] This auxiliary, however, was not sufficiently enantioselective to be effective in mismatched double asymmetric reactions - cases in which the stereochemical preferences dictated by the auxiliary and chiral aldehyde are dissonant.[21] Since C_2 symmetric diols had not been explored, we decided to focus our efforts on reagents incorporating this strategically significant symmetry element.[22]

The use of tartrate esters was an obvious place to start, especially since both enantiomers are readily available commercially and had already found widespread application in asymmetric synthesis (Figure 11) (e.g., Sharpless asymmetric epoxidation).[23,24] Reagents 36-38 are easily prepared and are reasonably enantioselective in reactions with achiral, unhindered aliphatic aldehydes (82-86% ee); typical results are given in Figure 12.[3c,h] Aromatic and α,β-unsaturated aldehydes, unfortunately, give lower levels of enantioselection (55-70% e.e.). It is also interesting to note that all other C_2 symmetric diols that we have examined (2,3-butanediol, 2,4-pentanediol, 1,2-diisopropylethanediol, hydrobenzoin, and mannitol diacetonide, among others) are relatively ineffective in comparison to the tartrate esters (see Table II).[25]

These chiral reagents are especially useful in the context of double

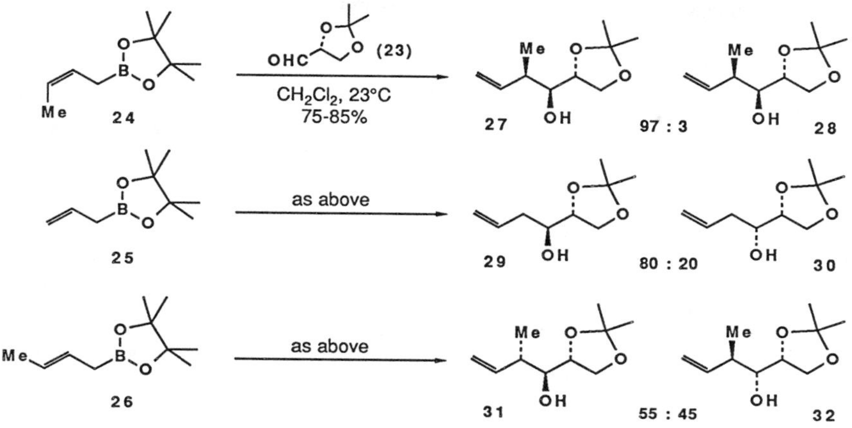

Figure 10. Reactions of allyl- and crotylboronates with
D-glyceraldehyde acetonide (23).

Table I. Representative Diastereofacial Selectivities (anti : syn)
in Reactions of Allyl Boronates and Chiral Aldehydes[a]

	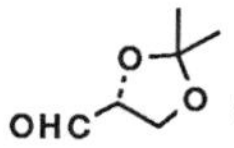 (23)	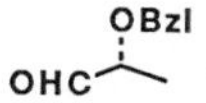 OBzl (34)	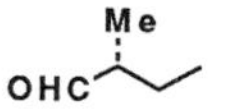 Me (35)
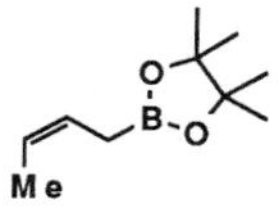 (24)	97 : 3	—	70 : 30
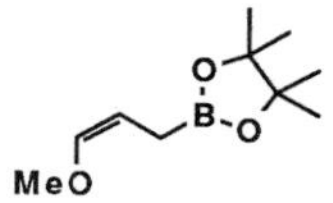 (14)	>95 : 5	82 : 18	—
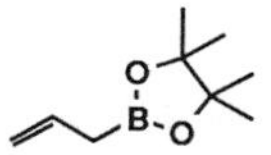 (25)	80 : 20	65 : 35	38 : 62
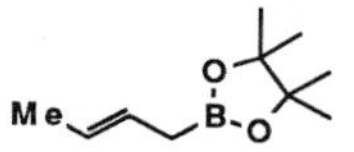 (26)	55 : 45	—	17 : 83
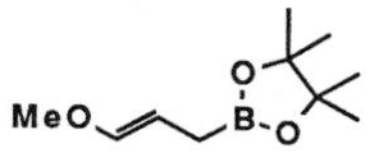 (33)	—	40 : 60	—

[a]The data represent the ratio of 4,5-anti to 4,5-syn carbonyl addition products;
see, for example, Figure 10.

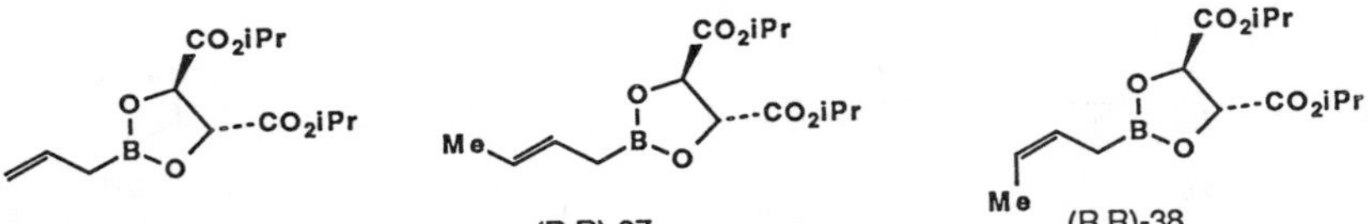

(R,R)-36 (R,R)-37 (R,R)-38

Figure 11. Tartrate allylboronates.

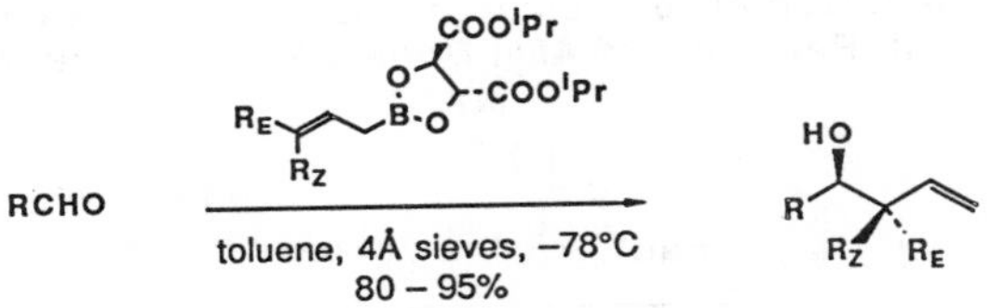

RCHO	Allyl	(E)–Crotyl*	(Z)–Crotyl*
n–C_9H_{19}CHO	86% e.e.	87% e.e.	82% e.e.
C_6H_{11}CHO	87%	88%	86%
t–C_4H_9CHO	86%	73%	70%
C_6H_5CHO (THF)	72%	67%	55%

*diastereoselectivity ≥97%

Figure 12. Products of the reactions of tartrate allylboronates with achiral, unhindered aliphatic aldehydes.

asymmetric synthesis.[3c,d,h,i] For example, whereas the reaction of D-glyceraldehyde acetonide (23) and pinacol allylboronate (25) provides the erythro diastereomer (29) as the major component of an 80:20 mixture (Figure 10), the reaction of 23 and (R,R)-36 provides this same product with up to 98:2 selectivity (matched case).[3c,26] When (S,S)-36 is used, however, the diastereoselectivity is reversed (mismatched combination) and the threo diastereomer 30 is the major component of a 92:8 mixture (Figure 13).[3d,26]

Thus, as this example clearly shows, reagent 36 (and 37 as well)[3c,h] is sufficiently enantioselective to control the stereochemical outcome of reactions with aldehydes that possess only modest intrinsic diastereofacial preferences. The consequences of this increased selectivity for organic synthesis are obvious. In the present case, compounds 29 and 30, which are synthetic precursors to 2-deoxyribose and 2-deoxylyxose,[17b] are each now easily prepared with excellent selectivity from readily available precursors.

A more striking example of the potential of these reagents in organic synthesis is provided by our recent synthesis of the C(19) - C(29) segment of the rifamycin S ansa chain (Figure 14).[3i] This synthesis pivots around four key C-C bond forming steps. The first (39 + 37) is a mismatched double asymmetric reaction that provides diastereomer 40 as the major component of an 88:11:1 mixture. The second (41 + 37), third (43 + 37) and fourth (45 + 36) proved to be matched double asymmetric reactions and provided 42, 44, and 46, respectively, with 98%, 95% and 91% diastereoselectivity. It is interesting to note that the minor diastereomers produced in steps 2 and 3 are the hydroxyl epimers of 42 and 44, and probably derive from reactions of the (Z)-crotyl reagent 38 that is a minor contaminant (3-5%) in the batches of (E)-crotyl reagent 37 used in this synthesis. Finally, with the exception of the first reaction leading to 40, the stereoselectivity is unoptimized: no effort has been made to "fine tune" any of the more advanced synthetic intermediates in order to enhance the diastereoselectivity of these C-C bond forming steps.

The synthesis of aldehyde 48 proceeds in 16 steps from (S)-39 in 15% yield and 75% stereoselectivity. The brevity, efficiency, and selectivity of this synthesis rivals alternative acyclic diastereoselective approaches to the rifamycin ansa chain, (see footnote 4 in reference 3i), thereby providing a clear testimony to the potential of the tartrate allylboronates as reagents for complex synthetic problems.

Additional applications of this methodology in the synthesis of carbohydrates will be discussed in subsequent sections in this chapter.

Synthesis of the AB Disaccharide Unit of Olivomycin A

Our strategy for synthesis of the oligosaccharide chains,[27,28] calls for 2,6-dideoxyhexoses or the corresponding glycals to serve as precursors for both α- and β-glycosidation reactions. If a selective β-glycosidation protocol can be developed, then in principle any structural isomer or analogue of the natural product can be assembled from a common set of monosaccharide precursors.[29]

Syntheses of each of the sugar residues in olivomycin A from commercially available carbohydrate precursors were known at the time our studies were initiated.[30,31] We elected not to synthesize these compounds via literature procedures, however, since we felt that totally synthetic methods might provide a more convenient and general solution,

Table II. C_2 Diols Used in the Allylboration Reaction[a]

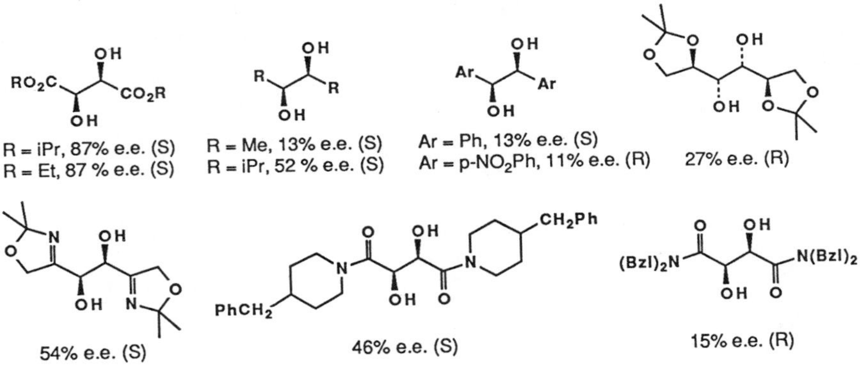

R = iPr, 87% e.e. (S)
R = Et, 87 % e.e. (S)

R = Me, 13% e.e. (S)
R = iPr, 52 % e.e. (S)

Ar = Ph, 13% e.e. (S)
Ar = p-NO$_2$Ph, 11% e.e. (R)

27% e.e. (R)

54% e.e. (S)

46% e.e. (S)

15% e.e. (R)

[a]Results obtained in reactions of the chiral allylboronates with cyclohexanecarboxaldehyde.

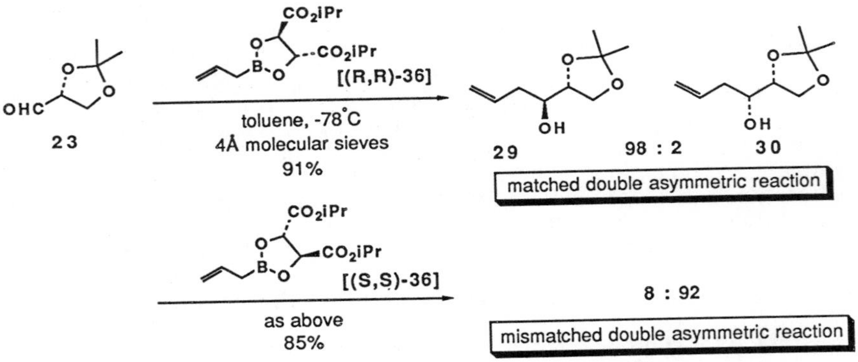

Figure 13. Reaction of D-glyceraldehyde acetonide with tartrate
allylboronate 36.

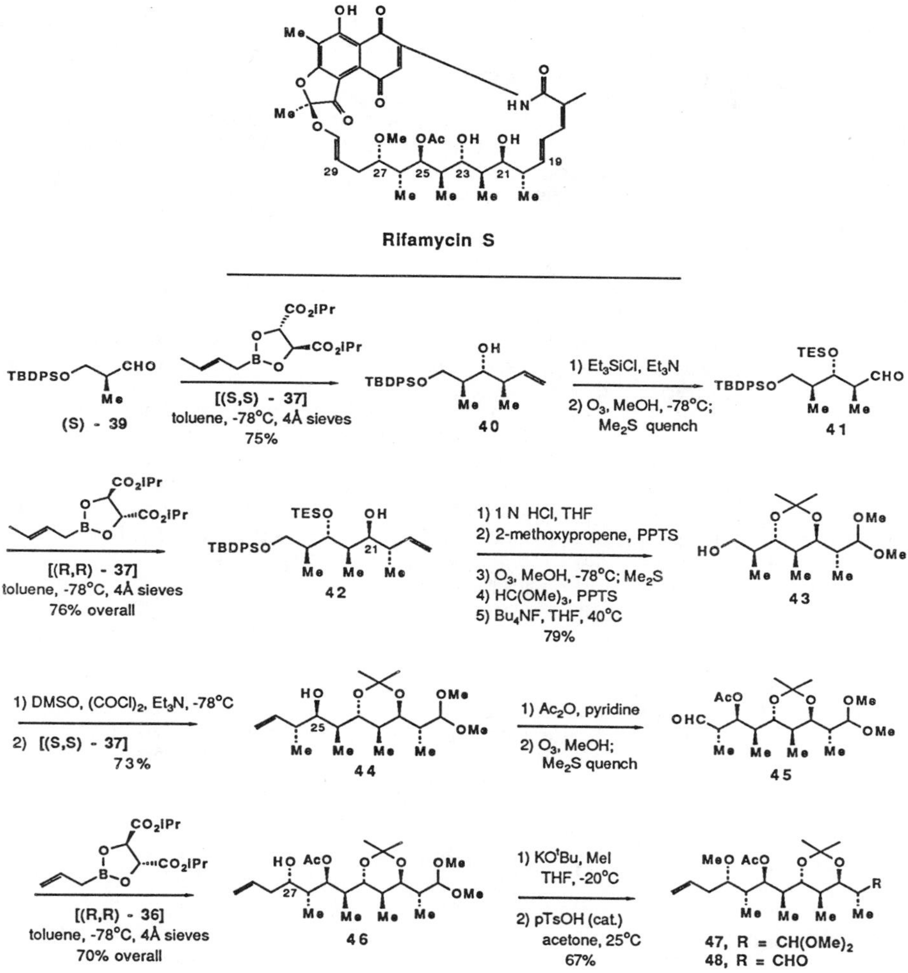

Figure 14. Synthesis of the C(19)–C(29) segment of the rifamycin S ansa chain.

particularly in the context of this program where the synthesis of structurally modified oligosaccharides was a long range goal.[29] We were also aware that several of these monosaccharides occur naturally (as glycosides of antibiotics) in both enantiomeric series,[30] and realized that a route involving asymmetric synthesis would provide the necessary stereochemical generality to achieve equal access to either antipodal series. Perhaps the most important influence on this decision, however, was the realization that acyclic stereoselective methods were becoming increasingly important in organic synthesis. We believed that methodology for synthesizing polyhydroxylated sugar-like acyclic systems could in fact compete with chiron-based strategies in many instances; our work on the synthesis of D-fucose derivative **3** for the olivin synthesis is but one example. Thus, we decided to embark on a program of monosaccharide synthesis from acyclic precursors as a means of developing methodology that would be useful to the organic chemist in a wide range of contexts.[1,32]

Epoxyalcohols prepared by the Sharpless kinetic resolution/ enantioselective epoxidation technology[33] served as the key synthetic intermediates in our initial studies on the synthesis of the olivomycin monosaccharides. Our most important contribution was the development of methodology for controlling the regioselectivity of nucleophilic substitution reactions of the epoxyalcohol intermediates.[34] Figure 15 summarizes two complimentary and highly regioselective procedures for substitution reactions with oxygen nucleophiles. The reaction of 2,3-epoxyalcohols with aqueous acid proceeds with very high selectivity for attack of water at the β-position, the epoxide carbon furthest away from the carbinol center. This mode of reactivity is illustrated in the digitoxose synthesis. In order for attack to occur at C_α, as required for the synthesis of olivose (**51**), it is necessary for the nucleophile to be delivered intramolecularly. We found phenylurethanes to be the best source of "tethered" oxygen nucleophiles, and that these neighboring group assisted reactions are best performed in the presence of Lewis acid catalysts such as Et_2AlCl.[34]

Although this approach has proved to be reasonably direct and efficient in the cases studied thus far, it suffers from several significant drawbacks: (i) because a resolution is involved, the maximum yield of useable chiral, non-racemic intermediates is 50%, and the separation of epoxyalcohol from the unreacted, kinetically resolved allylic alcohol is tedious, especially for large scale work; (ii) the generality of this method is restricted since the efficiency of the kinetic resolution (that is, the relative rate of epoxidation of the two allylic alcohol enantiomers) and the diastereoselectivity of the epoxidation step are poor for secondary (Z)-allylic alcohols,[33] an important class of substrates; (iii) the α-opening methodology is unattractive in cases where the intended role of the carbohydrate fragment is as an intermediate in subsequent reaction sequences. That is, the intrinsic differentiation of the C(4) and C(5) oxygen functionality in the epoxyalcohol substrate is lost in the course of the α-opening process (see **50** to **51**, Figure 15). This is undesirable since sugars with undifferentiated hydroxyl groups at C(3) and C(4) are produced; in fact, introduction of a suitable set of protecting groups into D-olivose (**51**) as the first step in studies on the synthesis of the olivomycin CDE trisaccharide has proven non-trivial.[35]

The important conclusion from an operational point of view is that if sugars are to be synthesized de novo, it is imperative that the method be direct, efficient, completely general, and provide access to intermediates in which all of the hydroxyl functionality is completely differentiated for use in subsequent synthetic schemes.

Our development of the tartrate ester modified allylboronates[3c,h] suggested to us that many of these problems could be avoided by using the reaction of a chiral aldehyde and a chiral allylboronate as a means of establishing the stereochemistry of the sugar backbone. This strategy has been used in our synthesis of the AB disaccharide unit of olivomycin A (Figures 16, 17).[3f]

Syntheses of monosaccharides **57** and **59** thus began with the reaction of aldehyde **53**[3a,15] and (S,S)-**36** that provided **54** in 93% yield and with 300:1 stereoselectivity. Diastereomer **54** was the major component of a 90:10 mixture when pinacol allylboronate (**25**) was used. The reaction of **53** and (S,S)-**36**, therefore, is a matched double asymmetric reaction. Benzylation of **54** provided **55** which was hydrolyzed by treatment with 4:1 HOAc - H$_2$O (98%). Ozonolysis of the resulting diol then provided 3-O-benzyl-2,6-dideoxy-D-*lyxo*-hexose (72%) as a mixture of pyranose and furanose anomers that was directly converted to the corresponding mixture of methyl glycosides by treatment with acidic methanol. This mixture was most conveniently separated following acylation. Thus, the desired α-pyranoside **56** was obtained as the major product in 36% yield along with an unseparated mixture of the β-pyranoside and the α,β-furanosides. The latter mixture was recycled three times ((i) MeOH, AcCl; (ii) Ac$_2$O, pyridine, DMAP; (iii) chromatographic separation) bringing the total yield of **56** to 71%.

Intermediate **56** served as precursor to both of the monosaccharide units in disaccharide **61**. The A ring sugar **57** was prepared in 84% yield by hydrogenation of **56** in EtOH over 10% Pd/C. Alternatively, treatment of **56** with powdered KOH in DMSO followed by excess CH$_3$I and catalytic 18-crown-6 gave **58** in 81% yield. This intermediate was then converted into thiosugar **59** as a mixture of anomers in 92% yield by using the method described by Hanessian.[36] Coupling of these two units (Figure 17) was smoothly accomplished by treatment of a mixture of **57** and **59** (1.1 equiv.) with NBS (1.2 equiv.) and 4Å molecular sieves in CH$_2$Cl$_2$.[37] Although a mixture of anomers was anticipated at the outset,[38] we were pleased to find that this method provided **60** in 61% yield as a >6:1 mixture in which the α,α-anomer predominated. The stereoselectivity was also independent of the anomeric composition of **59**. It is interesting to speculate that the excellent selectivity may be the consequence of neighboring group assistance as suggested at the bottom of Figure 17, since analogous glycosidations of 2-deoxyglucose derivatives, which have equatorial C(4)-alkoxy groups and are unable to form similar bridged structures, are substantially less selective. Finally, hydrogenation of **60** gave disaccharide **61**, the spectroscopic properties of which were in excellent agreement with literature values.[28b]

A General Synthetic Approach to Monosaccharides

The synthesis of disaccharide **61** is reasonably efficient (10 steps from **53**, 17% overall yield) and is readily amenable to scale up. Nevertheless, in contemplating extensions of this chemistry to the synthesis of differentially protected derivatives of D-olivose (e.g., **63**, Figure 18) needed for construction of the olivomycin CDE trisaccharide, it became apparent that this approach is unattractive because the required starting

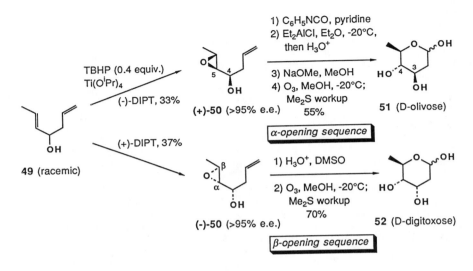

Figure 15. Reactions of epoxyalcohol intermediates in synthesis of olivomycin monosaccharides.

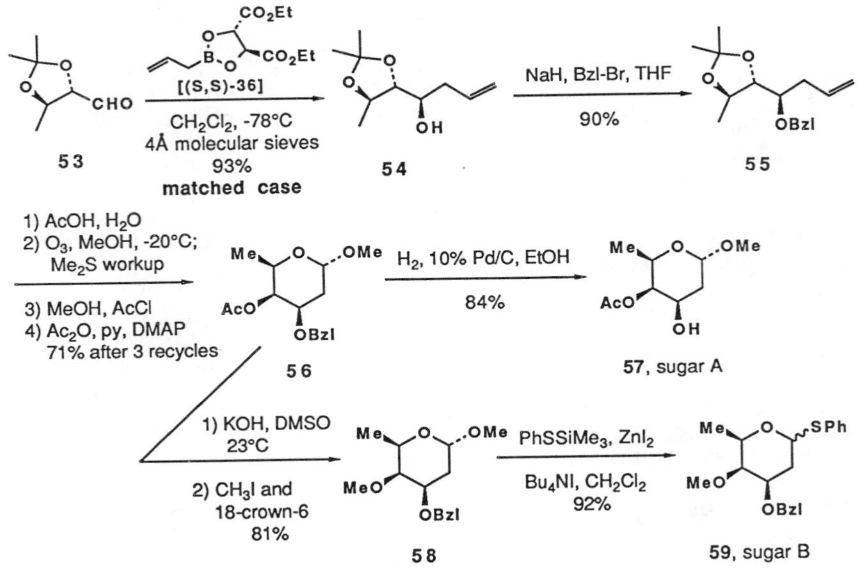

Figure 16. Synthesis of AB disaccharide unit of olivomycin A.

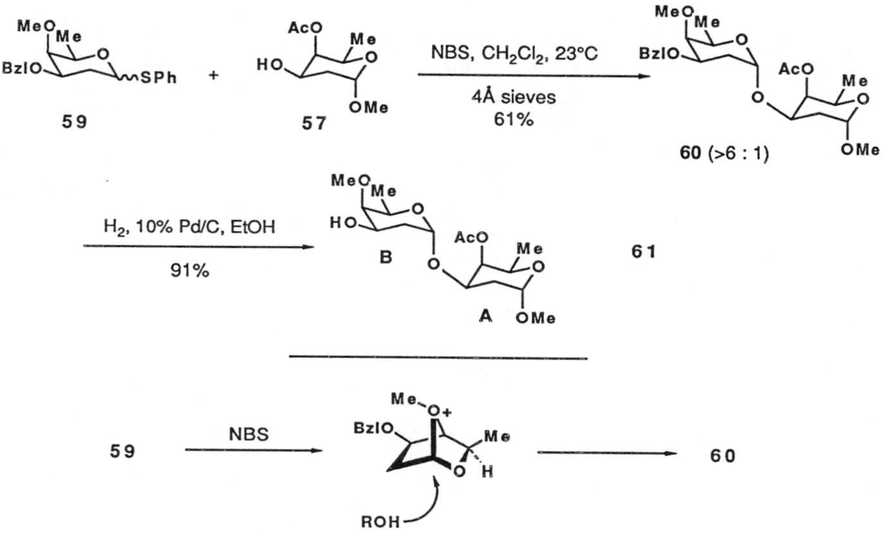

Figure 17. Coupling of A ring sugar **57** and thiosugar **59**.

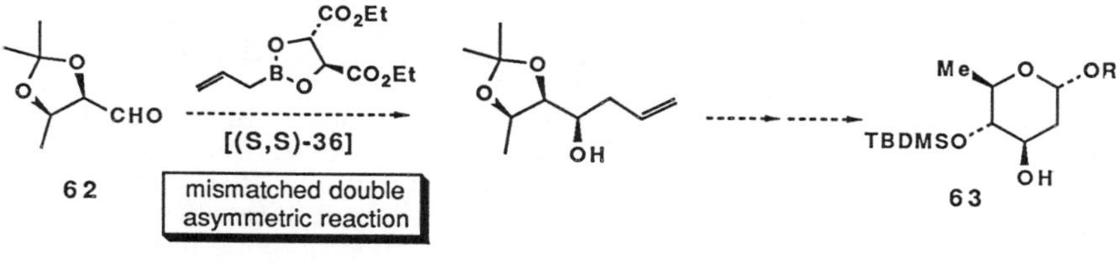

Figure 18. Synthesis of differentially protected D-olivose derivative
63.

material, 62, formally deriving from D-allo-threonine, is not readily accessible.[39]

This prompted us to begin exploratory studies of a more general synthetic approach to monosaccharides that would not rely on the accessibility of specific chiral pool precursors. This approach, outlined in Figure 19, relies on two asymmetric transformations: (i) the Sharpless asymmetric epoxidation[23] that can be used to prepare the epoxyallylic alcohol precursors to the indicated epoxyaldehydes; and (ii) the asymmetric allylboration reaction that presumably can be used to achieve diastereoface selection in the addition of allyl or γ-alkoxyallyl units to the epoxyaldehydes. Control of stereochemistry at C(3) relative to C(4) in 68 should be possible by selecting the appropriate reagent 66 or 67. Given the ability to rationally manipulate the epoxide functionality, all possible hexoses of either absolute configuration should be easily accessible. In addition, as long as β-epoxide opening reactions are employed, the intrinsic differentiation of the C(4) and C(5) oxygen functionality in 68 can be carried through to the target hexose. Finally, it was apparent that this monosaccharide synthesis, if successful, would be shorter and more practical than those based on iterative asymmetric epoxidation cycles.[40]

We began by studying the reactions of epoxyaldehydes 69 and 70 with both enantiomers of tartrate allylboronate 36 (Figure 20).[35,41] [Note that in this case diethyl tartrate was used as the auxiliary rather than diisopropyl tartrate as in all previous examples. These two readily available esters are used interchangeably in our laboratory.] The aldehydes were prepared by oxidation of the corresponding epoxyallylic alcohols with NaOAc buffered PCC (92-95% yield). When 69 was treated with achiral pinacol allylboronate (25), erythro epoxyalcohol 71 was produced as the major component of a 60:40 mixture. The reaction of 69 with (R,R)-36, therefore, constitutes a matched pair since the selectivity for 71 is increased to 96:4. Erythro epoxyalcohol 73 similarly is the major product (96:4) of a matched double asymmetric reaction of cis-epoxyaldehyde 70 and (S,S)-36. Thus, two of the four epoxyalcohol diastereomers are available with very good diastereoselectivity. The second pair of diastereomers, threo-epoxyalcohols 72 and 74, are available with lower selectivity (70-74: 30-26) via the mismatched double asymmetric reactions of 69 and 70 with (S,S)- and (R,R)-36, respectively. While we had hoped that the selectivity in these cases would be higher, these reactions may still be useful synthetically since two diastereomers are easily separated chromatographically.

An interesting aspect of this chemistry is that the enantiomeric purity of the two diastereomeric products are different in each of the reactions reported in Figure 20. This is a consequence of a kinetic resolution involving distinctly different pathways for the reaction of 36 with the two epoxyaldehyde enantiomers, both of which are present since the Sharpless epoxidation provides the epoxyalcohol precursors to 69 and 70 in only 95% and 90% e.e., respectively. For example, while the reaction of 70 with (S,S)-36 is a matched pair and leads preferentially to 73, the reaction of ent-70 with (S,S)-36 is a mismatched combination and leads preferentially to ent-74 (Figure 21). That is, the minor enantiomer of the epoxyaldehyde is converted preferentially to the minor product diastereomer, causing the enantiomeric purity of the major reaction product to be much greater than that of the epoxyaldehyde precursor, and the enantiomeric purity of the minor diastereomer to be significantly less so. This is most strikingly demonstrated by the reaction of 70 and (S,S)-

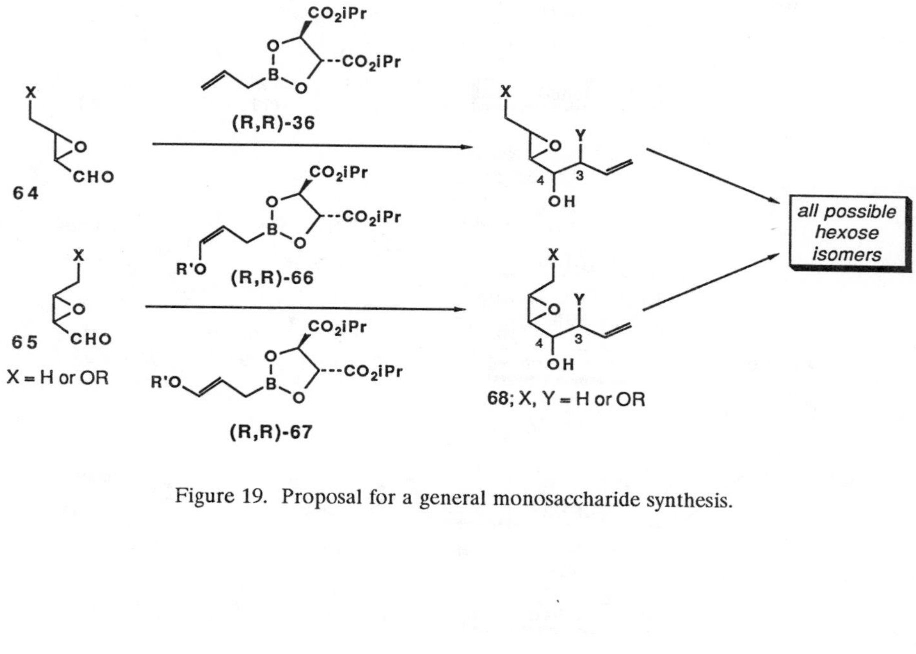

Figure 19. Proposal for a general monosaccharide synthesis.

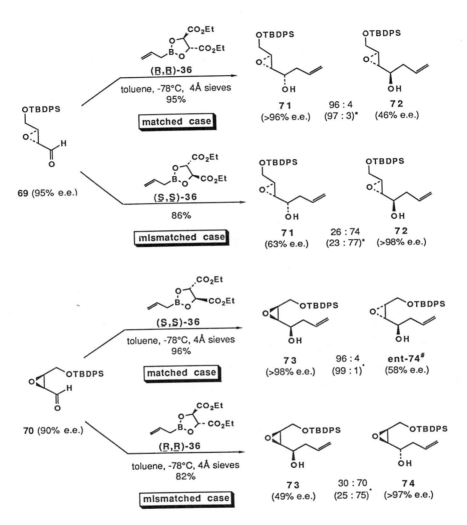

*Selectivity expected if epoxyaldehyde is 100% e.e.

#The major enantiomer of this product derives from the minor epoxyaldehyde enantiomer.

Figure 20. Reactions of epoxyaldehydes **69** and **70** with
enantiomers of tartrate allylboronate **36**.

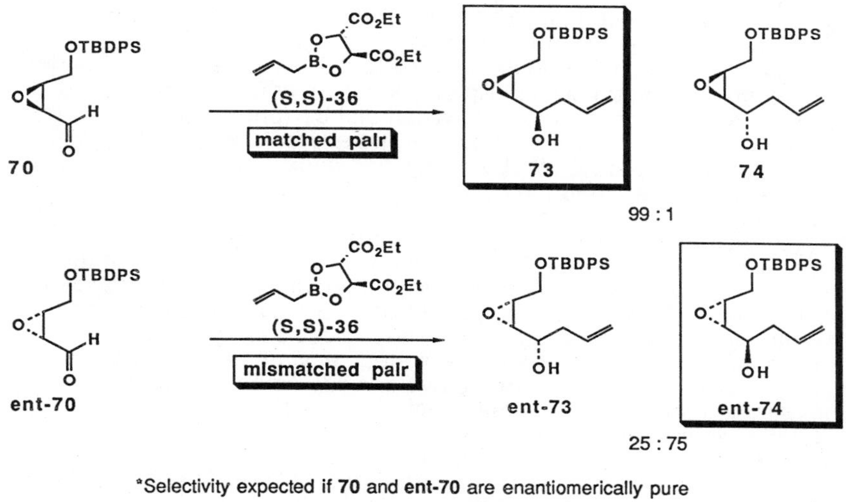

*Selectivity expected if **70** and **ent-70** are enantiomerically pure

Figure 21. Reaction of tartrate allylboronate **36** with epoxyaldehyde enantiomers of **70**.

36 where the *major* enantiomer of the minor reaction product **74** in fact derives from the *minor* enantiomer of **70**. This phenomenon has important ramifications in organic synthesis, since it clearly suggests that products with very high enantiomeric purity can be prepared by linking multiple double asymmetric transformations in a synthetic pathway. Similar observations have been made by Hoye[42] and Schreiber[43] in studies of asymmetric epoxidations of bisallylic alcohols.

The epoxyallylic alcohols prepared in this way are useful precursors to 2-deoxyhexoses or their immediate precursors. The flexibility of this approach is illustrated in Figure 22 by conversion of the two epoxyalcohol products of matched double asymmetric reactions (**71**, **73**) to precursors of all four 2-deoxyhexoses. The α-opening reactions of **71** (to **75**, and thence to 2-deoxy-L-glucose) and **73** (to **77**) involve the α-opening technology developed previously in our laboratory. As far as the β-opening reactions are concerned, two different methods have been employed. In the conversion of **71** to **76**, the silyl ether protecting group was first removed and then diol **79** was treated with NaOH under conditions where epoxide migration can occur (see Figure 23).[44] Monosubstituted epoxide **80** is the most reactive species present under these conditions, and nucleophilic attack occurs at C(7) of **80** to produce **75** with excellent regioselectivity. When diol **81** (prepared from **73**) was subjected to these conditions, however, tetrahydrofuran **84** and not tetraol **78** was the major product.[35] Evidently, two epoxide migration pathways are accessible to **81**, and the cyclization of **83** to **84** is faster than the intermolecular attack of hydroxide on **82**. We have subsequently found that tetrahydrofuran formation also competes to a limited extent (ca. 15%) in the alkaline hydrolysis of **79**. This problem has been solved in the case of **81** by using acidic hydrolysis conditions (Figure 22) which provided the desired tetraol **78** as major component of a 13:1 diastereomeric mixture (no tetrahydrofuran was produced.) The regioselectivity in this case presumably is dictated by the different steric environments at C(6) vs. C(5) since the electronic makeup of the two epoxide carbons should be comparable. It remains to be seen how general this acid hydrolysis will be with other epoxydiol substrates. We note in passing that the alkaline β-opening protocol has also been applied to the diols corresponding to **72** (10:1 regioselectivity; no tetrahydrofuran) and **74** (>20:1 regioselectivity; no tetrahydrofuran) and that efforts to optimize the β-opening sequence are continuing.

These results are strongly supportive of our initial hypothesis that the reactions of allylboronates and epoxyaldehydes may serve as the basis of an efficient approach to monosaccharides. Relatively little work, however, has been performed on the reactions of epoxyaldehydes and γ-alkoxyallylboronates, a transformation required to gain access to sugars in the 2-oxygenated series. One preliminary experiment designed to probe the intrinsic diastereoface selectivity of the epoxyaldehyde reaction partner is summarized in Figure 24. In this case, the reaction of epoxyaldehyde **85** with γ-methoxyallylboronate **14** provided a single major diastereomer (>6:1 selectivity) that has been tentatively assigned structure **86** by analogy to the anti diastereofacial selectivity exhibited by **14** in reactions with other α-oxygenated aldehydes (refer to Table I).

This example suggests that very high levels of selectivity will be realized in matched double asymmetric reactions involving chiral γ-alkoxyallylboronates of general structure **66** and **67** (Figure 19), but also foreshadows potential problems in applications of these reagents in

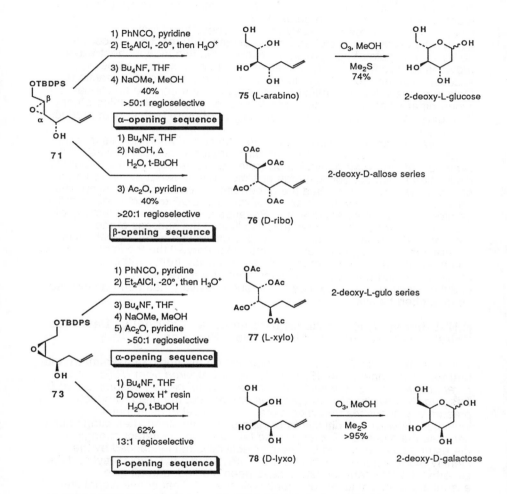

Figure 22. Conversion of epoxyalcohol products of matched double asymmetric reactions.

mismatched double asymmetric reactions with substrates like **69, 70**, and **85**. These reactions are not expected to be very selective since the mismatched reactions of **69** and **70** with allylboronate **36** are only marginally so, and at least with a (Z)-γ-alkoxyallylboronate like **14** the intrinsic diastereofacial preference of **85** will be a greater barrier to overcome. [Recall that the intrinsic diastereofacial preference of **69** in the reaction with pinacol allylboronate **25** was only 2:1]. Of course, in many instances it will not be necessary to have access to a highly diastereoselective mismatched double asymmetric reaction, since the ability to manipulate the epoxide functionality in two independent ways provides sufficient generality that all of the target hexoses can be accessed via the products of the matched double asymmetric reactions. Only in cases where it is necessary to maintain the intrinsic functional group differentiation in compounds like **86**, or when nucleophiles other than oxygen are used, will it be necessary to prepare both sets of product diastereomers.

We began these studies with the intention of applying this tandem asymmetric epoxidation/asymmetric allylboration sequence towards the synthesis of D-olivose derivative **63** (refer to Figure 18). As the foregoing discussion indicates, our research has moved somewhat away from this goal and we have not yet had the opportunity to undertake this synthesis. This, as well as the synthesis of the olivomycin CDE trisaccharide, remain as problems for future exploration. Because it is the enantioselectivity of the tartrate ester allylboronates that has limited the success of the mismatched double asymmetric reactions discussed here, as well as in several other cases published from our laboratory,[3h] the focus of our work on chiral allylboronate chemistry has shifted away from synthetic applications and towards the development of a more highly enantioselective chiral auxiliary. One such auxiliary has been developed, as described below.

N,N'-Dibenzyl-N,N'-ethylenetartramide, a Rationally Designed Chiral Auxiliary for the Allylboration Reaction

We begin with a discussion of our thoughts on the origin of asymmetry with tartrate allylboronates **36-38**.[3c,d] *Reagents prepared from (R,R)-tartrate invariably induce (S) configuration at the carbinol center, assuming that R has priority over the allyl group that is transferred;* the major product presumably arises via transition state A. The level of asymmetric induction, however, is difficult to explain by simple steric interactions alone because the aldehydic R group is too far removed to interact strongly with the ester substituents and because selectivity is not influenced by the identity of the ester group itself (Me, Et, iPr, adamantyl, cyclodecyl and 2,4-dimethyl-3-pentyl tartrate esters have been studied, and all give essentially identical levels of enantioselectivity). That conventional steric effects are probably not the dominant stereochemically determining factor is supported by our observation that the tartrate esters are substantially more enantioselective than any other C_2 symmetric diols examined to date (e.g., see Table II), many of which presumably do have a steric origin of enantioselection.[25]

These considerations prompted us to suggest early on that transition state **A** is favored as a consequence of n/n electronic repulsive interactions between the aldehydic oxygen atom and the β-face ester group that destabilizes **C** relative to **A**.[3c] These interactions (Figure 25) are

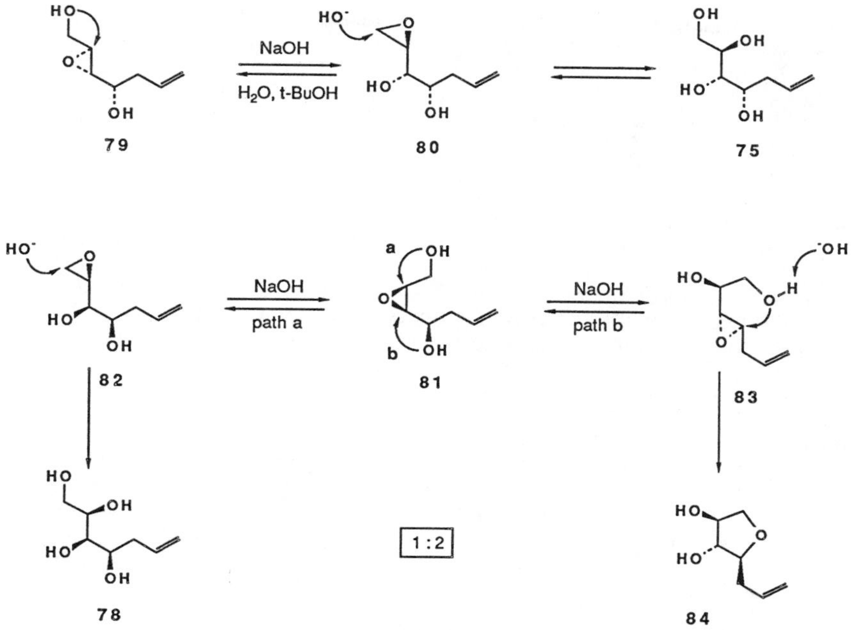

Figure 23. Treatment of diols with NaOH under conditions where epoxide migration can occur.

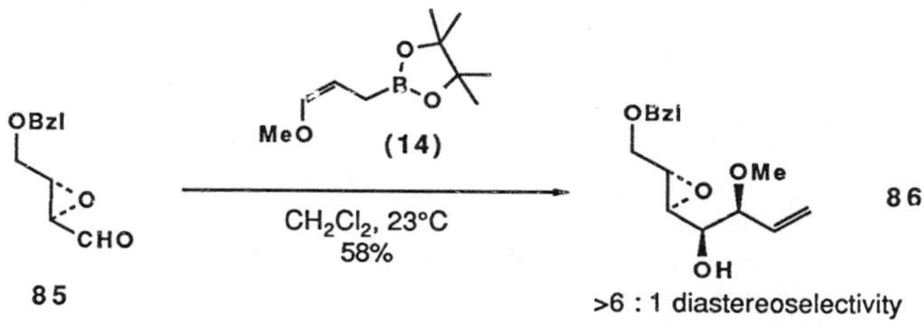

Figure 24. Reaction of epoxyaldehyde **85** with γ-methoxyallylboronate **14** to probe intrinsic diastereoface selectivity.

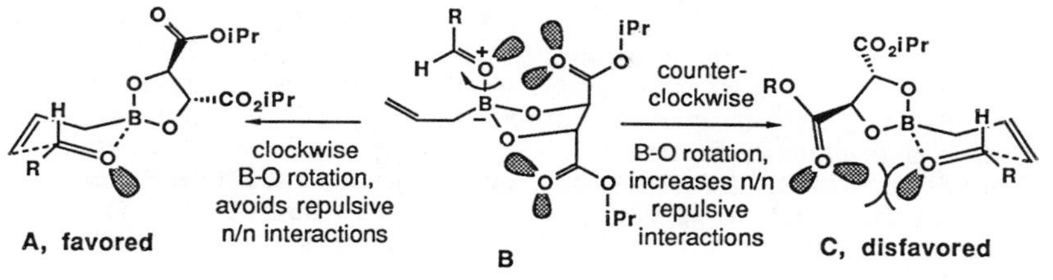

Figure 25. Origin of asymmetry.

possible since an easily accessible and frequently favored conformation of α-heteroatom substituted carbonyl systems is one in which the heteroatom and carbonyl are syn-coplanar.[45] Toluene appears to be particularly effective among nonpolar solvents in stabilizing this conformation[46] and, interestingly, also happens to be the solvent in which **36-38** generally display the best enantioselectivity.

For this mechanism to be correct, it is also necessary for the dioxaborolane to exist in conformation **B** with the two -CO_2iPr units pseudoaxial. In any other conformation of the dioxaborolane, or if other C-CO_2iPr bond rotational isomers are considered, the ester and aldehydic oxygen atoms are too far removed to interact. It should be noted further that reasonable transition states for C-C bond formation are not accessible if the aldehyde is symmetrically disposed with respect to the dioxaborolane system. Clockwise rotation about the B-O bond as indicated in **B** moves the aldehyde nonbonding lone pair away from the proximate ester carbonyl and leads to the favored transition state **A**. Rotation of the B-O bond in the reverse direction increases the n/n interactions and leads to disfavored transition state **C**.

These arguments imply that the aldehyde to boron complexation step (a Lewis acid/Lewis base reaction) is the critical enantioselectivity determining event, since conformation B most probably represents the ground state Lewis acid aldehyde complex. This conformation may be stabilized by a boron centered anomeric effect (n_O-σ^* interactions between the axial lone pairs of the ring oxygens and the B-O=CHR single bond).[47] The actual transition state for the allyl transfer probably occurs during a flipping motion of the dioxaborolane O-B-O unit that moves the allyl group towards a pseudoaxial position with development of two anti n_O-σ^* B-C interactions that facilitate cleavage of the B-C bond.

One further point is worthy of brief mention. While we have focused on lone pair/lone pair repulsive interactions that destabilize transition state **C**, it is conceivable that **A** is actually stabilized relative to **C** by a favorable charge-charge interaction between the ester carbonyl (δ^-) and the aldehydic carbonyl carbon (δ^+) owing to the proximity of these groups in **A**. While it is not yet possible to resolve the relative contributions of these distinct stereoelectronic effects, it is clear that our mechanistic proposal explains the experimental results only if the dioxaborolane and the C-CO_2iPr bonds exist in the conformations indicated in B. Any conformational infidelity at either site would be expected to lead to diminished enantioselectivity.

As a test of this hypothesis we decided to explore conformationally restricted auxiliaries such as **87** (Figure 26).[48] We recognized that as long as the tartrate unit is held within an eight membered ring, the critical conformational features discussed for **B** become structural constants in **D**. If our mechanism is correct, we expected reagent **88** to be substantially more enantioselective than the parent tartrate allylboronate **36**.[49]

Bislactam (R,R)-**87** was readily synthesized from benzylidene tartrate (**89**) and N,N'-dibenzylethylenediamine by a three step sequence in 40-42% overall yield (Figure 27). Interestingly, the Mukaiyama salt mediated[50] amidation-lactamization step proceeds in a preparatively useful yield (52-56%), while very poor results have been previously reported for the synthesis of eight membered lactams from ω-aminoacid precursors.[51]

Results obtained in the reactions of (R,R)-**88** with several representative achiral aldehydes are summarized in Table III. Also

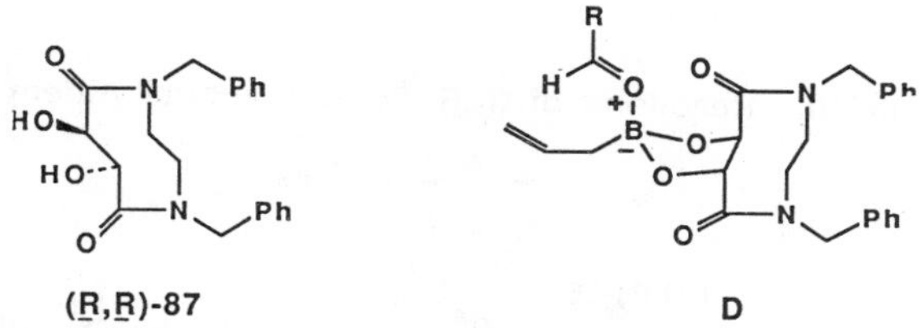

Figure 26. Conformationally restricted auxiliaries.

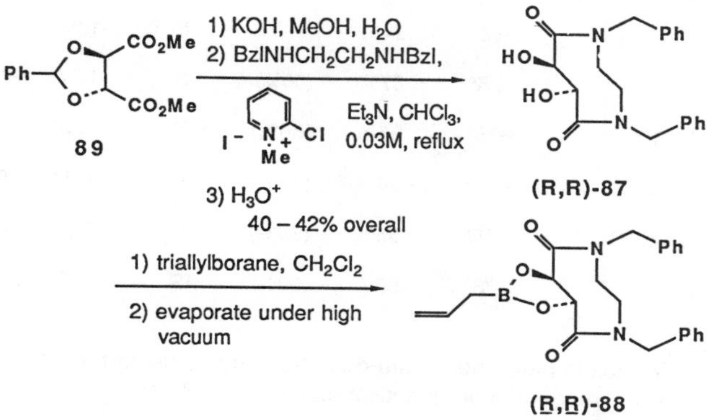

Figure 27. Synthesis of bislactam (R,R)-**87**.

Table III.   Reactions of (R,R)-88 and Achiral Aldehydes

RCHO	Temp	Selectivity[*]		Config	$\Delta\Delta G^{\ddagger}$ (kcal mol^{-1})[*]	
$C_6H_{11}CHO$	−78°C	97% e.e.	(87%)	S	-1.61	(-1.03)
"	−50°	94%	—	S	-1.53	—
"	+25°	87%	(50%)	S	-1.57	(-0.65)
$t\text{-}C_4H_9CHO$	−78°	96%	(86%)	S	-1.50	(-1.00)
$TBDPSOCH_2CH_2CH_2CHO$	−78°	94%	(84%)	R	-1.34	(-0.94)
$BzlOCH_2CHO$	−78°	85%	(60%)	S	-0.97	(-0.53)
C_6H_5CHO	−78°	85%	(60%)	S	-0.97	(-0.53)

[*]Values in parentheses are data obtained by using the parent
DIPT reagent under identical reaction conditions.

included are comparative reference data obtained in reactions with the parent tartrate allylboronate **36**. In every instance **88** greatly outperformed its predecessor. The reactions of **88** with cyclohexane-carboxaldehyde, pivaldehyde and 4-t-butyldimethylsilyloxybutanal proceed with 94-97% e.e., versus 84-87% e.e. with DIPT reagent **36**, a very significant improvement. Even benzyloxyacetaldehyde and benzaldehyde, which were very poor substrates for **36** (60% e.e.), now each give homoallyl alcohols with acceptable levels of enantioselection (85% e.e.). In energetic terms (see column with $\Delta\Delta G^{\neq}$ data), the new reagent is at least 50% more enantioselective on a case by case basis. Also significant are the observations that **88** is as selective in reactions at 25°C as is **36** at -78°C (entries 1,3), the $\Delta\Delta G^{\neq}$ of reactions of **88**, but not **36**, are independent of temperature within experimental error (entries 1-3), and the sense of asymmetric induction with **88** and **36** are the same.

The increased enantioselectivity of **88** is also apparent in reactions with chiral aldehydes (Figure 28). β–Alkoxypropionaldehydes **90** were relatively poor substrates when **36** was used.[3h] The best selectivity ever obtained for syn diastereomer **91** in the matched double asymmetric reactions was 89:11 [(S,S)-**36** and **90a**], whereas the best selectivity for anti diastereomer **92** was 87:13 [reaction of **90b** and (R,R)-**36**]. In contrast, the allylborations of **90a,b** with the new reagent **88** now proceed with up to 97:3 selectivity for either product diastereomer. Even more impressive results were obtained with glyceraldehyde acetonide (**23**): the matched double asymmetric reaction leading to **29** now proceeds with 300:1 diastereoselectivity, while the mismatched combination leading to **30** proceeds with 50:1 selectivity.

These data strongly support our original thesis regarding the origin of asymmetry with reagents **36-38**, and establish **88** as the most highly enantioselective allylmetal reagent yet devised.[21,52] It should be noted also that the increased enantioselectivity with **88** is not simply the consequence of the ester to lactam functional group modification, since a series of acyclic tartramides have been examined (e.g., bis-N,N-dibenzyl tartramide; see Table II), and their allylboron derivatives are *significantly* less enantioselective than even **36**. Consequently, these studies emphasize the important geometric relationships that must be present in the favored allylboration transition state and, further, suggest that the convergence of functional groups towards a metal center can be an exceedingly useful strategy for achieving a topological bias in the enantioselective functionalization of a carbonyl group.[53]

Although **88** is substantially more enantioselective than **36**, it is not, however, a superior reagent for organic synthesis. Compound **88** suffers from poor solubility in toluene especially at low temperatures, causing the reactions summarized in Table III and Figure 28 to be sluggish and require long reaction times for reasonable conversions.[48] We regard **88** to be a prototype of an improved auxiliary, and are actively striving to develop a reagent that combines the reactivity of **36** with the enantioselectivity and ease of preparation of **88**. If we are successful, then we will have an auxiliary that will greatly increase the utility of allylboronates in organic synthesis, especially in the context of mismatched double asymmetric reactions.

Future applications of allylboronates in the synthesis of carbohydrates along the lines suggested in Figure 19 will await the development of such an improved second generation reagent.[54]

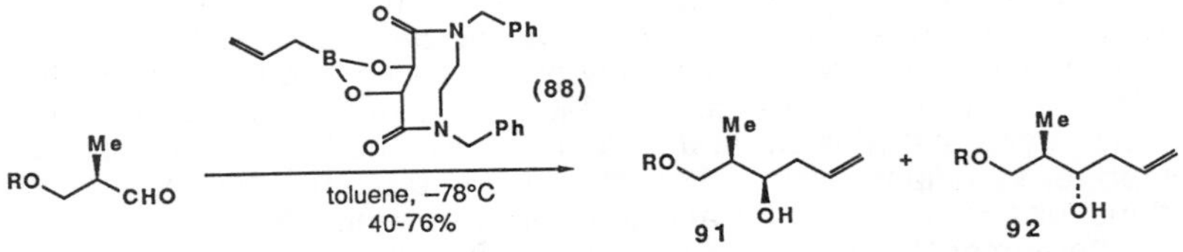

R	reagent		Selectivity*
90a	TBDMS	(S,S)	matched case	97 : 3	(89 : 11)
	"	(S,S), -50°	"	95 : 5	-
	"	(R,R)	mismatched	3 : 97	(19 : 81)
90b	TBDPS	(S,S)	matched case	95 : 5	(79 : 21)
	"	(R,R)	mismatched	3 : 97	(13 : 87)
	"	(R,R), -50°	"	5 : 95	-

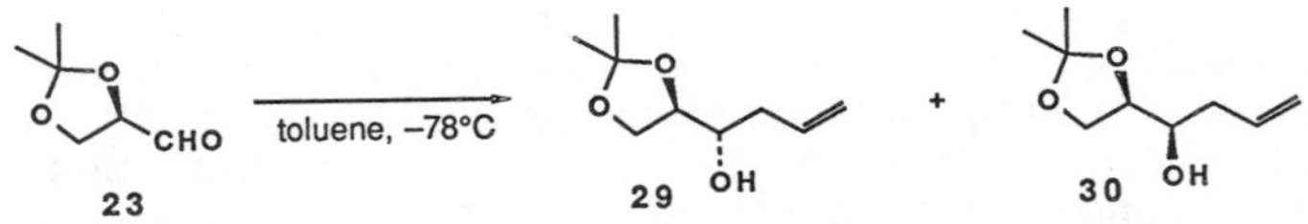

(R,R)	matched case	99.7 : 0.3	(98 : 2)*
(S,S)	mismatched	2 : 98	(8 : 92)

*Values in parentheses are data obtained by using the
parent DIPT reagent under identical reaction conditions

Figure 28. Reactions of **88** with chiral aldehydes.

Acknowledgments. I am indeed grateful to acknowledge the intellectual and experimental contributions of the many students and postdoctoral associates who participated in the research described in this presentation. They are, in alphabetical order: Michael A. Adam (now Dr.), Dr. Kaori Ando, Dr. Luca Banfi, Richard J. Brown (now Dr.), Dr. Ronald L. Halterman, Dr. David J. Harris, Lee K. Hoong, Dr. Brigitte Lesur, Michael R. Michaelides (now Dr.), Alan D. Palkowitz, Michelle A. J. Palmer, Julie A. Straub (now Dr.), and Alan E. Walts (now Dr.). Without the enormous contributions of these individuals there would be no story to tell. Finally, I would like to thank the National Institutes of Health (CA 29847, GM 26782, and AI 20779) for their continued and generous support of my research program.

Literature Cited

1. For recent reviews of syntheses of carbohydrates from noncarbohydrate precursors, see: (a) Danishefsky, S. J.; DeNinno, M. P. *Angew Chem., Int. Ed. Engl.* **1987**, 26, 15. (b) McGarvey, G. J.; Kimura, M.; Oh, T.; Williams, J. M. *J. Carbohydr. Chem.* **1984**, 3, 125. (c) Zamojski, A.; Banaszek, A.; Grynkiewicz, G. *Adv. Carbohydr. Chem. Biochem.* **1982**, 40, 1.

2. Reviews of the use of chiral pool precursors: (a) Hanessian, S.; "Total Synthesis of Natural Products: The 'Chiron' Approach;" Pergamon Press, Oxford, 1983. (b) Scott, J. W. in "Asymmetric Synthesis;" Morrison, J. O.; Scott, J. W., 5a.; Academic Press: New York, 1984; Vol. 4, p. 1.

3. (a) Roush, W. R.; Harris, D. J., Lesur, B. M. *Tetrahedron Lett.* **1983**, 24, 2227. (b) Roush, W. R.; Peseckis, S. M.; Walts, A. E. *J. Org. Chem.* **1984**, 49, 3429. (c) Roush, W. R.; Walts, A. E.; Hoong, L. K. *J. Am. Chem. Soc.* **1985**, 107, 8186. (d) Roush, W. R.; Halterman, R. L. *Ibid.* **1986**, 108, 294. (e) Roush, W. R.; Adam, M. A.; Walts, A. E.; Harris, D. J. *Ibid.* **1986,** 108, 3422. (f) Roush, W. R.; Straub, J. A. *Tetrahedron Lett.* **1986**, 27, 3349. (g) Roush, W. R.; Michaelides, M. R. *Ibid.* **1986**, 27, 3353. (h) Roush, W. R.; Palkowitz, A. D.; Palmer, M. A. J. *J. Org. Chem.* **1987**, 52, 316. (i) Roush, W. R.; Palkowitz, A. D. *J. Am. Chem. Soc.* **1987**, 109, 953. (j) Roush, W. R.; Coe, J. W. *Tetrahedron Lett.* **1987**, 28, 931.

4. For reviews, see: (a) Remers, W. A. "The Chemistry of Antitumor Antibiotics;" Wiley-Interscience: New York, 1979, Chapter 3. (b) Skarbeck, J. D.; Speedie, M. K. "Antitumor Compd. Nat. Origin: Chemistry and Biochemistry," Aszalos, A., Ed.; CRC Press, 1981, Chapter 5.

5. Roush, W. R.; Michaelides, M. R.; Tai, D. F.; Chong, W. K. M. *J. Am. Chem. Soc.*, **1987**, 109, 7575.

6. Bernet, B.; Vasella, A. *Helv. Chim. Acta.* **1979**, 62, 2411.

7. (a) Auge, C.; David, S.; Veyrieres, A. *J. Chem. Soc., Chem. Commun.* **1976**, 375. (b) Nashed, N. A. *Carbohydr. Res.* **1978**, 60, 200.

8. (a) Buchanan, J. G.; Edgar, A. R.; Power, M. J.; Theaker, P. D. *Carbohydr. Res.* **1974**, 38, C22. (b) Zhdanov, Y. A.; Alexeev, Y. E.; Alexeeva, V. G. *Adv. Carbohydr. Chem.* **1972**, 27, 227.

9. Zinner, H.; Ernst, B.; Kreienbring, F. *Chem. Ber.* **1962**, 821. Attempts to use milder conditions were less successful. Some cleavage of the benzyl ether was also observed.

10. 1-Alkoxyallyllithiums react with electrophiles preferentially at the γ position, so use of a metal additive (e.g., (RO)$_2$BX) to reverse the regioselectivity of the reaction would be necessary: (a) Evans, D. A.; Andrews, G. C.; Buckwalter, B. *J. Am. Chem. Soc.* **1974**, 96, 5560. (b) Still, W. C.; MacDonald, T. L. *Ibid.* **1974**, 96, 5561.

11. For a review of the diastereoselective addition of allylmetal compounds to aldehydes, see: Hoffmann, R. W. *Angew. Chem., Int. Ed. Engl.* **1982**, 21, 555.

12. (a) Hoffman, R. W.; Kemper, B. *Tetrahedron Lett.* **1982**, 23, 845; **1981**, 22, 5263. (b) Wuts, P. G. M.; Bigelow, S. S. *J. Org. Chem.* **1982**, 47, 2498. (c) Hoffmann, R. W.; Kemper, B.; Metternich, R.; Lehmeier, T. *Liebigs Ann. Chem.* **1985**, 2246.

13. For other γ-alkoxyallylmetal reagents, see: (a) Keck, G. E.; Abbott, D. E.; Wiley,   M. R. *Tetrahedron Lett.* **1987**, 28, 139. (b) Koreeda, M.; Tanaka, Y.; *Ibid.* **1987**, 38, 143. (c) Tamao, K.; Nakajo, E.; Ito, Y. *J. Org. Chem.* **1987**, 52, 957. (d) Yamamoto, Y.; Saito, Y.; Maruyama, K. *J. Organomet. Chem.* **1985**, 292, 311. (e) Yamaguchi, M.; Mukaiyama, T. *Chem. Lett.* **1982**, 237; **1981**, 1005; **1979**, 1279. (f) Koreeda, M.; Tanaka, Y. *J. Chem. Soc., Chem. Commun.* **1982**, 845.

14. Hoffmann, R W.; Zeiss, H. J.; Ladner, W.; Tabche, S. *Chem. Ber.* **1982**, 115, 2357.

15. Servi, S. *J. Org. Chem.* **1985**, 50, 5865.

16. (a) Mihelich, E. D.; Daniels, K.; Eickhoff, D. J. *J. Am. Chem. Soc.* **1981**, 103, 7690. (b) Sharpless, K. B.; Verhoeven, T. R. *Aldrichim. Acta* **1979**, 12, 63.

17. (a) Hoffman, R. W.; Weidmann, U. *Chem. Ber.* **1985**, 118, 3966. (b) Hoffmann, R. W.; Endesfelder, A.; Zeiss, H. J. *Carbohydr. Res.* **1983**, 123, 320. (c) Wuts,   P. G. M.; Bigelow, S. S. *J. Org. Chem.* **1983**, 48, 3489.

18. Dimethyl allylboronates (e.g., **17, 18**) are sensitive to hydrolysis, and the allylboronic acids are unstable in the presence of O$_2$. The pinacol and tartrate esters, however, are quite stable and many have been purified by distillation. We routinely monitor the isomeric purity of the crotyl reagents by capillary GC analysis (ref 3e).

19. For a review, see: Masamune, S.; Choy, W.; Petersen, J. S.; Sita, L. R. *Angew. Chem., Int. Ed. Engl.* **1985**, 24, 1.

20. (a) Herold, T.; Schrott, U.; Hoffman, R. W.; Schnelle, G.; Ladner, W.; Steinbach, K. *Chem. Ber.* **1981**, 114, 359. (b) Hoffmann, R. W.; Herold, T. *Ibid.* **1981**, 114, 375.

21. For leading references to other classes of chiral allylmetal compounds, see: (a) Hoffman, R. W.; Landmann, B. *Chem. Ber.* **1986**, 119, 2013. (b) Hoffmann, R. W.; Dresely, S. *Angew. Chem., Int. Ed. Engl.* **1986**, 25, 189. (c) Ditrich, K.; Bube, T.; Sturmer, R.; Hoffmann, R. W. *Ibid.* **1986**, 25, 1028. (d) Jadhav, P. K.; Bhat, K. S.; Perumal, P. T.; Brown, H. C. *J. Org. Chem.* **1986**, 51, 432. (e) Brown, H. C.; Bhat, K. S. *J. Am. Chem. Soc.* **1986**, 108, 5919. (f) Brown, H. C.; Bhat, K. S.; Randad, R. S. *J. Org. Chem.* **1987**, 52, 3701, 319. (g) Midland, M. M.; Preston, S. B. *J. Am. Chem. Soc.* **1982**, 104, 2330. (h) Jephcote, V. J.; Pratt, A. J.; Thomas, E. J. *J. Chem. Soc., Chem Commun.* **1984**, 800. (i) Mukaiyama, T.; Minowa, N.; Oriyama, T.; Narasaka, K. *Chem. Lett.* **1986**, 97. (j) Hayashi, T.; Konishi, M.; Ito, H.; Kumada, M. *J. Am. Chem. Soc.*

1982, 104, 4962. (k) Roder, H.; Helmchen, G.; Peters, E. M.; Peters, K.; von Schnering, H. G. *Angew. Chem., Int. Ed. Engl.* **1984**, 23, 898.

22. An important consequence of the use of a C_2 symmetric auxiliary is that the number of competing transition states is reduced from four (as in the case of Hoffmann's bornandiol reagents) to two, thereby increasing the probability that a single, selective pathway will be found.

23. For recent reviews, see: (a) Rossiter, B. E. in "Asymmetric Synthesis"; Morrison, J. D.; Ed.; Academic Press: New York, 1985; Vol. 5, p. 193. (b) Finn, M. G.; Sharpless, K. B. *Ibid.* Vol. 5, p. 247.

24. Use of tartrate esters as chiral auxiliaries in the asymmetric reactions of allenyl boronic acid also have been reported: Ikeda, N.; Arai, I.; Yamamoto, H. *J. Am. Chem. Soc.* **1986**, 108, 483; Haruta, R.; Ishiguro, M.; Ikeda, N.; Yamamoto, H. *Ibid.* **1982**, 104, 7667.

25. Roush, W. R.; Banfi, L., unpublished research results.

26. Roush, W. R.; Palmer, M. A. J., unpublished, optimized data.

27. (a) Thiem, J.; Meyer, B. *J. Chem. Soc. Perkin II* **1979**, 1331. (b) Thiem, J.; Meyer, B. *Tetrahedron* **1981**, 37, 551. (c) Thiem, J.; Schneider, G. *Angew. Chem., Int. Ed. Engl.* **1983**, 22, 58.

28. (a) Thiem, J.; Gerken, M. *J. Org. Chem.* **1985**, 50, 954. (b) Thiem, J.; Meyer, B. *Chem. Ber.* **1980**, 113, 3058. (c) Thiem, J.; Gerken, M. *J. Carbohydr. Chem.* **1982-83**, 1, 229. (d) Thiem, J.; Elvers, *J. Chem. Ber.* **1981**, 114, 1442; **1980**, 113, 2049. (e) Thiem, J.; Meyer, B. *Ibid.* **1980**, 113, 3067. (f) Thiem, J.; Gerken, M.; Snatzke, G. *Liebigs Ann. Chem.* **1983**, 448.

29. Available structure activity correlation data, reviewed in reference 4, clearly indicates that the oligosaccharide fragments have a marked effect on the biological properties of the aureolic acid antibiotics. A strategy for developing an improved therapeutic agent, therefore, might involve the synthesis of analogues with structurally modified oligosaccharide chains.

30. (a) Horton, D.; Wander, J. D. in "The Carbohydrates", 2nd ed.; Academic Press: New York, 1980, Vol 1B, p. 643. (b) Williams, N. R.; Wander, J. D. *Ibid*, p. 761. (c) Butterworth, R. F.; Hanessian, S. *Adv. Carbohydr. Chem. Biochem.* **1971**, 26, 279.

31. The following list is not intended to be inclusive, but rather only to provide leading references to the synthesis of the olivomycin monosaccharides from carbohydrate precursors. (a) D-olivose (residues C, D): Durette, P. L. *Synthesis* **1980**, 1037. Stanek, J., Jr.; Marek, M.; Jary, J. *Carbohydr. Res.* **1978**, 64, 315. (b) D-Oliose (sugar A) and olivomose (sugar B): Cheung, T. M.; Horton, D.; Weckerle, W. *Carbohydr. Res.* **1977**, 58, 139. Garegg, P. J.; Norberg, T. *Acta Chem. Scand. Ser. B.* **1975**, 29, 205. Brimacombe, J. S.; Portsmouth, D. *Carbohydr. Res.* **1965**, 1, 128. (c) L-olivomycose (sugar E): Thiem, J.; Elvers, *J. Chem. Ber.* **1979**, 112, 818. Williams, E. H.; Szarek, W. A.; Jones, J. K. N. *Can. J. Chem.* **1969**, 47, 4467.

32. For a general review of acyclic diastereoselective synthesis, see: Bartlett, P. A. *Tetrahedron* **1980**, 36, 2.

33. Martin, V. S.; Woodard, S. S.; Katsuki, T.; Yamada, Y.; Ikeda, M.; Sharpless, K. B. *J. Am. Chem. Soc.* **1981**, 103, 6237.

34. (a) Roush, W. R.; Brown, R. J.; DiMare, M. *J. Org. Chem.* **1983**, 48, 5083. (b) Roush, W. R.; Brown, R. J. *Ibid.* **1983**, 48, 5093. (c) Roush,

W. R.; Hagadorn, S. M. *Carbohydr. Res.* **1985**, 136, 187. (d) Roush, W. R.; Adam, M. A. *J. Org. Chem.* **1985**, 50, 3752.

35. Straub, J. A., Ph. D. Thesis, MIT, Cambridge, MA 1987.
36. Hanessian, S.; Guindon, Y. *Carbohydr. Res.* **1980**, 86, C3.
37. Nicolaou, K. C.; Seitz, S. P.; Papahatjis, D. P. *J. Am. Chem. Soc.* **1983**, 105, 2430.
38. For leading references to other glycosidation methods involving thio sugars, see reference 17c and: (a) Hanessian, S.; Bacquet, C.; Lehong, N. *Carbohydr. Res.* **1980**, 80, C17. (b) Ferrier, R. J.; Hay, R. W.; Vethaviyasar, N. *Ibid.* **1973**, 27, 55.
39. The (S,S)-enantiomer of **62**, but not **62** itself, is available from cinnamaldehyde via an asymmetric hydroxylation reaction involving fermenting Baker's yeast: Fronza, G.; Fuganti, C.; Grasselli, P. *J. Chem. Soc., Chem. Commun.* **1980**, 442.
40. (a) Ko, S. Y.; Lee, A. W. M.; Masamune, S.; Reed, L. A., III.; Sharpless, K. B.; Walker, F. J. *Science* **1983**, 220, 949. (b) Minami, N.; Ko, S. S.; Kishi, Y. *J. Am. Chem. Soc.* **1982**, 104, 1109.
41. For other recent studies of diastereoselective nucleophilic additions to epoxyaldehydes, see: (a) Howe, G. P.; Wang, S.; Proctor, G. *Tetrahedron Lett.* **1987**, 28, 2629. (b) Takeda, Y.; Matsumoto, T.; Sato, F. *J. Org. Chem.* **1986**, 51, 4728.
42. (a) Hoye, T. R.; Suhadolnik, J. C. *J. Am. Chem. Soc.* **1985**, 107, 5312. (b) Hoye, T. R., Suhadolnik, J. C. *Tetrahedron* **1986**, 42, 2855.
43. Schreiber, S. L.; Schreiber, T. S.; Smith, D. B. *J. Am. Chem. Soc.* **1987**, 109, 1525.
44. (a) Payne, G. B. *J. Org. Chem.* **1962**, 27, 3819. (b) Katsuki, T.; Lee, A. W. M.; Ma, P.; Martin, V. S.; Masamune, S.; Sharpless, K. B.; Tuddenham, D.; Walker, F. J. *J. Org. Chem.* **1982**, 47, 1373.
45. (a) Kroon, J. in "Molecular Structure and Biological Activity," Griffin, J. F. and Duax, W. L. , ed.; Elsevier Biomedical: New York, 1982, p. 151. (b) Karabatsos, G. J.; Fenoglio, D. J. *Topics Stereochem.* **1970**, 5, 167.
46. (a) Karabatsos, G. J.; Fenoglio, D. J.; Lande, S. S. *J. Am. Chem Soc.* **1969**, 91, 3572. (b) Karabatsos, G. J.; Fenoglio, D. J. *Ibid.* **1969**, 91, 3577. (c) Brown, T. L. *Spectrochim Acta* **1962**, 18, 1615.
47. For a previous example of a boron centered anomeric effect, see: Shiner, C. S.; Garner, C. M.; Haltiwanger, R. C. *J. Am. Chem. Soc.* **1985**, 107, 7167.
48. Roush, W. R.; Banfi, L. *J. Am. Chem. Soc.* **1988**, 110, 3979.
49. Eight-membered dilactones have also been considered, but so far we have been unable to synthesize these materials.
50. (a) Mukaiyama, T.; Usui, M.; Saigo, K. *Chem. Lett.* **1976**, 49. (b) Bald, E.; Saigo, K.; Mukaiyama, T. *Ibid.* **1975**, 1163.
51. (a) Steliou, K.; Poupart, M.A. *J. Am. Chem. Soc.* **1983**, 105, 7130. (b) Collum, D. B.; Chen, S. C.; Ganem, B. *J. Org. Chem.* **1978**, 43, 4394.
52. Reagent **88** also ranks among the most highly enantioselective chiral acetate aldol enolate equivalents: (a) Braun, M. *Angew. Chem., Int. Ed. Engl.* **1987**, 26, 24, and literature cited therein. (b) Masamune, S.; Sato, T.; Kim, B. M.; Wollmann, T. A. *J. Am. Chem. Soc.* **1986**, 108, 8279.
53. The significance of convergent functional groups on the design of molecular receptors and clefts has been described: (a) Rebek, J., Jr.;

Askew, B.; Ballester, P.; Doa, M. *J. Am. Chem. Soc.* **1987**, 109, 4119. (b) Rebek, J., Jr.; Askew, B.; Nemeth, D.; Parris, K. *Ibid.* **1987**, 109, 2432. (c) Rebek, J., Jr.; Askew, B.; Killoran, M.; Nemeth, D.; Lin, F.-T. *Ibid.* **1987**, 109, 2426.
54. This manuscript is based on a lecture presented in the "Stereoselective Synthesis of Carbohydrates from Acyclic Precursors" Symposium at the 194th American Chemical Society National Meeting, New Orleans, September 2, 1987. This research is also discussed in Roush, W. R., "Strategies and Tactics in Organic Synthesis," Vol. 2.; Lindberg, T., ed.; Academic Press: New York, 1988.

RECEIVED October 14, 1988

Chapter 15

A Fresh Approach to the Synthesis of Carbohydrates

Teruaki Mukaiyama

Department of Chemistry, Faculty of Science, University of Tokyo, Hongo, Bunkyo-ku, Tokyo 113, Japan

Various monosaccharides and glycosides are successfully synthesized by the application of new and highly stereoselective reactions developed in our laboratory.

In the area of synthetic organic chemistry, various tools for diastereoselective carbon-carbon bond formation in acyclic systems have recently been developed. Furthermore, the asymmetric synthesis of chiral compounds in high optical purity is achieved without the use of enzymes. Based on these developments, much attention has been focused on the synthesis of carbohydrates in recent years. In general, carbohydrates consist of various monosaccharides linked by glycosyl bonds and, therefore, possess many chiral centers. Thus, the following goals exist as challenging problems in the synthesis of carbohydrates; (a) to prepare various monosaccharides with correct stereochemistry and (b) to make glycosyl bonds stereoselectively between sugar derivatives.

This article summarizes a variety of novel and useful methodologies developed in our laboratory for the synthesis of carbohydrates.

(A) STEREOSELECTIVE SYNTHESIS OF MONOSACCHARIDES

Various monosaccharides are usually synthesized starting from the readily available sugars, such as glucose, galactose, etc. In recent years, useful methods for the synthesis of various sugars by stereoselective carbon-carbon bond formation between simple organic molecules have been much sought after. For this purpose, several new reactions using common metal salts such as potassium enolates, cadmium salts, zinc enolates, boron enolates etc. were developed (I,II). Next, novel divalent tin species mediated carbon-carbon bond forming reactions were explored and a convenient synthesis of a variety of monosaccharides has been achieved by use of these reactions (III,IV). Furthermore, a 4-carbon building block, a key starting material for sugar synthesis, was devised and prepared from L- or D-tartaric acid (V).

0097-6156/89/0386-0278$06.00/0

I. Stereoselective Additions to 2,3-0-Isopropylidene-D- or L-glyceraldehyde

The cadmium salt of the 2-allyloxybenzimidazole derivative **1** reacts
with various aldehydes to afford adducts **2** with high regio- and
stereoselectivities. The adducts **2** are subsequently transformed
into trans-vinyloxiranes **3** (1).

D- and L-Ribose are synthesized starting from 2,3-0-isopropylidene-D-
and L-glyceraldehyde, respectively, by the application of this reaction (2).

Zinc enolate **4**, prepared from acetylene ether pyridine 1-oxide,
mercuric chloride, and zinc, adds to aldehydes to form α-chloro-β-
hydroxy esters **5** in good yields (3). Subsequent treatment with base
gives trans-epoxyesters, one of which **6** is converted to 2-amino-2-
deoxy-D-ribose stereoselectively in good yields (4).

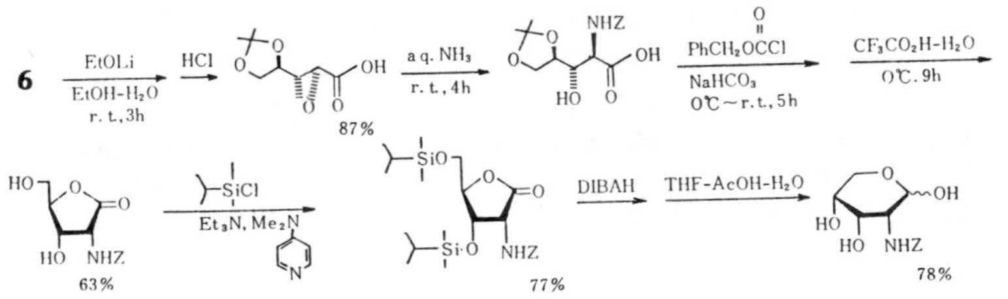

6 $\xrightarrow[\substack{EtOH-H_2O \\ r.t.,3h}]{EtOLi}$ HCl $\cdots$ (87%) $\xrightarrow[r.t.,4h]{aq.\ NH_3}$ $\cdots$ $\xrightarrow[\substack{NaHCO_3 \\ 0°C-r.t.,5h}]{PhCH_2OCCl}$ $\xrightarrow[0°C.9h]{CF_3CO_2H-H_2O}$

(63%) $\xrightarrow[Et_3N,\ Me_2N]{\ \ SiCl\ \ }$ (77%) $\xrightarrow{DIBAH}$ $\xrightarrow{THF-AcOH-H_2O}$ (78%)

Replacement of pyridine 1-oxide by borinic acid in the above reaction leads to formation of vinyloxyborane intermediate **7** which further reacts with aldehydes to give the aldol products **8** and **9**. 2-Deoxy-D-ribose is prepared efficiently by application of this reaction (5).

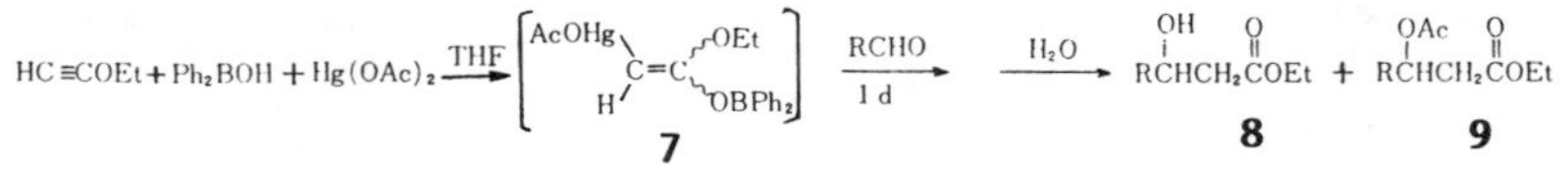

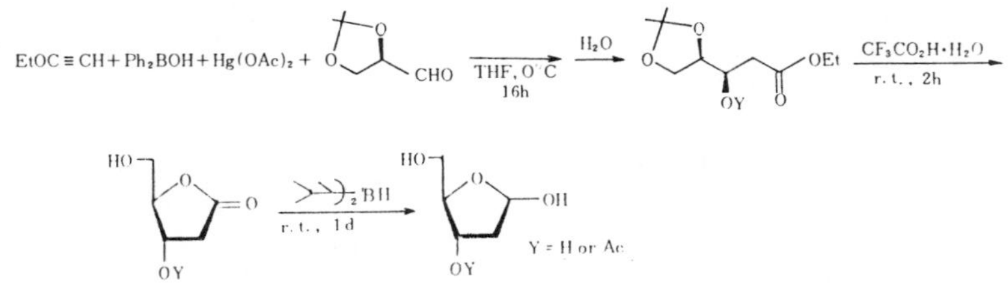

$HC\equiv COEt + Ph_2BOH + Hg(OAc)_2 \xrightarrow{THF}$ $\begin{bmatrix} AcOHg \\ \\ H \end{bmatrix} C=C \begin{matrix} OEt \\ \\ OBPh_2 \end{matrix}$ **7** $\xrightarrow[1\ d]{RCHO}$ $\xrightarrow{H_2O}$ $\underset{\textbf{8}}{RCHCH_2COEt} + \underset{\textbf{9}}{RCHCH_2COEt}$

(with OH O and OAc O substituents respectively)

$EtOC\equiv CH + Ph_2BOH + Hg(OAc)_2 +$ $\cdots CHO$ $\xrightarrow[16h]{THF,\ 0°C}$ $\xrightarrow{H_2O}$ $\cdots OEt$ $\xrightarrow[r.t.,\ 2h]{CF_3CO_2H-H_2O}$

$\xrightarrow[r.t.,\ 1d]{\ \ \ \ \ _2BH}$ Y = H or Ac

Stereocontrolled addition of 2-furyl anion to 2,3-O-isopropylidene-D-glyceraldehyde can be achieved by the addition of some metal salts. In the presence of zinc iodide, 2-furyllithium undergoes stereoselective addition to afford the _anti_ adduct **10** with high selectivity. This adduct can be transformed into D-ribulose through standard sequences (6).

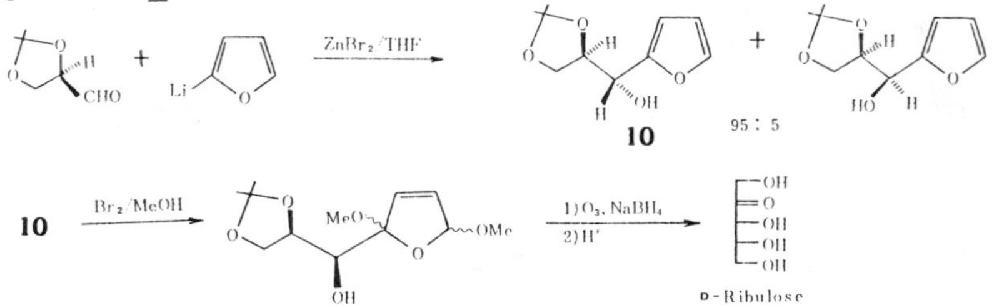

$\cdots CHO$ + $\cdots$ Li $\xrightarrow{ZnBr_2\ THF}$ **10** + $\cdots$ 95 : 5

10 $\xrightarrow{Br_2/MeOH}$ $\cdots OMe$ $\xrightarrow[2)H^+]{1)O_3,\ NaBH_4}$ $\begin{matrix} OH \\ O \\ OH \\ OH \\ OH \end{matrix}$ D-Ribulose

II. Asymmetric Aldol Reaction by the Use of Optically Active Imines

The potassium enolate prepared from atrolactic acid derivative **11** undergoes aldol reaction to give **12** in a highly stereoselective manner. Successive acid treatment gives _syn_-amino acid **13** in good yield (7).

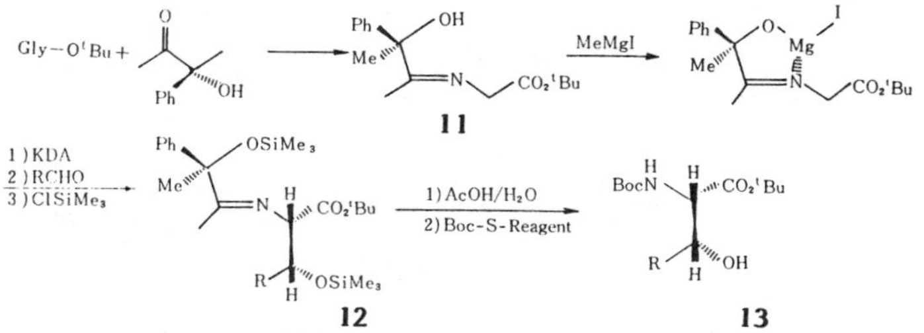

2-Acetamido-2-deoxy-D-arabinose **14** and 2-acetamino-2-deoxy-D-ribose **15** are prepared from (R)- and (S)-atrolactic acid derivatives respectively using the above reaction (8).

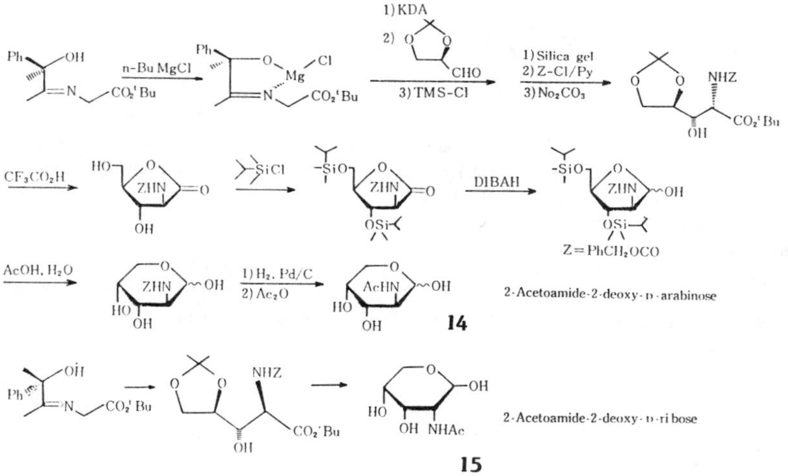

III. New and Useful Synthetic Reactions by the Use of Stannous Fluoride or Metallic Tin

Allyltin difluoroiodide, formed in situ by the oxidative addition of stannous fluoride to allyl iodide, is found to react with carbonyl compounds to give the corresponding homoallylic alcohols in excellent yields under mild reaction conditions (9).

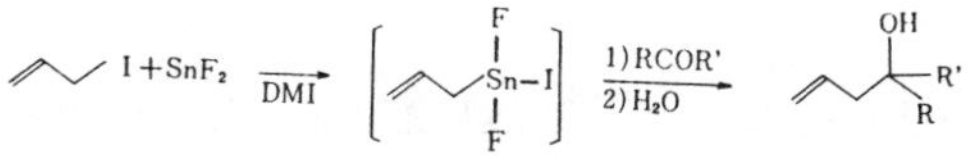

This reaction is applied to the synthesis of 2-deoxy-D-ribose (10).

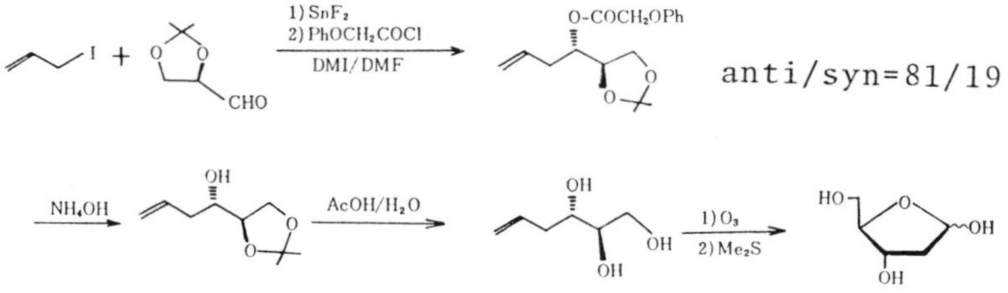

anti/syn=81/19

2,2,2-Tribromoethanol derivatives are also obtained by the reaction of carbon tetrabromide with aldehydes in the presence of stannous fluoride. 2,3-Di-O-acetyl-D-erythronolactone 16 is synthesized as shown below (11).

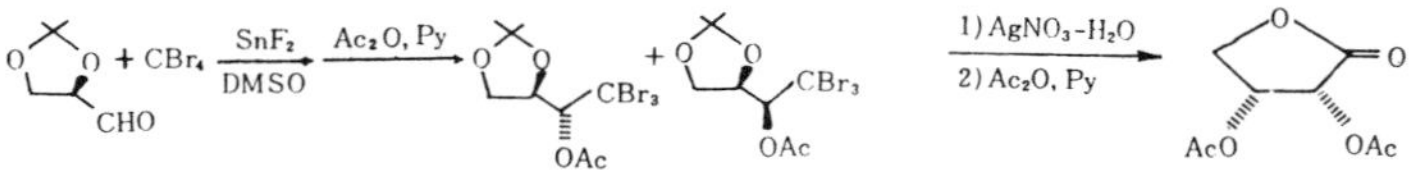

3 : 1   **16**

Metallic tin, Sn(0), is even more effectively employed. For example, in the presence of Sn(0), allyl bromide and α-halocarbonyl compounds afford nucleophilic organometallic species, which add to aldehydes in good yields to give homoallylic alcohols (12) and β-hydroxycarbonyl compounds (13,14) respectively. α-Diketones could be reduced by activated Sn(0), to give tin(II) enediolates which in turn undergo aldol reaction to form α,β-dihydroxyketones (15,16). This reaction was successfully applied to a stereoselective synthesis of methyl D-glucosaminate (17).

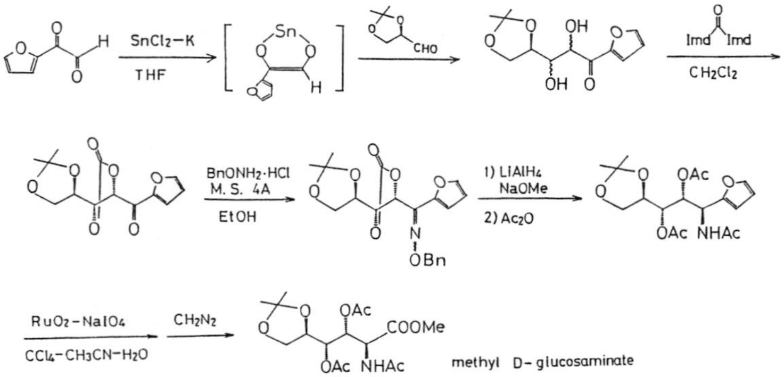

IV. Stereoselective Aldol Reaction by the Use of Stannous Enolates

The aldol reaction is one of the most fundamental and useful syn-thetic methods in organic synthesis and boron enolates are known (18) to be the most efficient intermediates in view of their mild reaction conditions and high stereoselectivity. Recently, stannous enolates

were found to have the prominent features shown above; stannous
enolates generated from β-bromoketones and Sn(0) react with aldehydes
in a highly regio- and stereoselective manner. Also, a convenient
method for the generation of stannous enolates was newly developed;
the stannous enolates are generated by treatment of ketones with
stannous triflate in the presence of N-ethylpiperidine and the
stannous enolates thus formed show high reactivity as well as high
stereoselectivity (19,20), and the cross aldol reaction between two
different ketones is also realizable in good yields (21).

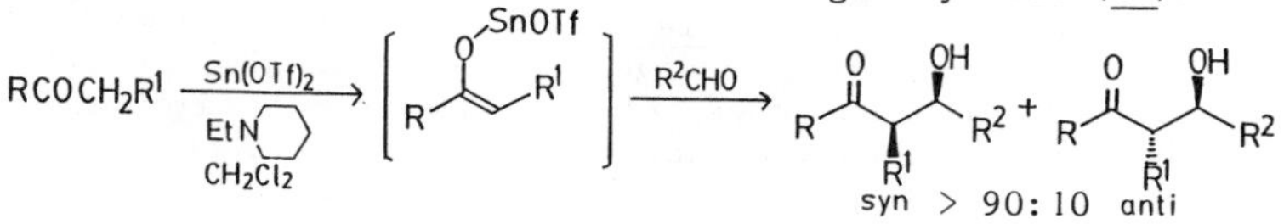

Various cis-α,β-epoxycarbonyl compounds are stereoselectively prepared via α-bromo-β-hydroxycarbonyl compounds by application of this
Sn(II) mediated aldol reaction to α-bromocarbonyl compounds (22).
Also this reaction is employed for the stereoselective synthesis of
2-amino-2-deoxy-D-arabinitol.

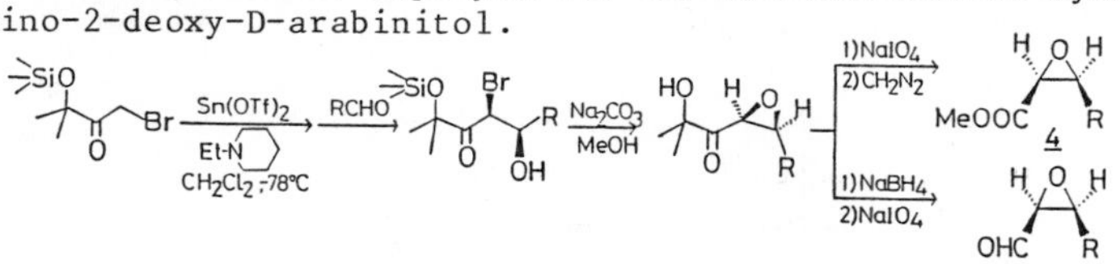

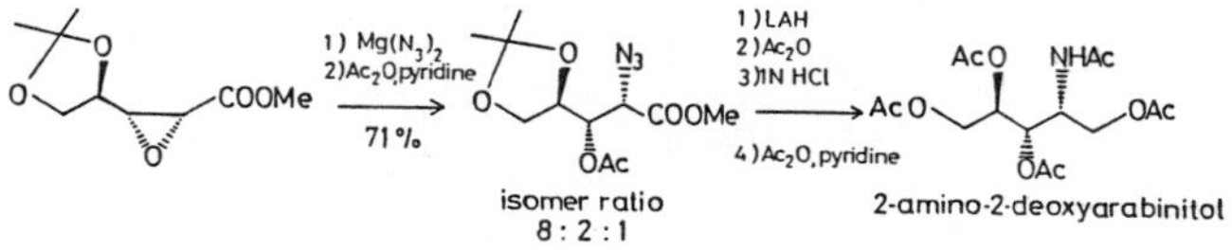

A simple synthesis of the branched chain sugar, 2-C-methyl-D,L-
lyxofuranoside **17** has been achieved by using the tin(II) enolate of a
1,3-dihydroxy-2-propanone derivative and methyl pyruvate (23).

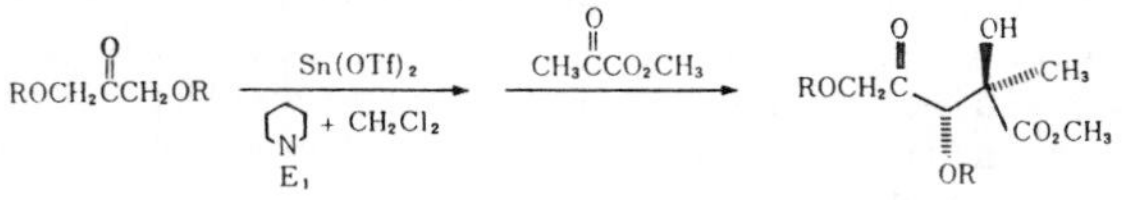

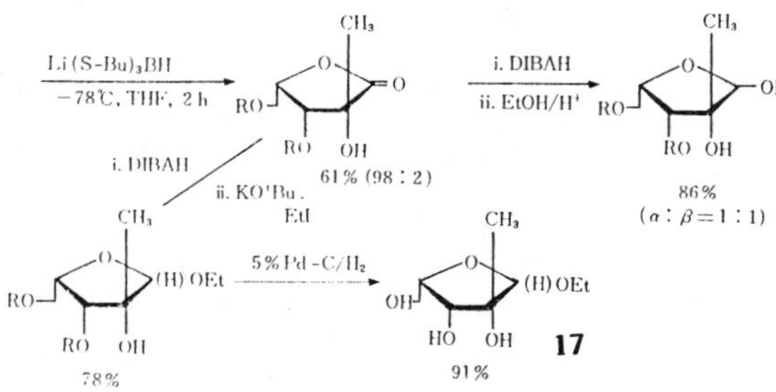

Although several asymmetric aldol reactions have been reported recently, chiral auxiliary groups are usually attached to the ketone equivalent molecules in these reactions (24). Previously, no example existed of an aldol-type reaction where two achiral carbonyl compounds are used for constructing a chiral molecule with the aid of a ligand. It had already been shown that chiral diamines derived from (S)-proline are efficient ligands in certain asymmetric reactions (25,26). Then, based on the considerations that divalent tin has vacant d orbitals and ability to accept a bidentate ligand, enantioselective aldol reactions via divalent tin enolates with chiral diamines were explored. A highly enantioselective cross aldol reaction between aromatic ketones (27,28) or 3-acetylthiazolidine-2-thione (29) and various aldehydes has been achieved, in which chiral diamines derived from (S)-proline work very effectively as ligands. This is the first example for the formation of cross aldols in high optical purity starting from two achiral carbonyl compounds by employing chiral diamines as chelating agents.

These compounds derived from 3-acetylthiazolidine-2-thione are very versatile chiral materials, capable of being transformed into various synthetic intermediates as previously demonstrated (30). Furthermore, in the stannous enolate mediated aldol-type reactions of 3-(2-benzyloxyacetyl)thiazolidine-2-thione, the stereochemical course of the reaction is dramatically altered by the addition of TMEDA as a ligand. High asymmetric induction is also achieved by the addition of a chiral diamine derived from (S)-proline (31).

Thus the characteristic features of tin(II) enolates enable the stereoselective synthesis of aldol products even from two different

ketones.   The combination of stannous triflate and N-ethylpiperidine
provides an easy approach to tin(II) enolates, whereas tin(IV)
enolates have been prepared through relatively laborious multi-step
procedures.   Enantioselective aldol reaction effected by chiral diamines also enhances the utility of tin(II) enolates in organic
synthesis.

V. New 4-Carbon Building Block for Sugar Synthesis

A number of sugar derivatives are successfully synthesized by employing the newly devised stereoselective carbon-carbon bond forming
reactions starting from 2,3-O-isopropylidene-D- or L-glyceraldehyde.
Recently, syntheses of monosaccharides starting from simple molecules
were reported by Masamune, Sharpless, Kishi, Danishefsky, and other
groups (32,33).   In most of these approaches, a carbon framework is
extended by a Wittig-type reaction and a generation of chiral centers
is achieved by the Sharpless asymmetric epoxidation.   In contrast,
our approach mentioned in the foregoing chapters is based on the
stereoselective carbon-carbon bond formation, that is, the stereocontrolled carbon chain extention.
Next, a development of a new and potentially useful 4-carbon building
block for the L-sugars was undertaken, and it was found 4-0-benzyl-
2,3-0-isopropylidene-L-threose 18 is readily prepared starting from
L-tartaric acid (34).

Bn=CH₂Ph

2-Deoxy-L-galactose 19, 3-amino-2,3-dideoxy-L-xylo-hexose 20 and L-
diginose 21 are synthesized conveniently from the aldehyde 18 (35).

Further, convenient syntheses of the two fragments of Polyoxin J,
deoxypolyoxin C and 5-0-carbamoylpolyoxamic acid, have been achieved
in a highly stereoselective manner starting from the same chiral
building block, 4-0-benzyl-2,3-0-isopropylidene-L-threose, leading to
a formal total synthesis of Polyoxin J (36).

(B)   STEREOSELECTIVE GLYCOSYLATION REACTIONS

Besides stereoselective synthesis of various monosaccharides, stereoselective reaction for the preparation of glycosides is an important
problem in the synthetic field of carbohydrate chemistry. However,
the classical methods, which require the assistance of heavy metal
salts or drastic reaction conditions, are still employed by and large
in the synthesis of such compounds. Taking these disadvantages into
consideration, new glycosylation reactions, which proceed under mild
reaction conditions with high selectivity, have been developed and
exploited.

I. New Approaches for the Synthesis of Glucosides. An Alkoxylation
of Silyl Enol Ethers.

α-Alkoxyketones and α-ketoacetals are prepared in good yields by the
reaction of silyl enol ethers with alkyl hypochlorites in the
presence of palladium(0) catalyst (37).

The reaction is applied to the synthesis of α-glucoside as shown in
the following equation.

II. A Cyclization of Hydroxy Enol Ethers

A stereoselective cyclization of (Z)-(2R,3R,4R)-6-cyclohexyloxy-1,3,4-tribenzyloxy-5-hexen-2-ol promoted by mercuric trifluoroacetate or PhSeCl is successfully achieved (38). 2-Deoxy-α-hexopyranoside derivative is obtained almost exclusively by the treatment of **22** with mercuric trifluoroacetate followed by reductive work up, while a predominant formation of the β-anomer is achieved by the reaction of **22** with PhSeCl, and the successive deselenylation.

N-Iodosuccinimide also promotes the cyclization to give the α-glucoside derivative, and the disaccharide derivative is synthesized with high stereoselectivity starting from a hydroxy vinylether (39).

III. The Glycosyl Fluoride Method

To achieve higher stereoselectivity, the combination of various sugar derivatives and activating reagents was examined. As a result, it was found in 1981 that α-glucosides are prepared with high stereoselectivity by the reaction of β-glucosyl fluorides with various hydroxy compounds in the presence of stannous chloride and silver perchlorate, or stannous chloride and trityl perchlorate (40). The use of fluoro-sugars has been extensively studied by many research groups (41-43) after this result was published.

$$\alpha/\beta = 88/12 - 81/19$$

IV. Trityl Perchlorate Mediated Glycosylation

Recently it was found that the aldol reaction of silyl enol ethers
with acetals or aldehydes is effectively promoted by a catalytic
amount of trityl perchlorate to give the corresponding aldols in good
yields (44,45). Polymer-bound trityl perchlorate also successfully
catalyzed the aldol reaction (46).

The trityl perchlorate mediated reaction of 1-0-acylsugars with alcohols is successfully carried out at 0 °C and the corresponding α-
glycosides are stereoselectively produced (47).

$$\alpha/\beta = 96/4 - 92/8$$

Similarly, C-glycosylation reaction of 1-0-acylsugars with silylated
carbon nucleophiles, such as silyl enol ethers, proceeded to give the
corresponding C-glycosides in good yields in the presence of a
catalytic amount of trityl perchlorate (48). This reaction was also
catalyzed by ploymer-bound trityl perchlorate (Mukaiyama, T.;
Kobayashi, S. Carbohydrate Research, in press.)

$$\alpha/\beta = 96/4$$

Literature Cited

1) Yamaguchi, M.; Mukaiyama, T. Chem. Lett. 1979, 1279.
2) Yamaguchi, M.; Mukaiyama, T. Chem. Lett. 1981, 1005.
3) Mukaiyama, T.; Murakami, M. Chem. Lett. 1981, 1129.
4) Murakami, M.; Mukaiyama, T. Chem. Lett. 1982, 1271.
5) Murakami, M.; Mukaiyama, T. Chem. Lett. 1982, 241.
6) Suzuki, K.; Yuki, Y.; Mukaiyama, T. Chem. Lett. 1981, 1529.
7) Nakatsuka, T.; Miwa, T.; Mukaiyama, T. Chem. Lett. 1981, 279.
8) Mukaiyama, T.; Miwa, T.; Nakatsuka, T. Chem. Lett. 1982, 145.
9) Mukaiyama, T.; Harada, T.; Shoda, S. Chem. Lett. 1980, 1507.

10) Harada, T.; Mukaiyama, T. Chem. Lett. 1981, 1109.
11) Mukaiyama, T.; Yamaguchi, M.; Kato, J. Chem. Lett. 1981, 1505.
12) Mukaiyama, T.; Harada, T. Chem. Lett. 1981, 1527.
13) Harada, T.; Mukaiyama, T. Chem. Lett. 1982, 467.
14) Harada, T.; Mukaiyama, T. Chem. Lett. 1982, 161.
15) Mukaiyama, T.; Kato, J.; Yamaguchi, M. Chem. Lett. 1982, 1291.
16) Mukaiyama, T.; Tsuzuki, R. Kato, J. Chem. Lett. 1983, 1825.
17) Mukaiyama, T.; Tsuzuki, R.; Kato, J. Chem. Lett. 1985, 837.
18) Mukaiyama, T; Inomata, K.; Muraki, M. J. Am. Chem. Soc. 1973, 95 967.
19) Mukaiyama, T. Org. React. 1982, 28, 203, and references cited therein.
20) Mukaiyama, T.; Stevens, R. W.; Iwasawa, N. Chem. Lett. 1982, 353.
21) Stevens, R. W.; Iwasawa, N.; Mukaiyama, T. Chem. Lett. 1982, 1459.
22) Mukaiyama, T.; Yura, T.; Iwasawa, N. Chem. Lett. 1985, 809.
23) Stevens, R. W.; Mukaiyama, T. Chem. Lett. 1983, 595.
24) Heathcock, C. H. In Asymmetric Synthesis; Morrison, J. D., Ed.; Academic Press: Orland, FL, 1984; Vol 3, pp 165-206.
25) Mukaiyama, T.; Asami, M.; Hanna, J.; Kobayashi, S. Chem. Lett. 1977, 783.
26) Mukaiyama, T.; Soai, K.; Sato, T.; Shimizu, H.; Suzuki, K. J. Am. Chem. Soc. 1979, 101, 1455.
27) Iwasawa, N.; Mukaiyama, T. Chem. Lett. 1982, 1441.
28) Stevens, R. W.; Mukaiyama, T. Chem. Lett. 1983, 1799.
29) Iwasawa, N.; Mukaiyama, T. Chem. Lett. 1983, 297.
30) Mukaiyama, T.; Iwasawa, N. Chem. Lett. 1982, 1903.
31) Mukaiyama, T.; Iwasawa, N. Chem. Lett. 1984, 753.
32) Pfenninger, A. Synthesis 1986, 89 and references cited therein.
33) Danishefsky, S.; Kobayashi, S.; Kerwin, J. F., Jr. J. Org. Chem. 1982, 47, 1981.
34) Mukaiyama, T.; Suzuki, K.; Yamada, T. Chem. Lett. 1982, 929.
35) Mukaiyama, T.; Yamada, T.; Suzuki, K. Chem. Lett. 1983, 5.
36) Tabusa, F.; Yamada, T.; Suzuki, K.; Mukaiyama, T. Chem. Lett. 1984, 405.
37) Nakatsuka, T.; Mukaiyama, T. Chem. Lett. 1982, 369.
38) Suzuki, K.; Mukaiyama, T. Chem. Lett. 1982, 683.
39) Suzuki, K.; Mukaiyama, T. Chem. Lett. 1982, 1525.
40) Mukaiyama, T.; Murai, Y.; Shoda, S. Chem. Lett. 1981, 431.
41) Araki, Y.; Watanabe, K.; Kuan, F.-H.; Itoh, K.; Kobayashi, N.; Ishido, Y. Carbohydr. Res. 1984, 127, C5.
42) Hashimoto, S.; Hayashi, M.; Noyori, R. Tatrahedron Lett. 1984, 25, 1379.
43) Dolle, R. E.; Nicolaou, K. C. J. Am. Chem. Soc. 1985, 107, 1691, 1695.
44) Mukaiyama, T.; Kobayashi, S.; Murakami, M. Chem. Lett. 1984, 1759.
45) Mukaiyama, T.; Kobayashi, S.; Murakami, M. Chem. Lett. 1985, 447.
46) Mukaiyama, T.; Iwakiri, H. Chem. Lett. 1985, 1363.
47) Mukaiyama, T.; Kobayashi, S.; Shoda, S. Chem. Lett. 1984, 907.
48) Mukaiyama, T.; Kobayashi, S.; Shoda, S. Chem. Lett. 1984, 1529.

RECEIVED May 31, 1988

Chapter 16

Asymmetric Reactions Toward the Synthesis of Carbohydrates

Koichi Narasaka

Department of Chemistry, Faculty of Science, University of Tokyo, Hongo, Bunkyo-ku, Tokyo 113, Japan

By the use of chiral oxazolidines derived from a chiral norephedrine and methyl ketones, an asymmetric aldol reaction proceeds in a highly enantioselective manner. In the case of ethyl or α-methoxy ketones, the corresponding anti aldol products were obtained with high diastereo- and enantioselectivities. A chiral titanium reagent, generated from dichlorodiisopropoxytitanium and a chiral 1,4-diol, prepared from dimethyl tartrate, is found to be an effective catalyst for asymmetric Diels-Alder reactions. 1,3-Oxazolidin-2-one derivatives of α,β-unsaturated carboxylic acids react with dienes in the presence of a 10% molar equivalents of the chiral titanium reagent, giving the adducts in high optical purity. The chiral titanium reagent is also applied to the hydrocyanation of aldehydes successfully.

Asymmetric Aldol Reaction

The development of enantioselective aldol reactions has been widely studied in conjunction with the synthesis of natural products. Highly enantioselective aldol reactions have been achieved by employing chiral enolates of ethyl ketones and propionic acid derivatives.(1) On the other hand, achieving high asymmetric induction in the asymmetric aldol reaction of methyl ketones is still a problem.(2)

With this in mind, the asymmetric aldol reaction of chiral oxazolidines 1, prepared from chiral norephedrine and methyl ketones was investigated. The general pathway of this asymmetric aldol reaction is outlined in the following scheme. It would be expected that the treatment of a chiral oxazolidine with 2 molar amounts of lithium diisopropylamide (LDA) would generate a lithiated enamine, which would be converted to the cyclic metallo-enamine 2 by the addition of a metal salt. As a rigid 5-membered chelate is formed in the cyclic metallo-enamine 2, high asymmetric induction would be expected in the successive reaction with an aldehyde through a [4.3.0]-bicyclic transition state 3. Removal of the chiral auxiliary from the adduct 4, performed by acid treatment, will give a corresponding aldol product 5.

0097–6156/89/0386–0290$06.00/0

Based on this assumption, the asymmetric aldol reaction of chiral 1,3-oxazolidines **1** of methyl ketones was examined. It was found that the corresponding aldol products were obtained in good optical purity when divalent tin chloride was used as an additive metal salt.

The typical experimental procedure is as follows: The oxazolidine **1a** was prepared from (+)-norephedrine and acetone, and purified by distillation. To a THF (2 mL) solution of LDA (2.1 mmol) was added a THF solution of **1a** (1 mmol) at 0 °C and the mixture was stirred for 2 hr. Then a THF (4 mL) solution of SnCl$_2$ (1.05 mmol) was added at 0 °C and stirred for 30 min. To this mixture was added a THF (2 mL) solution of 2,2-dimethyl-propanal(1.2 mmol). After stirring for 20 min at 0 °C, the mixture was quenched with aq. 4% NaHCO$_3$. The crude reaction mixture was treated with acetone in the presence of a catalytic amount of BF$_3$ OEt$_2$ to remove the chiral auxiliary. The crude products were purified by chromatographic procedure to give the corresponding β-hydroxy ketone **5a** with 86% ee.

In Table 1, the yields and the optical purities of the aldol products between some methyl ketones and aldehydes are summarized. It is apparent that the aldol products of methyl ketones were obtained in good to excellent optical purity.(3)

Furthermore, this reaction was applied to the aldol reaction of 3-pentanone. When the chiral oxazolidine **6** was prepared from 3-pentanone and the aldol reaction was carried out by the same procedure, the α,β-anti β-hydroxy ketones **7** were produced predominantly over the syn-isomer **8** with excellent optical purities.(5) (See Table 2.)

Table 1. Asymmetric Aldol Reaction of Methyl Ketones

$RCOCH_3$	R'CHO	Yield/%	Optical Purity/%ee
$PhCOCH_3$	$Ph(CH_2)_2CHO$	68	70[a]
	n-PrCHO	69	76[a]
	$c-C_6H_{11}CHO$	64	77[a]
	t-BuCHO	66	93[b]
$t-BuCOCH_3$	$Ph(CH_2)_2CHO$	64	85[a]
	$c-C_6H_{11}CHO$	54	84[a]
	t-BuCHO	56	95[b]

a) Determined by ^{1}H NMR or ^{19}F NMR measurement of its MTPA
   ester(4) in the presence of $Eu(fod)_3$.
b) Determined by ^{1}H NMR measurement in the presence of
   $Eu(hfc)_3$.

Table 2. Asymmetric Aldol Reaction of Ethyl Ketones

entry	RCHO	Yield/%	anti:syn[a]	Optical Purity/%ee
1	$Ph(CH_2)_2CHO$	77	7 : 1	92[b]
2	$c-C_6H_{11}CHO$	75	9 : 1	92[b]
3	t-BuCHO	56	6 : 1	95[c]

a) The relative stereochemistry in entry 1 and 2 is assigned
   by the coupling constant between the protons on C-4 and C-
   5.(6) In the cases of entry 2 and 3, it was also determined
   by the chemical shift of the methyl carbon on C-4.(7)
b) Determined by ^{1}H and ^{19}F NMR measurement of its MTPA
   ester in the presence of $Eu(fod)_3$.
c) Determined by ^{1}H NMR measurement in the presence of
   $Eu(hfc)_3$.

In order to determine the absolute stereochemistry of the anti-aldol **7**, a diastereoisomer of the insect pheromone, (3S,4R)-4-methylheptan-3-ol **11** was synthesized from the (+)-oxazolidine **6**. The reaction of (+)-**6** and propanal gave aldols **9a** and **10a** in 61% yield (anti:syn = 7.3:1). They were converted to the corresponding acetates **9b,10b** and separated by column chromatography. The optical purity of the anti-acetate **9b** was determined as 95% ee using a chiral shift reagent. The anti-acetate **9b** was then converted to the acetate of the desired isomer of the pheromone **11** as shown below. The absolute configuration (3S,4R) and the optical purity (95% ee) were proved by comparison with those of the literature.(8) Highly enantioselective synthesis of anti-aldols has remained as a formidable synthetic problem. The present asymmetric aldol reaction affords a useful method for their preparation.

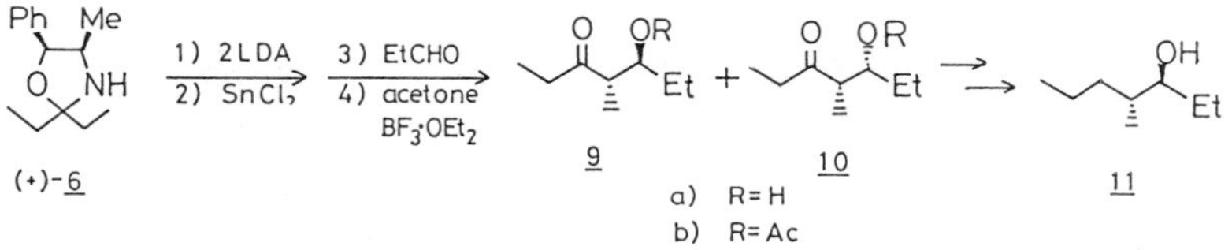

For the preparation of optically active polyhydroxy compounds, such as synthetic intermediates of monosaccharides, the asymmetric aldol reaction of 1,3-dimethoxy-2-propane (Table 3) by

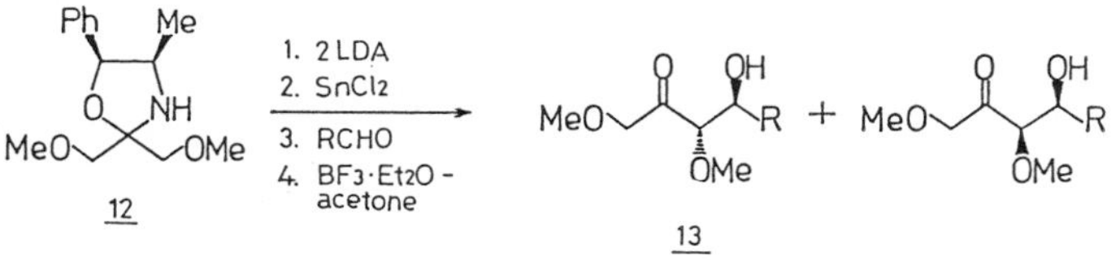

Table 3. Asymmetric Aldol Reaction of α-Methoxy Ketone

entry	RCHO	Yield/%	anti:syn[a]	Optical Purity/%ee
1	Ph(CH$_2$)$_2$CHO	72	7 : 2	93[b]
2	n-C$_9$H$_{19}$CHO	74	9 : 1	95[b]
3	c-C$_6$H$_{11}$CHO	64	7 : 1	95[c]
4	t-BuCHO	45	10 : 0	95[c]

a)The relative stereochemistry is assigned by the coupling constant between the protons on C-3 and C-4 in entry 1 and 2, and in the cases of entry 3 and 4 the stereochemistry is assigned by the coupling constant between the protons on C-3 and C-4 of their acetates.(6)
b)Determined by ^{1}H NMR measurement of its acetate in the presence of Eu(hfc)$_3$.
c)Determined by ^{1}H NMR measurement in the presence of Eu(hfc)$_3$.

this methodology was examined. As in the case of 3-pentanone, the reaction of a chiral oxazolidine **12** with aldehydes afforded predominantly the corresponding anti-aldol adducts **13**. Quite high enantioselectivity was observed and in particular one enantiomer was obtained almost exclusively in the reaction with secondary and tertiary aldehydes.(9)

The L-lyxose derivative **14** was prepared from the anti-adduct **13a** of 3-phenylpropanal as outlined below, and the absolute configuration was determined by comparison with the optical rotation of an authentic sample(10) derived from D-lyxose.

Relative configurations of A and B are determined by coupling constant of ^{1}H NMR.

As mentioned, the present methodology affords an effective means for the synthesis of optically active aldols. A high level of asymmetric induction was observed with a wide range of ketones such as methyl, ethyl and α-methoxy. Furthermore, when (1S,2R)-norephedrine is employed as a chiral auxiliary, it is always the si-face of aldehydes that is attacked by the tin(II) enamine.

Asymmetric Diels-Alder Reaction

In the above asymmetric aldol reaction, the introduction and the removal of the chiral auxiliary are carried out by simple procedures, and high asymmetric induction is achieved even at ice-bath temperature. However, at least a stoichiometric amount of a chiral auxiliary is required in such a stereo-differentiating reaction (chiral auxiliary is attached to the reactant.).

In order to perform asymmetric reactions more efficiently, it is desirable that a large amount of an optically active compound should be produced with only a small investment of chiral auxiliary in a catalytic process. This use of asymmetric catalysts is an area of investigation that has developed rapidly in the last two decades, from which very efficient methods have been developed. Most of them, however, are concerned with asymmetric functional group transformations, and little has been done for the construction of optically active carbon skeletons.(11)

In organic synthesis, Lewis acids have been applied to a wide variety of carbon-carbon bond forming reactions. Therefore, it is expected that if we could design an effective chiral Lewis acid, various carbon-carbon bond forming reactions could be well controlled in an enantioselective manner. As it is well known that titanium reagents can be applied to a variety of carbon-carbon bond forming reactions as Lewis acids, the exploration of chiral titanium reagents was initially investigated.

One of the efficient method for constructing carbohydrates is the Diels-Alder reaction, which gives a variety of important synthetic intermediates for the synthesis of cyclitol derivatives. We therefore were interested in the development of asymmetric Diels-Alder reaction by the use of chiral titanium reagents.

Recent progress in this area has led to the development of various, highly selective, asymmetric reactions by employing chiral dienes and dienophiles in the presence of Lewis acids.(12) On the other hand, little work has been done on asymmetric Diels-Alder reactions promoted by chiral Lewis acids.(13)

Firstly, we modified the α,β-unsaturated acids to be used as dienophiles by converting them into the corresponding 1,3-oxazolidin-2-one (abbreviated as oxazolidone) derivatives 15. This modification is based on the consideration that such bidentate dienophiles would form rigid complexes with a chiral Lewis acid, resulting in high reactivity and a good level of π-facial selectivity during the cycloaddition reaction.

As the chiral Lewis acids, cyclic dialkoxydichlorotitaniums(IV) were chosen and prepared in situ from various chiral 1,2- or 1,4-diols and dichlorodiisopropoxytitanium(IV) according to the alkoxy exchange method.(14)

$$*(R) \begin{array}{c} \text{—OH} \\ \text{—OH} \end{array} + TiCl_2(O\text{-}^iPr)_2 \longrightarrow *(R) \begin{array}{c} \text{—O} \\ \text{—O} \end{array} TiCl_2 + 2^iPrOH$$

The reaction of 3-crotonoyloxazolidone 15a and cyclopentadiene was examined in toluene in the presence of an equimolar amount of various chiral alkoxy titanium(IV) derivatives. It was found that the corresponding endo-adduct 18a was obtained in 55% ee when the chiral 2,3-O-isopropylidene-1,1,4,4-tetraphenylbutanetetraol 17a(15) was introduced as a chiral auxiliary. In this titanium catalyst, the conformation of

the 7-membered ring of the alkoxy titanium is thought to be
important in controlling the enantioselectivity along with the
conformation of the 5-membered acetal ring. The reaction was,
therefore, examined further by using 1,1,4,4-
tetraphenylbutanetetraol derivatives which have various
substituents on the acetal center, and the titanium reagent

derived from 2,3-O-phenylethylidene derivative **17b** was found to
promote the asymmetric cycloaddition reaction in a selective
manner yielding **18a** with 75% optical purity. Moreover, by the
use of 2 molar amounts of the titanium reagent, the product **18b**
was obtained in 92% ee.(15)
      The (R)-(+)-chiral 1,4-diol **17b** was easily prepared from L-
(+)-dimethyl tartrate. Dimethyl tartrate was converted to the
corresponding phenylethylidene derivative by treatment with 1,1-
dimethoxy-1-phenylethane and cat. p-toluenesulfonic acid in
refluxing benzene, followed by conversion to the diol **17b** with
excess phenylmagnesium bromide. The diol was purified by column
chromatography on silica gel (hexane : ethyl acetate = 5 : 1),
and recrystalization from a mixture of hexane and 2-propanol.
The crystals were obtained as a adduct of **17b** and 2-propanol (mp
111-114 °C). The azeotropic removal of 2-propanol with benzene
afforded the diol **17b** as a white amorphous solid.
      Since high enantioselectivity was achieved by employing two
molar equivalents of the chiral titanium reagent generated from
17b, the asymmetric Diels-Alder reaction of various oxazolidone
derivatives of α,β-unsaturated acids **15** and cyclopentadiene was
studied. The results are listed in Table 4. With the exception
of the acryloyl derivative **15b**, various dienophiles reacted with
cyclopentadiene to give the endo-adducts **18** in high optical
purity.

Table 4. Asymmetric Diels-Alder Reaction of 15 with
          Cyclopentadiene

R	React. Temp./ °C	Yield/%	endo:exo[a]	Optical Purity/%ee[b]
Me	-15	93	90:10	92[c] (2S,3R)
H	-78	69	86:14	38[c] (2S)
Ph	0	97	92: 8	81[d]
n-Pr	-15	82	90:10	90[e]
CH₃CH=CH	rt	77	92: 8	82[e]

a) These isomers were separated by silica gel chromatography.
b) Those of endo isomers.
c) The products were converted to the corresponding benzyl esters
   by Evans' procedure,(16) and the absolute configuration and the
   optical purity were determined by the optical rotation.(12d)
d) The product was reduced to an alcohol with lithium aluminum
   hydride, and the optical purity was determined by ^{19}F NMR
   analysis of the corresponding chiral MTPA ester.(4)
e) The products were reduced to alcohols with lithium aluminum
   hydride, and the optical purity was determined by HPLC analysis
   of the corresponding Pirkle's carbamates.(17)

     High enantioselectivity was also attained upon additon of
the chiral titanium reagent to the reaction of acyclic dienes
with oxazolidones of 2-butenoic acid 15a and fumaric acid 19.
The present titanium reagent was noted to exhibit a wide
applicability to the asymmetric Diels-Alder reaction of various
prochiral dienes and dienophiles.

46 % y., 92 % e.e.

86% y., 85 % e.e.

Next we investigated the effect of altering the ratio of the
titanium reagent to dienophiles. It was found, however, the use
of two molar equivalents of the chiral titanium reagent is
indispensable for a high degree enantioselection. That is, when
a catalytic amount (10-17%) of the titanium reagent is used in
the reaction of **15a** and cyclopentadiene, the endo-adduct **18a** is
obtained in high yield but in low enantioselectivity.

$n = 2$ (−)−92% e.e.
$n = 0.17$ (+)− 9 % e.e.

To improve this procedure a highly enantioselective process
employing only a truly catalytic amount of the chiral Lewis acid
was desirable. It was found that the presence of Molecular
Sieves 4A (MS 4A) in the reaction mixture enhances the
enantioselectivity in the chiral titanium-catalyzed reaction.(18)
The chiral titanium alkoxide was prepared by the alkoxy
exchanging method, removing isopropanol azeotropically from the
refluxing toluene solution. However, the same high level of
asymmetric induction was observed by the use of the chiral
titanium species generated in situ by mixing the chiral 1,4-diol
17b and dichlorodiisopropoxytitanium in toluene at room
temperature in the presence of MS 4A. For example, the chiral
catalyst was prepared from a 10.5% molar amount (to the
dienophile) of the 1,4-diol **17b** and a 10% molar amount of
dichlorodiisopropoxytitanium in toluene at room temperature, and
then powdered MS 4A, 3-crotonoyloxazolidone **15a** and
cyclopentadiene were added successively at 0 °C. After being
stirred for 24 hr, the endo-adduct **18a** was produced in 91% ee
with the same absolute configuration (2$\underline{S}$,3$\underline{R}$) as that observed in
the original procedure.

As can be seen in Table 5, various oxazolidones of acrylic
acid derivatives **15** react with cyclopentadiene to afford the
endo-adducts **18** in good to high (64-91%) enantioselectivity by
the combined use of a catalytic amount of the chiral titanium
reagent and MS 4A.

Compared to the previous procedure in which 2 molar
equivalents of the chiral titanium was employed, almost the same
level of enantioselectivity was attained in the Diels-Alder
reaction of **15a**, and the optical purity of the cycloadduct **18b**
was improved significantly.

On the other hand, the reaction of the oxazolidone
derivative of fumaric acid **19** and butadiene was found to proceed
in poor enantioselectivity as compared with the reaction carried
out in the presence of excess amounts of the chiral titanium
reagent, and the cyclohexenedicarboxylic acid derivative **20** was
obtained in 32-45% ee. In order to achieve wide applicability
for the catalytic procedure, the reaction conditions were

Table 5. Asymmetric Diels-Alder Reaction Using Catalytic Amount of Titanium Reagent

R	React. Temp./°C	endo:exo[a]	Yield/%	Optical Purity/%ee[b]
Me	0	92: 8	87	91 (2S,3R)
H	-40	96: 4	93	64 (2S)
n-Pr	0	91: 9	79	72
Ph	rt	88:12	72	64

a)These isomers were separated by silica gel chromatography.
b)Determined by the same procedures described in the Table 4.

examined in detail. The reaction of **19** and butadiene was examined in various solvents in the presence of a 10% molar equivalents of the chiral titanium alkoxide and powdered MS 4A. The enantioselectivity displayed by the reactions in various solvents are summarized in Table 6. It was noted that the enantioselectivity is influenced strongly by the solvent, and that alkyl substituted benzenes are very suitable solvents for the present reaction. The enantioselectivity is dependent on the number of methyl groups on the benzene ring and the optical purity of the adduct **20** was greatly increased in the order of toluene, xylenes and trimethylbenzenes. Furthermore, rather high enantioselectivity was attained by employing hexylbenzene as the solvent, and the trans-4-cyclohexene-1,2-dicarboxylic acid derivative **20** was obtained in 98% yield with 85% ee.(19)

Table 6.   Solvent Effect on the Enantioselectivity

Solvent	Optical Purity/%ee[a]	Solvent	Optical Purity/%ee[a]
benzene	22	CCl$_4$	58
toluene	32-45	1,2,3-TMB[b]	77
o-xylene	67	1,3,5-TMB[b]	81
p-xylene	68	1,3,5-TIPB[c]	85
cumene	74	hexylbenzene	85

a) Determined by the NMR analysis of the corresponding dimethyl
   ester (Mg(OMe)$_2$ using chiral shift reagent Eu(hfc)$_3$.
b) TMB=trimethylbenzene.
c) TIPB=triisopropylbenzene.

The generality of the solvent effect on the enantioselectivity was examined in the following examples using 1,3,5-
trimethylbenzene (1,3,5-TMB) as the common solvent (under
unoptimized reaction conditions).

The reaction of **19** with isoprene was also found to proceed
smoothly in 1,3,5-TMB to afford the 4-methylcyclohexene-1,2-
dicarboxylic acid derivative **21** in 92% ee.

99% yield, 92% ee.
(in toluene, 61% e.e.)   **21**

An example of the solvent effect is also seen in the
reaction of 3-acryloyl-1,3-oxazolidin-2-one (**15b**) which did not
give sufficient asymmetric induction by the previous method.(18)
The reaction of **15b** with butadiene in 1,3,5-TMB gives the 3-
cyclohexenecarboxylic acid derivative **22** in 77% optical purity.

62% yield, 77% e.e.
(in toluene, 67% e.e.)

The reactions of various oxazolidones of β-substituted
acrylic acids with cyclopentadiene were found to proceed with
high asymmetric induction by using 1,3,5-trimethylbenzene as the

solvent as compared with the results of the reactions in toluene, and the corresponding endo-adducts **18** are prepared in 80-90% enantioselectivity.

Table 7. Reaction of **15** with Cyclopentadiene

R	Yield/%	endo:exo[a]	Optical Purity/%ee[b]
Me	90	91: 9	91
Ph	97	92: 8	82
n-Pr	75	91: 9	75

a)These isomers were separated by silica gel chromatography.
b)Determined by the procedures described in Table 4.

As shown by Table 7 above, the chiral titanium catalyst-MS 4A system is widely applicable to the reactions of a variety of dienophiles and dienes when a suitable alkyl substituted benzene is employed as a solvent, and synthetically important Diels-Alder adducts are prepared in high enantioselectivity by the present catalytic process.

Asymmetric Hydrocyanation Reaction
    The chiral titanium reagent preparerd in situ from the chiral 1,4-diol and dichlorodiisopropoxytitanium is expected to be applicable to various reactions catalyzed by Lewis acids. We, therefore, investigated the asymmetric synthesis of cyanohydrins from aldehydes and cyanotrimethylsilane employing the chiral titanium reagent.(20)
    Asymmetric synthesis of cyanohydrins is an important process in organic synthesis because cyanohydrins can be easily converted into a variety of valuable synthetic intermediates such as α-hydroxy acids, α-hydroxy ketones, and β-amino alcohols. Optically active cyanohydrins are obtained in good selectivity by the nucleophilic attack of cyanating reagents to chiral acetals.(21) However, the chiral auxiliaries are destroyed, and not recovered. In catalytic processes with chiral boryl compounds,(22) D-oxynitrilase,(23) and synthetic peptides,(24) the optical purities of the resulting cyanohydrins are generally not sufficient.
    Firstly, we examined the asymmetric hydrocyanation of 3-phenylpropanal. When 3-phenylpropanal was treated with cyanotrimethylsilane using the chiral titanium alkoxide prepared from dichlorodiisopropoxytitanium and the chiral 1,4-diol **17b** in toluene at room temperature, only a trace amount of the

cyanohydrin **23a** was generated after 2 days. On the other hand, by the addition of MS 4A to this solution, the reaction proceeded smoothly at -65 °C to give 2-hydroxy-4-phenylbutanenitrile, which was isolated in 89% yield with 74% ee. The hydrocyanation of primary, secondary and aryl aldehydes was examined with cyanotrimethylsilane to give the optically active cyanohydrins in good optical purity. In particular, benzaldehyde is converted into (R)-mandelonitrile **23b** in 96% ee. (See Table 8.)

Table 8. Asymmetric Hydrocyanation of Aldehydes with Cyanotrimethylsilane

R	Reaction Time/h	Yield/%	Optical Purity/%ee[a]
Ph	12	79	96 (R)[b]
PhCH$_2$	12	66	77 (R)[b]
PhCH$_2$CH$_2$	12	89	74 (R)[c]
c-C$_6$H$_{11}$	48	77	68[d]
C$_8$H$_{17}$	24	66	76[d]

a) Optical purities were determined by HPLC or ^{19}F NMR analysis of the corresponding MTPA esters.
b) The absolute configurations were determined by the comparison of the optical rotations with those of the literature.(25)
c) The absolute configuration was determined after hydrolysis to the known 2-hydroxy-4-phenylbutanoic acid.(26)
d) The absolute configurations were not determined.

    The optical purity of these cyanohydrins was influenced largely by the reaction temperature and was very low at room temperature. Furthermore, when benzaldehyde was treated with cyanotrimethylsilane in the presence of the chiral titanium reagent and MS 4A at room temperature, followed by stirring the reaction mixture at -65 °C for 1 day, the optical purity of mandelonitrile **23b** was ca. 10%. These results indicated that the cyanohydrins were produced as kinetic products in this catalytic process.
    As chiral cyanohydrins are important synthetic intermediates for the preparation of chiral amino acids, carbohydrates, and so on, this process would be a useful asymmetric one-carbon homologation procedure widely employable for organic synthesis.

Conclusion
    As mentioned, the asymmetric reactions useful for the

preparation of carbohydrates have been developed. By application
of the asymmetric aldol reaction, poly-hydroxy compounds are
prepared in high diastereo- and enantioselectivities. The chiral
titanium reagent generated in situ from the chiral 1,4-diol and
dichlorodiisopropoxytitanium is found to be effective for
promoting asymmetric hydrocyanation of aldehydes, and is
successfully used for preparation of the optically active
cyanohydrins from aryl aldehydes. The chiral titanium reagent
also realized the asymmetric Diels-Alder reaction, which affords
cyclohexenedicarboxylic acid and norbornenecarboxylic acid
derivatives in a highly enantioselective manner by a convenient
catalytic process.

Literature Cited

1. Reviews; Evans, D. A.; Nelson, J. V.; Taber, T. R.;
   "Stereoselective Aldol Condensations," in "Topics in
   Stereochemistry," Allinger, N. L.; Eliel, E. L.; Wielen, S. H.
   Eds.; John Wiley and Sons, Inc., New York, 1982, Vol. 13, p. 1;
   Heathcock, C. H. "The Aldol Addition Reaction," in "Asymmetric
   Synthesis," Morrison, J. D. Ed.; Academic Press, Inc., New
   York, 1984, Vol. 3, p. 111.
2. a) Seebach, D.; Ehrig, V.; Teschner, M. Justus Liebigs Ann.
   Chem. 1976, 1357; b) Eichenauer, H.; Friedrich, E.; Lutz, W.;
   Enders, D. Angew. Chem., Int. Ed. Engl. 1978, 17, 206; c)
   Sugasawa, T.; Toyoda, T.; Tetrahedron Lett. 1979, 1423; d)
   Heathcock, C. H.; White, C. T. J. Am. Chem. Soc. 1979, 101,
   7076; e) Evans, D. A.; Taber, T. R. Tetrahedron Lett. 1980, 21,
   4675; f) Heathcock, C. H.; White, C. T.; Morrison, J. J.;
   VanDerveer, D. J. Org. Chem. 1981, 46, 1296; g) Heathcock, C.
   H.; Pirrung, M. C.; Lampe, J.; Buse, C. T.; Young, S. D. J.
   Org. Chem. 1981, 46, 2290; h) Braun, M. Angew. Chem., Int. Ed.
   Engl. 1987, 26, 24; i) Braun, M.; Devant, R. Tetrahedron lett.
   1984, 25, 5031; j) Iwasawa, N.; Mukaiyama, T. Chem. Lett. 1983,
   297; k) Nagao, Y.; Hagiwara, Y.; Kumagai, T.; Ochiai, M.;
   Inoue, T.; Hashimoto, K.; Fujita, E. J. Org. Chem. 1986, 51,
   2391.
3. Narasaka, K.; Miwa, T.; Hayashi, H.; Ohta, M. Chem. Lett. 1984,
   1399.
4. Dale, J. A.; Dull, D. L.; Mosher, H. S. J. Org. Chem. 1969, 34,
   2543.
5. Narasaka, K.; Miwa, T. Chem. Lett. 1985, 1217.
6. a) Stiles, M.; Winkler, R.; Chang, Y.; Traynor, L. J. Am. Chem.
   Soc. 1964, 86, 3337; b) House, H. O.; Crumrine, D. S.;
   Teranishi, A. Y.; Olmstead, H. D. J. Am. Chem. Soc. 1973, 95,
   3310.
7. Heathcock, C. H.; Pirrung, M. C.; Sohn, J. E. J. Org. Chem.
   1979, 44, 4294.
8. Mori, K. Tetrahedron 1977, 33, 289.
9. Narasaka, K.; Yasuda, H. unpublished results.
10. Wolfrom, M. L.; Moody, F. B. J. Am. Chem. Soc. 1940, 62, 3465.
11. "Asymmetric Synthesis," Morrison, J. D. Ed; Academic Press,
    Inc., New York, 1985, Vol 5.
12. a) Paquette, L. A. "Asymmetric Synthesis," Morrison, J. D. Ed;

Academic Press, Inc., New York, 1984, Vol. 3, Chap.4; b)
Oppolzer, W. Angew. Chem., Int. Ed. Engl. 1984, 23, 876; c)
Oppolzer, W.; Chapuis,C.; Bernardinelli, G. Helv. Chim. Acta
1984, 67, 1397; d) Evans, D. A.; Chapman, K. T.; Bisaha, J. J.
Am. Chem. Soc. 1984, 106, 4261; e) Trost, B. M.; O'Krongly, D.;
Belletire, J. L. J. Am. Chem. Soc. 1980, 102, 7595; f)
Helmchen, G.; Karge, R.; Weetman, J. "Modern Synthetic Methods,"
Scheff23ord, R. Ed; Springer-Verlag, Berlin, 1986, Vol. 4, p. 261;
g) Masamune, S.; Reed, L. A.; Davis, J. T.; Choy, W. J. Org.
Chem. 1983, 48, 4441; h) Kelly, T. R.; Whiting, A.;
Chandrakumar, N. S. J. Am. Chem. Soc. 1986, 108, 3510.

13.a) Hashimoto, S.; Komeshima, N.; Koga, K. J. Chem. Soc., Chem.
Commun. 1979, 437; b) Bednarski, M.; Maring, C.; Danishefsky,
S. Tetrahedron Lett. 1983, 24, 3451; c) Chapuis, C.;
Jurczak, J. Helv. Chim. Acta 1987, 70, 436; Seebach, D.; Beck,
A. K.; Imwinkelried, R.; Roggo, S.; Wonnacott, A. Helv. Chim.
Acta 1987, 70, 954; Maruoka, K.; Ito, T.; Shirasaka, T.;
Yamamoto, H. J. Am. Chem. Soc. 1988, 110, 310.

14.Seebach, D. "Modern Synthetic Methods," Schefford, R. Ed; Otto
Salle Verlag and Verlag Sauerlander, Frankfurt am Main, 1983,
Vol. 3, p. 217.

15.Narasaka, K.; Inoue, M.; Okada, N. Chem. Lett. 1986, 1109.

16.Evans, D. A.; Ennis, M. D.; Mathre, D. J. J. Am. Chem. Soc.
1982, 104, 1737.

17.Pirkle, W. H.; Hoekstra, M. S. J. Org. Chem. 1974, 39, 3904.

18.Narasaka, K.; Inoue, M.; Yamada, T. Chem. Lett. 1986, 1967.

19.Narasaka, K.; Inoue, M.; Yamada, T.; Sugimori, J.; Iwasawa, N.
Chem. Lett. 1987,2409.

20.Narasaka, K.; Yamada, T.; Minamikawa, H. Chem.Lett. 1987, 2073.

21.a) Elliott, J. D.; Choi, V. M. F.; Johnson, W. S. J. Org. Chem.
1983, 48, 2294; b) Choi, V. M. F.; Elliott, J. D.; Johnson, W.
S. Tetrahedron Lett. 1984, 25, 591.

22.Reetz, M. T.; Kunisch F.; Heitmann, P. Tetrahdron Lett. 1986,
27, 4721.

23.Becker, W.; Freund, F.; Pfeil, E. Angew. Chem. 1965, 77, 1139.

24.a) Oku, J.; Ito, N.; Inoue, S. Makromol. Chem. 1982, 183, 579;
b) Asada, S.; Kobayashi, Y.; Inoue, S. Makromol. Chem. 1985,
186, 1755; c) Kobayashi, Y.; Asada, S; Watanabe, I.; Hayashi,
H.; Motoo, Y.; Inoue, S. Bull. Chem. Soc. Jpn. 1986, 59, 893.

25.Gountzos, H.; Jackson, W. R.; Harrington, K. J. Aust. J. Chem.
1986, 39, 1135.

26.Yanagisawa, H.; Ishihara, S.; Ando, A.; Kanazaki, T.; Miyamoto,
S.; Koike, H.; Iijima, Y.; Oizumi, K.; Matsushita, Y.; Hata, T.
J. Med. Chem. 1987, 30, 1984.

RECEIVED August 30, 1988

Chapter 17

Microbially Aided Synthesis of Carbohydrates

Claudio Fuganti

Centro di Studio sulle Sostanze Organiche Naturali del Consiglio Nazionale delle Ricerche, Dipartimento di Chimica del Politecnico di Milano, Piazza Leonardo da Vinci 32, 20133 Milan, Italy

The (2$\underline{S}$,3$\underline{R}$) methyldiols (4), generated in a multienzymatic process in fermenting baker's yeast from α-position unsaturated aromatic aldehydes, are used as starting materials in the synthesis of deoxy- and deoxy aminosugars of the L-series, including the L-daunosamine and L-vancosamine derivatives 17 and 34.

Our interest in the synthesis of deoxy- and deoxy-amino sugars of the L-series originates from studies (1) on the products obtained by the action of fermenting baker's yeasts on such aromatic, α, β-unsaturated aldehydes as cinnamaldehyde (1a) and α-methylcinnamaldehyde (1b). The transformation proceeds as indicated in Eq. 1.

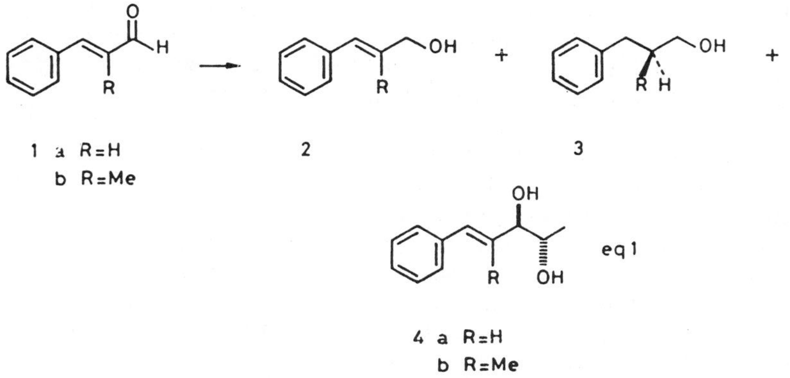

1 a R=H   2   3
  b R=Me

4 a R=H
  b R=Me

eq 1

While the production of 2 and 3 from 1 falls amongst the known capacities of baker's yeast, the formation, in yields of 20-25 % of 4, containing two additional carbon atoms with respect to the precursor aldehyde and two adjacent chiral centres of the type $R^2 \cdot R^1 CHOH$, is new and quite fruitful from the synthetic viewpoint (2).

0097–6156/89/0386–0305$06.00/0

The generation of the (2$\underline{S}$,3$\underline{R}$) diol 4 from 1 is the consequence of a multienzymic process involving two distinct chemical operations (2): (i) Addition of a C_2 unit equivalent to acetaldehyde onto the $\underline{si}$ face of the carbonyl carbon atom of the unsaturated aldehyde to form a (3$\underline{R}$) α-hydroxyketone, in an acyloin of type condensation, and (ii) reduction of the latter intermediate on the $\underline{re}$ face of the carbonyl group to give rise to the diol actually isolated (Eq. 2).

The foregoing optically active diols show some structural features that render them useful starting materials. (a) The (2$\underline{S}$,3$\underline{R}$) absolute configurations of the diols 4 matches that at position 5 and 4 of 6-deoxy sugars of the L-series such as L-amicetose (5) and L-olivomycose (6), which have been indeed prepared from 4a and 4b , respectively, in work designed to establish their absolute configurations (3). (b) The double bond of the diols may be stereospecifically functionalized, the stereochemistry of the process being dictated by the stereochemistry of the adjacent allylic center. (c) Once the double bond has been saturated, it is possible to functionalize the derived products regioselectively at the benzylic position. (d) In saturated products degradation of the benzene ring with O_3 affords 6-deoxy--hexonic acids, bearing all of the chiral centers present in the side-chain of the parent compound. Thus, D(-)-$\underline{allo}$-muscarine (7) (4) and the enantiomeric forms of γ-hexanolide (8) (5) have been prepared from 4a by taking advantage of the foregoing properties.
(e) More importantly, from suitably protected forms of aforementioned diols (4) the action of ozone leads to carbonyl compunds; when isopropylidene derivatives are used, the products have in the α and β positions two chiral oxygen substituents incorporated in a cyclic acetal framework. Furthermore, the C_4 and C_5 $\underline{erythro}$ carbonyl compounds 9 and 10 so obtained may be converted into their epimers 11 and 12, having the $\underline{threo}$ stereochemistry, upon base treatment. We may thus consider the diols 4, readly accesible by yeast treatment of the aldehydes 1, as carbohydrate-like chiral synthons that may be used, in some instances, as convenient alternatives to natural carbohydrates as starting materials for the synthesis of natural products containing in their framework relatively few carbon atoms that are chiral through oxygen susbtitution. Key intermediates in the synthesis based on the

chiral diols 4 are thus the C_4 aldehydes 9 and 11 and the C_5 methyl ketones 10 and 12.

However, the number of relatively small, highly functionalized chiral synthons derived from 9-12 has been increased considerably, taking advantage of current knowledge of methods for stereocontrolled chain-elongation through nucleophilic addition onto the sp^2 carbon atom of 9-12 or of their phenylsulfenimino derivatives.

Synthesis Of Deoxy Amino Sugars Of L-Series

The most significant application of 9-12 is in the synthesis of amino-deoxy sugars of the L series, such as L-daunosamine, its 3-C- methyl analog (L-vancosamine) and their configurational isomers. At the end of a series of experiments, two different strategies have been assessed for preparation of the four configurational isomers of 3-amino—2,3,6-trideoxy-L-hexoses from the erythro and threo C_4 aldehydes 9 and 11, depending upon the relative stereochemistry as positions 3 and 4 of the target amino sugar. For the 3,4-underline{erythro} lyxo and ribo isomers

(17 and 18), the C_4-$\underline{N}$-phenylsulfenimines 13 and 14, prepared from 11 and 9 by the action of NH_3, diphenyl disulfide, and $AgNO_3$, have been used as key intermediates ($\underline{6}$). Addition of diallyzinc into 13 and 14 is highly selective and affords 15 and 16, with 4,5-$\underline{erythro}$ stereochemistry.

13   14

15   16

17   18

Products 15 and 16 contain all of the chiral centers and the required functionalities for direct conversion into $\underline{N}$-protected L—daunosamine (17) and L-ristosamine (18). This is achieved by $\underline{N}$-protection and removal of a C_1 unit by ozonolysis of the terminal vinyl group. Using allylmagnesium bromide instead of diallylzinc, the Felkin-Ahn adducts 15 and 16 are obtained from 13 and 14 in 55:45 and 30:70 ratios, respectively, as the 4,5-$\underline{threo}$ diastereoisomers.

Construction from 9 and 11 of the 3,4-$\underline{threo}$ $\underline{arabino}$ and $\underline{xylo}$ isomers 23 and 24, respectively, requires a different approach. Instead of the C_4 + N $\longrightarrow C_4$-N; C_4-N + C_3 $\longrightarrow C_7$-N; C_7-N $\longrightarrow C_6$-N + C, sequence, the C_4 + C_2 $\longrightarrow C_6$; C_6 + N $\longrightarrow C_6$-N one is used to elaborate the C_6-N chiral framework of 23 and 24. Thus, the C_6, α,β-unsaturated esters 19 and 20 are obtained from 9 and 11. The latter add ammonia stereoselectively (7) to give the intermediates 21 and 22, with the correct stereochemistry for conversion into 23 and 24 by simple acid hydrolysis, N-protection, and reduction by DIBAH. A feature common to the two pathways leading from 9 and 11 to L-daunosamine 17, L-ristosamine 18, L-acosamine 23, and to the L-xylo isomers 24 (all in the N-protected form) is that no manipulation of the chiral centers is required once the C_6 and C_7 intermediates have been obtained in order to assess the correct stereochemistry.

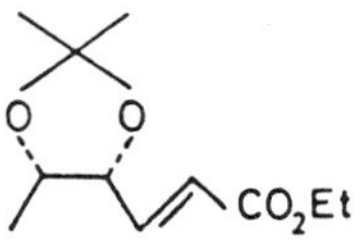

19

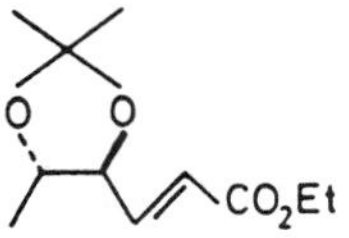

20

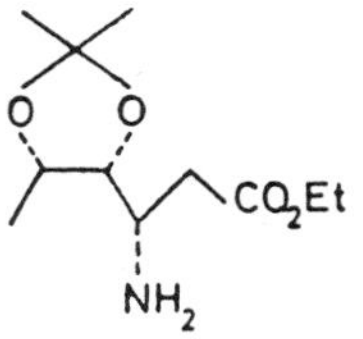

21

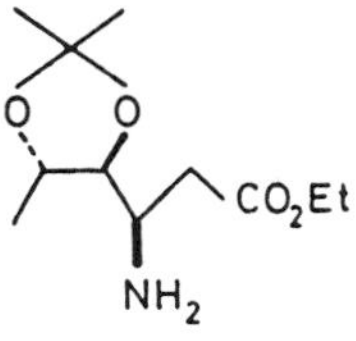

22

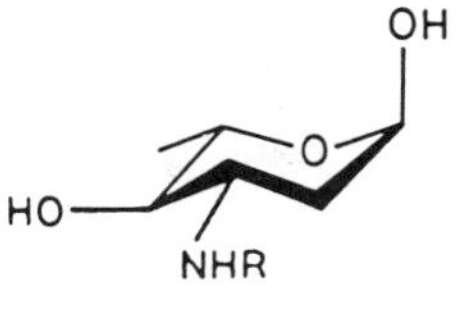

23

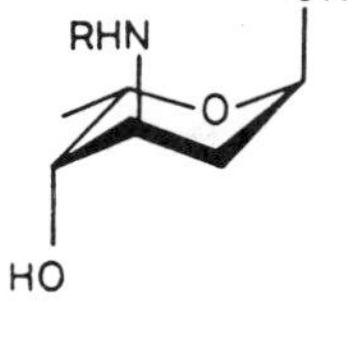

24

Attempted direct extension of the two foregoing methodologies to synthesis of the 3-C-methyl analogs of 17, 18, 23 and 24 starting from the diastereoisomeric C_5 methyl ketones 10 and 12 was ineffective, as ammonia does not add across the triply substituted double bond of the 3-C-methyl analogs of the esters 19 and 20. However, the phenyl-sulfenimines 25 and 26 add allylmagnesium bromide and diallylzinc with different stereochemistry (8), so that it is possible to have eventual access to the four configurational isomers of 3-amino-2,3,6-trideoxy--3-C-methyl-3-L-hexose.

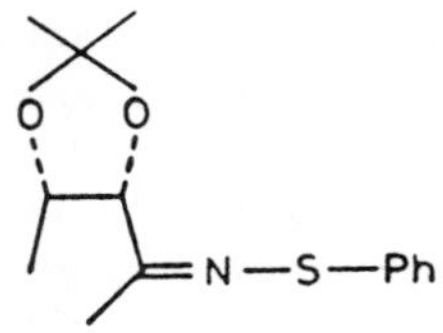

25

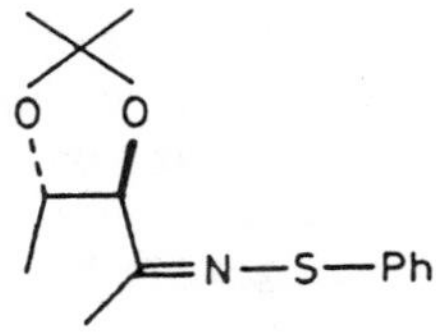

26

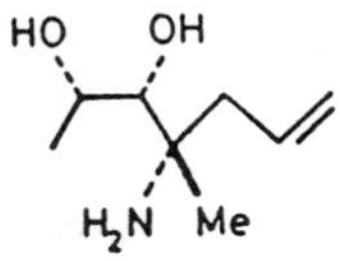

27

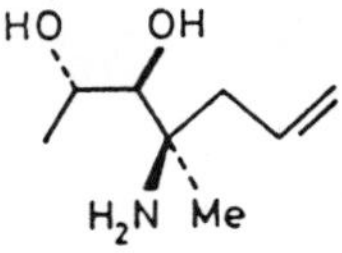

28

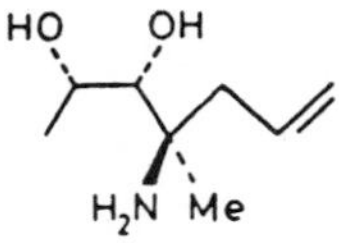

29

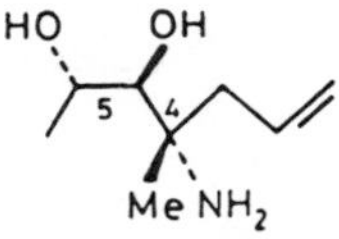

30

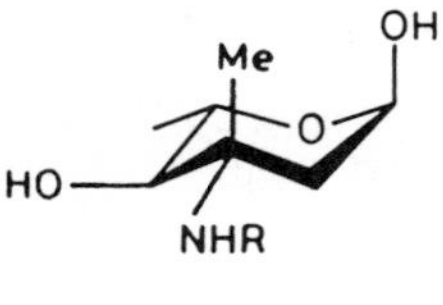

31

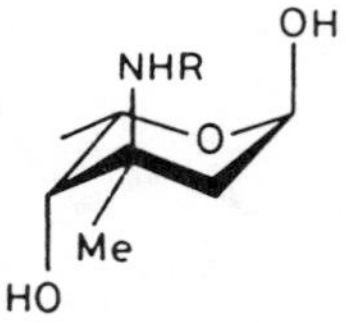

32

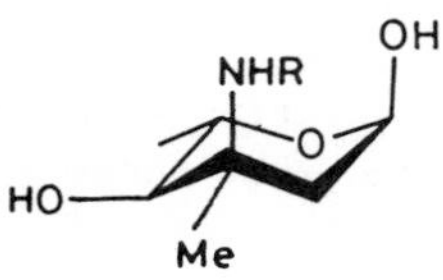

33

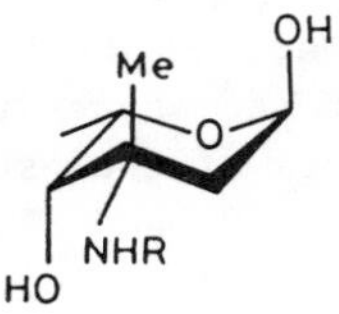

34

Thus, allylmagnesium bromide affords from 25 and 26 the 4,5-<u>threo</u> adducts 27 and 28 in   95:5 ratio with the 4,5-<u>erythro</u> materials 29 and 30.   The latter products,   formed according to Felkin's model, are obtained in   7:3 ratio with 27 and 28 when diallylzinc is added to 25 and   26.   The set 27—30,   as before,´ on <u>N</u>-protection and ozonolysis, affords L-vancosamine 34 and its configurational isomers 31—33.
The   methyl ketone 10 behaves at variance with 25 and 26   with   allyl-magnesium bromide and diallylzinc; both reagents give rise exclusively to   the 4,5-<u>erythro</u> adduct 35,   subsequently converted into L-mycarose (36) (<u>9</u>).   Similarly, the aldehydes 9 and 11 react with diallylzinc to afford   almost exclusively products 37 and   38   respectively,   whereas allylmagnesium bromide adds to 37 and 38 to give. 30—35 % of the 4,5--<u>threo</u>   adducts.   These materials,   on acid hydrolysis and ozonolysis, give rise to 39,   40, 41 and 42, the four configurational isomers of a 2,3,6-trideoxy-L-hexose.   The C_7 adduct 37 constitutes a flexible car-bohydrate-like   starting   material for the synthesis of a   variety   of chiral products. An application of 37 in the carbohydrate field is the synthesis of 4-amino-2,4,6-trideoxy-L-<u>lyxo</u> hexose (45) (<u>10</u>).

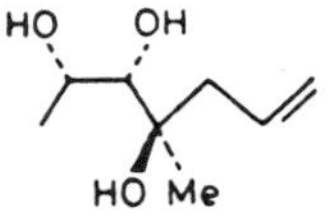

35

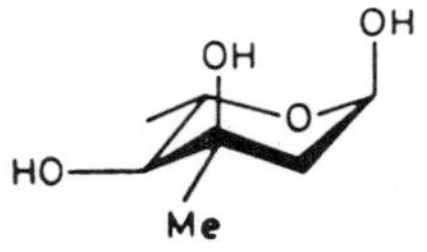

36

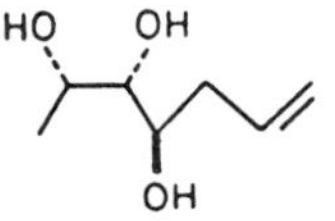

37

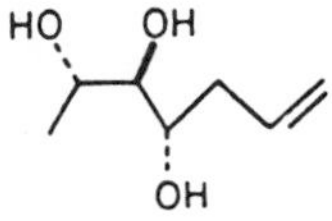

38

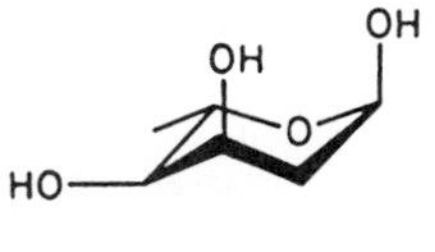

39

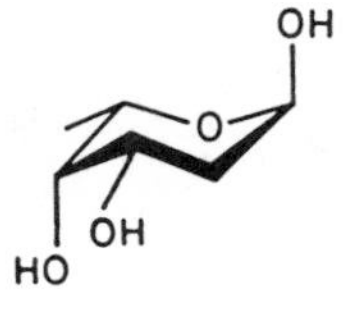

40

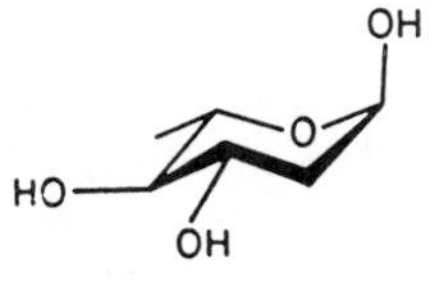

41

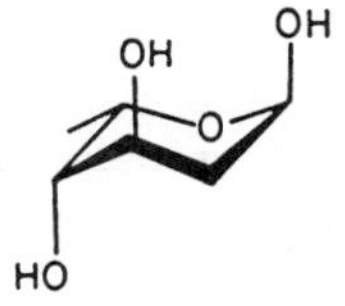

42

    The synthesis proceeds from 37 through the intermediacy of 43 and
44.   The   glycoside derived from 45 and adriamycinone shows   antitumor
activity similar to that of the isomer adriamycin, but exhibits a much
faster cellular uptake ($\underline{11}$).   Also,   the C_4 aldehyde 11 may be used as
an   alternative   to L-rhamnose in the synthesis   of   L-daunosamine   17
based   on   regioselective double-bond   functionalization   ($\underline{12}$).   Thus,
methyl 2,3,6-trideoxy-α-L-$\underline{threo}$-hex-2-enopyranoside (46), a key intermediate   in   the aforementioned procedure,   is obtained from   11   upon
reaction with $Ph_3P=CHCHOCH_2CH_2O$, followed by controlled methanolysis.

43   44   45

46

Synthesis Of Deoxy Sugars Of L-And D-Series

In   addition to 9—12,   several useful chiral carbonyl compounds   have
been   obtained from the diols obtained by yeast treatment of the   corresponding α-hydroxyketones. As a part of a study ($\underline{2}$) on the substrate
specificity of the multienzymic conversion shown in Eq.   2,   a serie of
racemic α-hydroxyketones has been prepared and submitted to the yeast
treatment. The reduction process is stereospecific, but depending upon

the structure of the substrate, as the result of enzymic kinetic resolution, <u>anti</u> diols are obtained as prevalent products or <u>anti-syn</u>
mixtures. Thus, from ketones of the type 47, the <u>anti</u> diols 48 are
prepared. From 49, the two <u>anti</u> and <u>syn</u> diols 50 and 51 are obtained.
The latter material, as its isopropylidene acetal, affords on ozonolysis the C_5 <u>threo</u> methyl ketone 52, the enantiomer of 12, from which
the D-enantiomer of L-vancosamine (34) becomes accessible.

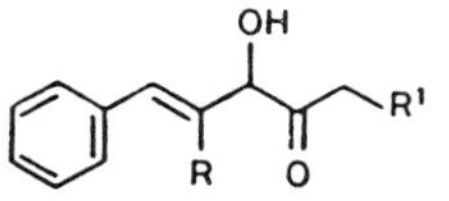

47 R=H,Me ; R¹=H,Me

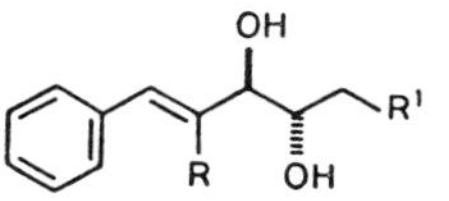

48 R=H,Me ; R¹=H,Me

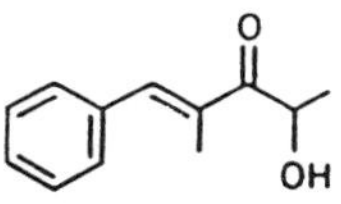

49

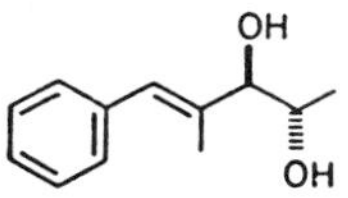

50

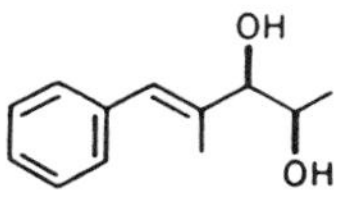

51

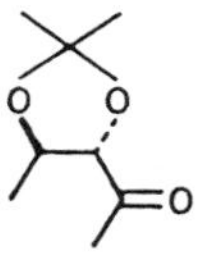

52

Furthermore, yeast treatment of the α-acetoxy ketone 53, bearing'
two oxygen substituents (13) at α and β, affords the carbinol 54 in 20
% yield, somewhat less than 10 % of the (2R) diastereoisomer, and 70 %
of recovered starting material. From the carbinol 54, crystalline 55
is obtained, which may be converted through suitable manipulation of
the protecting groups and ozonolysis, into 4-deoxy-D-<u>talo</u>-hexose 56.
The minor diastereoisomer similarly affords 4-deoxy-D-<u>xylo</u>-hexose 57.
Thus, in the yeast treatment of 53, as the results of the enzymic
kinetic resolution, the (2<u>S</u>, 4<u>S</u>, 5<u>R</u>) diol 55, a carbohydrate-like
chiral synthon, is accessible out of eight possible isomers.
Similar carbohydrate-like chiral synthons are obtained by yeast reduction of the α-acetoxyketones 58 and 59 (14), bearing in 1,6 and 1,5
relationship, respectively, two masked carbonyl functions accessible
chemoselectively with different reagents. The mode of reduction of 58
and 59 is, however, opposite. Whereas yeast treatment of 58 affords

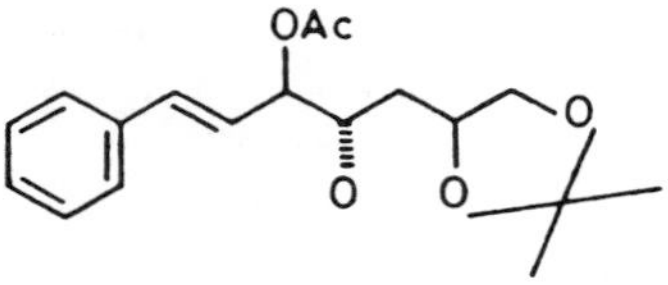

53

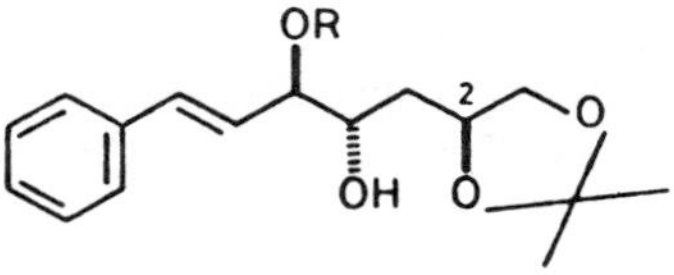

54 R=COCH₃

55 R=H

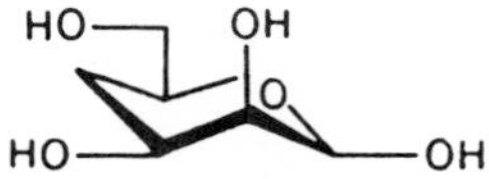

56 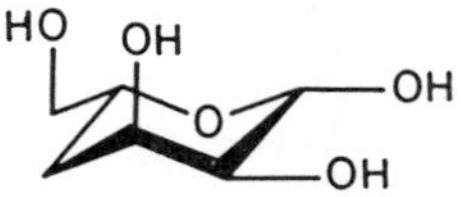   57

the carbinol 60, in optically pure form (as shown by its conversion by
basic hydrolysis, methanolysis, ozone treatment, and NaBH₄ reduction,
into the methyl-2,3-dideoxy-D-glycopyranoside (61)), compound 59 (4)
gives rise to a carbinol containing   70 % of the (3R, 4S) enantiomer
62. The latter material, on basic hydrolysis, yields a diol, which on
fractional crystallization, yields the optically pure (3R, 4S) product
63, from which 2-deoxy-L-erythro-pentose (64) is obtained. It is
probable that the reversal of the mode of reduction of 58 and 59
arises through due participation, during the bioconversion, of several
enzymes acting with opposite stereochemistry. Thus, the acceptability
of a non-natural substrate by a synthetically useful enzymes strongly
depends on even subtle structural modifications of substrates.

In conclusion, the direct synthetic access to L-daunosamine,
L-vancosamine and their configurational isomers reported here, as
compared to the existing methods to the same class of derivatives
(15), shows the revelance to organic synthesis of enzymic methods
leading from accessible substrates to such highly functionalized,
stereochemically rich compounds as the diol 4, using commercially
available, inexpensive baker's yeast. The products prepared by baker's-yeast reduction of synthetic α-acetoxyketones, leading to deoxy-
-D-hexoses and to deoxy-L-penthoses, appear of less general applicability, but might constitute an alternative to natural carbohydrates as
advanced intermediates in the synthesis of particular natural
products.

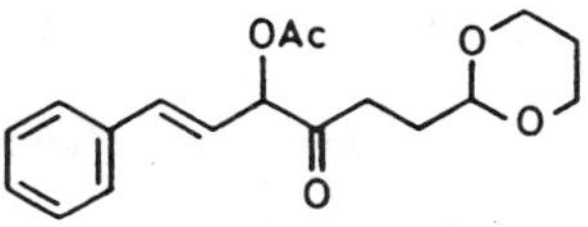

58

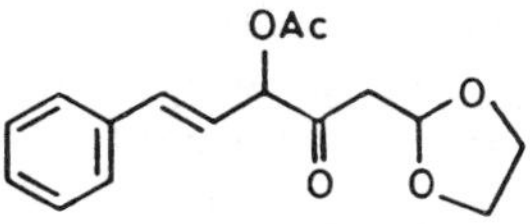

59

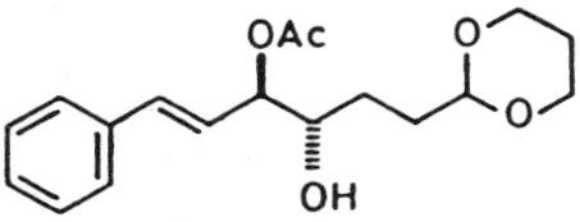

60

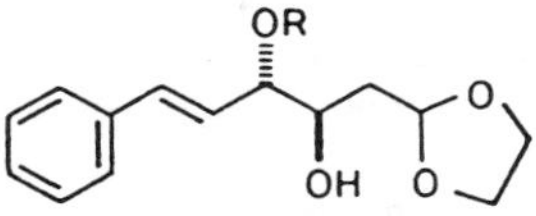

62 R=COCH₃

63 R=H

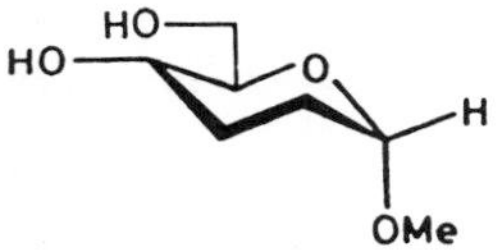

61

64

References

1. Fuganti, C.; Grasselli, P. Chem. Ind. (London) 1977, 983.

2. Fuganti, C.; Grasselli, P.; Servi, S.; Spreafico, F.; Zirotti, C.; Casati, P. J.Org.Chem. 1984, 49, 4087.

3. Fuganti, C.; Grasselli, P. J.C.S.Chem.Commun. 1978, 299.

4. Fronza, G.; Fuganti, C.; Grasselli, P. Tetrahedron Lett. 1978, 3941.

5. Bernardi, R.; Fuganti, C.; Grasselli, P.; Marinoni, G. Synthesis 1980, 50

6. Fuganti, C.; Grasselli, P.; Pedrocchi-Fantoni, G. J.Org.Chem., 1983, 48, 909.

7. Dyong, I.; Bendlin, H.Chem. Ber., 1978, 1677. Fronza, G.; Fuganti, C.; Grasselli, P. and Marinoni, G. Tetrahedron Lett., 1979, 883. Fronza, G.; Fuganti, C. and Grasselli, P., J.C.S: Chem. Commun., 1980, 442.

8. Fronza, G.; Fuganti, C.; Grasselli, P. and Pedrocchi-Fantoni, G.
   Tetrahedron Lett., 1981, 5073.
   Fronza, G.; Fuganti, C.; Grasselli, P.; Pedrocchi-Fantoni, G.
   J.Carbohydr. Chem. 1983, 225

9. Fuganti, C.; Grasselli, P.; Pedrocchi-Fantoni, G. Tetrahedron
   Lett., 1982, 4143.

10. Fronza, G.; Fuganti, C.; Grasselli, P.; Pedrocchi-Fantoni,
    G.Carbohydr. Res. 1985, 136, 115.

11. Penco, S. personal communication

12. Servi, S.; J. Org. Chem., 1985, 50, 5865.

13. Fronza, G.; Fuganti, C.; Grasselli, P.; Servi, S. J. Org.
    Chem., 1987, 52, 2086.

14. Fronza, S.; Fuganti, C.; Grasselli, P.; Servi, S. Tetrahedron
    Lett., 1985, 40, 4961

15. Hauser, F. M.; Ellenberger, S.R. Chem. Rev., 1986.86, 35

RECEIVED September 19, 1988

Chapter 18

Enzyme-Catalyzed Synthesis of Carbohydrates

C.-H. Wong, D. G. Drueckhammer, J. R. Durrwachter, B. Lacher,
C. J. Chauvet, Y.-F. Wang, H. M. Sweers, G. L. Smith, L. J.-S. Yang, and
W. J. Hennen

Department of Chemistry, Texas A&M University, College Station,
TX 77843

Several enzymatic procedures have been
developed for the synthesis of carbohydrates
from acyclic precursors. Aldolases appear to
be useful catalysts for the construction of
sugars through asymmeteric C-C bond formation.
2-deoxy-KDO, 2-deoxy-2-fluoro-KDO, 9-$\underline{O}$-acetyl
sialic acid and several unusual sugars were
prepared by a combined chemical and enzymatic
approach. Alcohol dehydrogenases and lipases
have been used in the preparation of chiral
furans, hydroxyaldehydes, and glycerol
acetonide which are useful as building blocks
in carbohydrate synthesis.

Recent developments in the enzymatic synthesis of
carbohydrates can be classified into four approaches: 1)
asymmetric C-C bond formation catalyzed by aldolases (1-
10); 2) enzymatic synthesis of carbohydrate synthons (10-
11); 3) asymmetric glycosidic formation catalyzed by
glycosidases (12-17) and glycosyl transferases (18-23);
and 4) regioselective transformations of sugars and
derivatives (24-25). These enzymatic transformations are
stereoselective and carried out under mild conditions
with minimum protection of functional groups. They hold
promise in preparative carbohydrate synthesis. In
connection with this book, we focus on the first two
approaches.

Aldol Condensations

FDP Aldolase. There have been approximately 15 aldolases
isolated, each of which catalyzes a distinct aldol
reaction (25-26). The aldolases which have been studied
the most as synthetic catalysts are fructose-1,6-

0097–6156/89/0386–0317$06.00/0
© 1989 American Chemical Society

diphosphate aldolase (FDP aldolase, EC 4.1.1.13) from
rabbit muscle (1-4) and N-acetylneuraminic acid aldolase
(NANA aldolase, EC 4.1.3.3.) from clostridia (5). From
the results we and other groups have obtained so far, it
appears that the enzyme FDP aldolase is quite specific
for the aldol donor dihydroxyacetone phosphate (DHAP),
but will accept a variety of aldehydes as the acceptor
(1-4). The stereochemistry of C-C bond formation is not
affected by the substituents at the carbon center next to
the aldehyde group. The α-substituents, however, do
affect the rate of C-C bond formation. Using the
structural representation of the aldehyde acceptor shown
in Figure 1, an aldehyde with R being the largest and R'
the second largest group at the chiral center would allow
the reaction to take place much faster in relation to its
enantiomer. However, under totally reversible, catalytic
conditions, the product distribution is thermodynamically
controlled.

Typical aldol reactions can be carried out on a 10-
100 mmol scale with free or immobilized enzyme. After
removal of the phosphate group from the aldol product by
acid- or phosphatase-catalyzed hydrolysis, a free ketose
can be prepared. Several unusual ketoses with different
substituents at R and R' have been prepared. These
include deoxysugars, fluorosugars, O-alkyl sugars (1-3)
and other higher monosaccharides (4). These ketoses, 3,
may be further transferred to the corresponding aldoses 4
upon treatment with the enzyme glucose isomerase (EC
5.3.1.5) (3), an industrial enzyme used in the
manufacture of high fructose corn syrup. This combined
aldolase/isomerase catalysis provides a new route to a
variety of sugars and derivatives which are useful for
other applications (2-3).

The substrate DHAP required in these enzymatic
syntheses can be prepared chemically (27) or
enzymatically (1-4). It can also be generated in situ
from FDP (1-4). Potential problems with the use of DHAP
in these procedures include the stability of DHAP in
solution ($t_{1/2}$ ~ 15 h at pH 7.5) and the the non-trivial
preparation of DHAP. We have found that a mixture of
dihydroxyacetone and a small amount of inorganic arsenate
can be used to replace DHAP in the aldol reactions in
which the inorganic arsenate is recycled about 50 times
during the reactions (3). This eliminates the need for
the preparation of DHAP. From the results of kinetic and
competition studies, we conclude that the reactions
proceed through dihydroxyacetone arsenate monoester which
is analogous to DHAP and is accepted by the enzyme as a
substrate (Figure 2). The C-C bond formation is thus
completely determined by the enzyme which only accepts
the arsenate monoester as a substrate, although the
reversible formation of arsenate esters in aqueous
solution is nonselective and several enzymatically
inactive arsenate species may exist in solution. Under

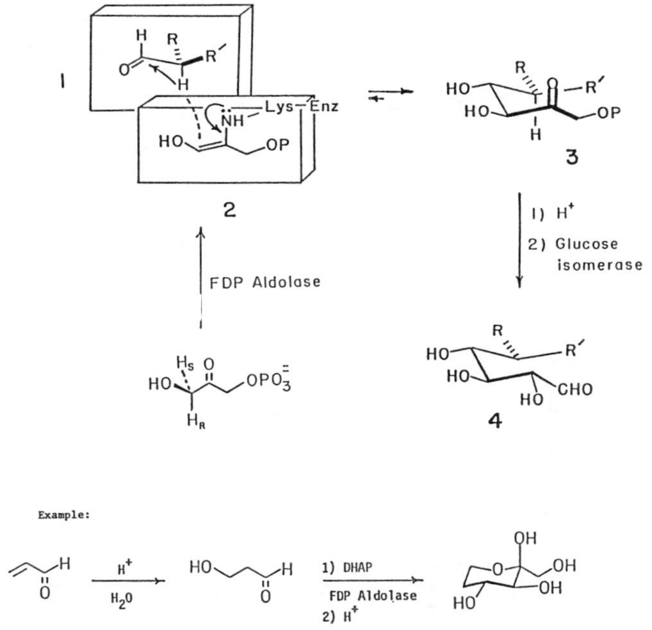

Figure 1. Synthesis of usual and unusual sugars using FDP aldolase and glucose isomerase as catalysts.

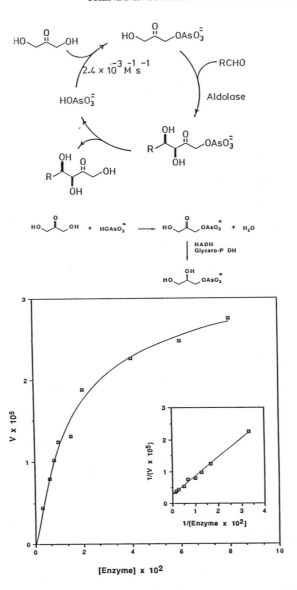

Figure 2.   Mechanism of dihydroxyacetone/arsenate
reaction with FDP aldolase.   Both dihydroxyacetone
and inorganic arsenate are not the inhibitor of the
aldolase reactions.   The rate constant for the
arsenate ester formation is determined enzymatically
(a plot of 1/v vs 1/E gives a non-zero intercept
which is attributed to the rate at infinite enzyme
concentration and that rate corresponds to the rate
of nonenzymatic formation of the arsenate ester).

the reaction conditions, the arsenate formation is the rate determining step and the aldol reaction step is virtually irreversible. This is contrary to the normal reversible aldol reactions with DHAP where the C-C bond formation is rate determining. These results were supported by the fact that a high diastereoselectivity was observed in the reaction of DHAP with an excess of a racemic aldehyde, whereas in the reaction with dihydroxyacetone/arsenate (DHA/As) no significant selectivity was observed (Figure 3). In an attempt to use inorganic vanadate instead of arsenate in the aldolase reactions no appreciable aldol condensation was observed, presumably due to the undesired redox reaction between DHA and inorganic vanadate.

KDO-8 Phosphate Synthetase. KDO (3-deoxy-α-D-manno-2-octulopyranosonic acid) is a component of capsular polysaccharides (k-antigens) and lipopolysaccharides (LPS, also known as endotoxins) found in Gram-negative bacteria (28). These KDO-containing membrane components are unique to gram-negative bacteria and are crucial to the survival of the organism. Since KDO is not found in mammalian tissues, it has become a primary target for investigation by medicinal and synthetic chemists (28-31) for the development of antibacterial agents effective against Gram-negative bacteria. β-2-Deoxy KDO (32) and the 8-dipeptidyl derivative of 8-amino-2,8-dideoxy-β-KDO (33) have been shown to be potent inhibitors of the enzyme CMP-KDO synthetase which utilizes the β-pyranose form of KDO as substrate (34). Unlike 2-deoxy-β-KDO which cannot penetrate intact bacteria, the glycopeptide analog can be transported through the cell membrane and hydrolyzed by a peptidase inside the cell to release the inhibitor (33). Several C-glycosidic derivatives of KDO have recently been prepared from protected KDO (35).

We have been successfully in immobilizing the enzyme KDO-8-phosphate synthetase which was isolated from E. coli B cells (ATCC 11303) (36). About 400 U of the enzyme can be produced from 500 g of wet cells. The use of this enzyme preparation is illustrated in Figure 4. With the use of a system of co-immobilized enzymes on polyacrylamide gels prepared according to previously described procedures (37), KDO-8-phosphate was prepared on a 10-50 g scale starting with D-arabinose and phosphoenol pyruvate (38). After removal of the phosphate moiety by acid- or phosphatase-catalyzed hydrolysis, the product was converted chemically to 2-deoxy-β-KDO and 2-deoxy-2-fluoro KDO (see Figure 5). The procedure for the introduction of the F group was similar to that reported previously (39). After completing the work, we noticed that KDO-8-phosphate was also prepared by Whitesides' group in a similar way with three enzymes enclosed in a dialysis membrane (40).

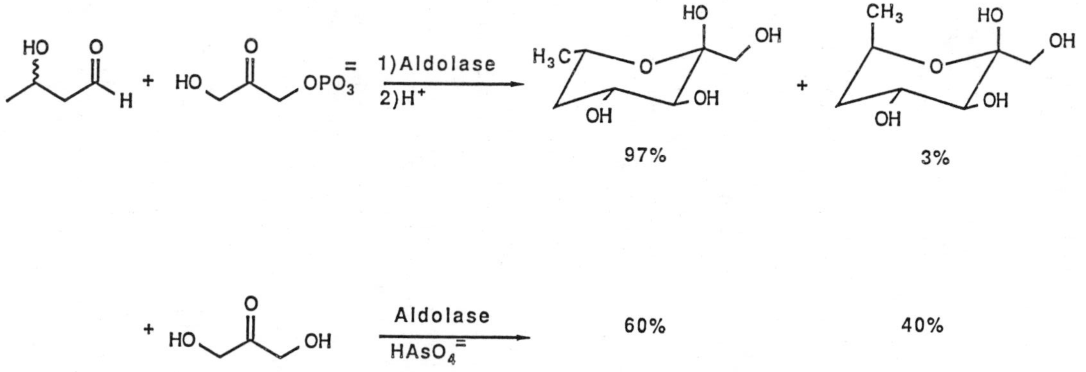

Figure 3. Selectivity of the FDP-aldolase reactions using DHAP vs. dihydroxyacetone/arsenate as a substrate. In the former case, the more stable sugar is obtained due to the reversible nature of the reaction. In the later case, both sugars were obtained in nearly equal amounts, because the reaction was found to be virtually irreversible and the formation of the arsenate ester was rate limiting.

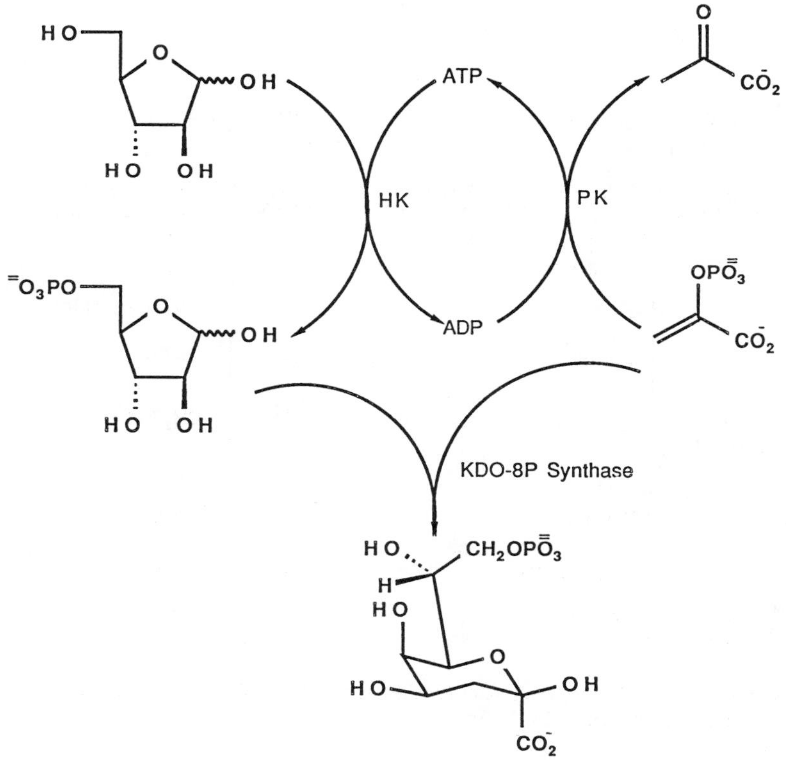

Figure 4. Synthesis of KDO-8-phosphate. HK, hexokinase; PK, pyruvate kinase.

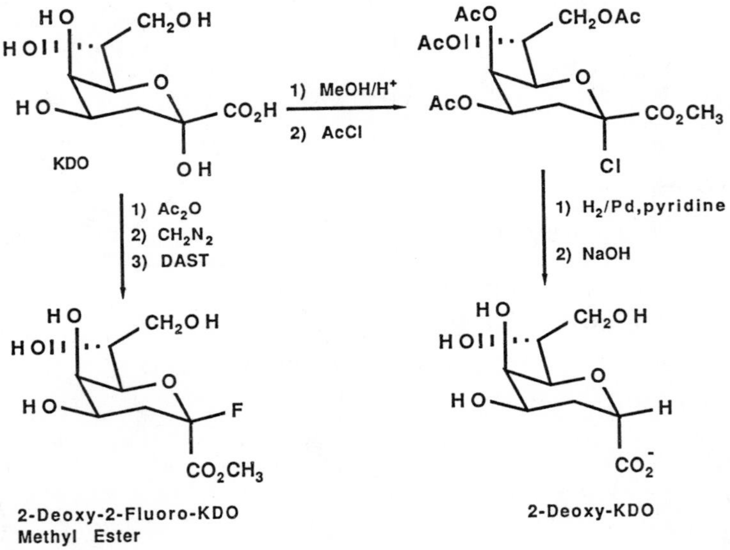

Figure 5.   Chemical conversion of KDO to 2-deoxy-β-KDO and 2-deoxy-2-fluoro-KDO

<u>N-Acetylneuraminic Acid Aldolase</u>. A new procedure has also been developed for the synthesis of 9-<u>O</u>-acetyl-<u>N</u>-acetylneuraminic acid using the aldolase catalyzed reaction methodology. This compound is an unusual sialic acid found in a number of tumor cells and influenza virus C glycoproteins (<u>41</u>). The aldol acceptor, 6-<u>O</u>-acetyl-D-mannosamine was prepared in 70% isolated yield from isopropenyl acetate and N-acetyl-D-mannosamine catalyzed by protease N from <u>Bacillus</u> <u>subtilis</u> (from Amano). The 6-<u>O</u>-acetyl hexose was previously prepared by a complicated chemical procedure (<u>42</u>). The target molecule was obtained in 90% yield via the condensation of the 6-<u>O</u>-acetyl sugar and pyruvate catalyzed by NANA aldolase (Figure 6). With similar procedures applied to KDO, 2-deoxy-NANA and 2-deoxy-2-fluoro-NANA were prepared from NANA.

Carbohydrate Synthons

Enzymatic reduction of carbonyl compounds and enzymatic enantioselective transformation of racemic or meso alcohols (<u>25</u>,<u>43</u>) are two methodologies that have proven to be beneficial in the preparation of optically active hydroxyl compounds, key chiral building blocks used in carbohydrate and natural product syntheses (<u>44</u>-<u>45</u>). Our interest in this area is to develop enzymatic routes to optically active glycerol and furan derivatives, and hydroxyaldehydes.

<u>Chiral Furans</u>. The alkyl furyl carbinol **5** is a useful building block for the synthesis of deoxy-L-sugars (<u>46</u>) (Figure 7). This synthon can be prepared from acyl furans via enzymatic reductions catalyzed by the alcohol dehydrogenase from <u>Thermoanaerobium brockii</u>, or from an esterase-catalyzed kinetic resolution of the racemic alcohol (<u>11</u>). Both approaches provide the chiral building block in > 90% ee using quite straightforward procedures on g scales (Figure 8).

<u>Chiral Hydroxyaldehydes</u>. Similarly, optically active lactaldehyde and α-hydroxybutyraldehyde, both useful aldol acceptors, can be prepared enzymatically (<u>10</u>) according to the scheme shown in Figure 9.

<u>Chiral Glycerol Derivatives</u>.
<u>Irreversible Transesterification</u>. A new preparation of chiral glycerol acetonide (2,2-dimethyl-1,3-dioxolane-4-methanol) involving an enantioselective hydrolysis of 2-<u>O</u>-benzylcerol diacetate to the (<u>R</u>)-monoacetate catalyzed by a lipoprotein lipase (<u>47</u>) has recently been developed. In an effort to prepare the (<u>S</u>)-enantiomer, we have used the aforementioned irreversible transesterification reaction using isopropenyl acetate as an acylating reagent, which upon reaction gives acetone as a

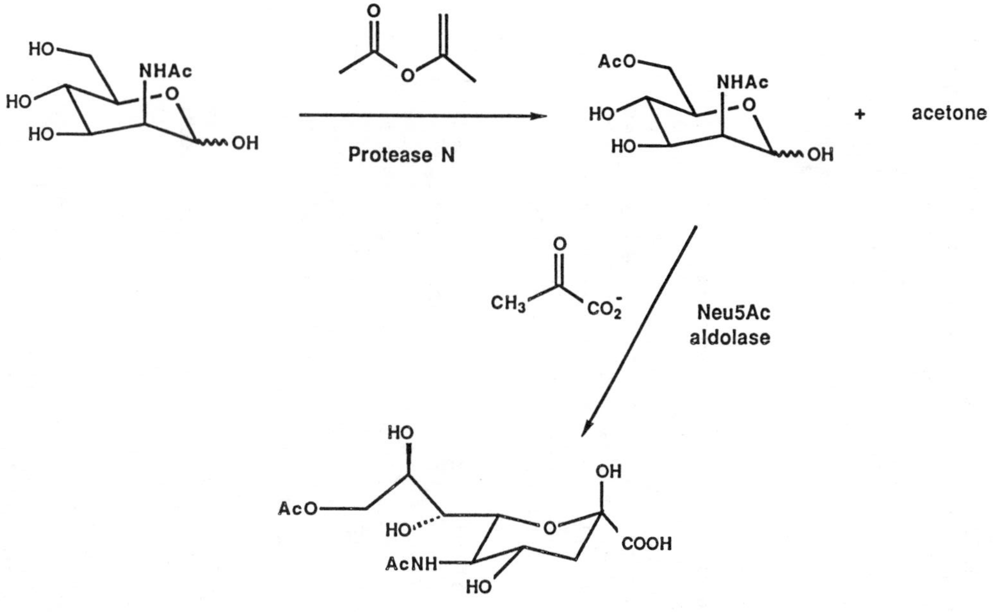

Neu5,9Ac$_2$

Figure 6. Synthesis of 9-O-acetyl-N-acetylneuraminic acid. The aldol acceptor was prepared from N-acetylmannosamine and isopropenyl acetate in DMF catalyzed by protease N obtained from Amano. The aldol condensation was carried out by using N-acetylneuraminic acid aldolase as catalyst.

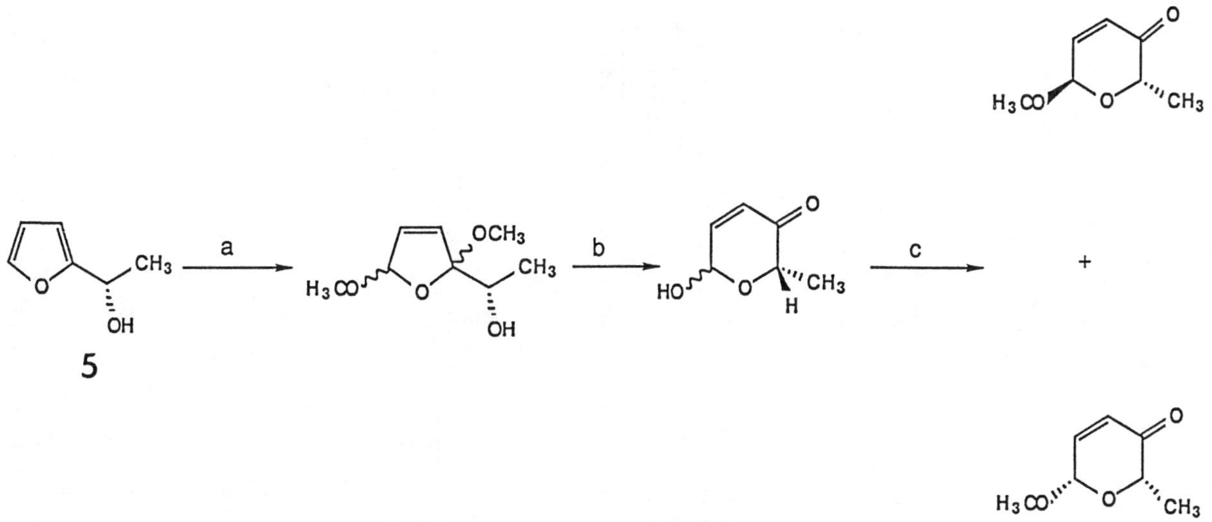

Figure 7. Conversion of (S)-furyl methyl carbinol to a mixture of α- and β-L-dihydropyranone. a, Br$_2$/MeOH; b, H$^+$; c, HC(OMe)$_3$/SnCl$_4$.

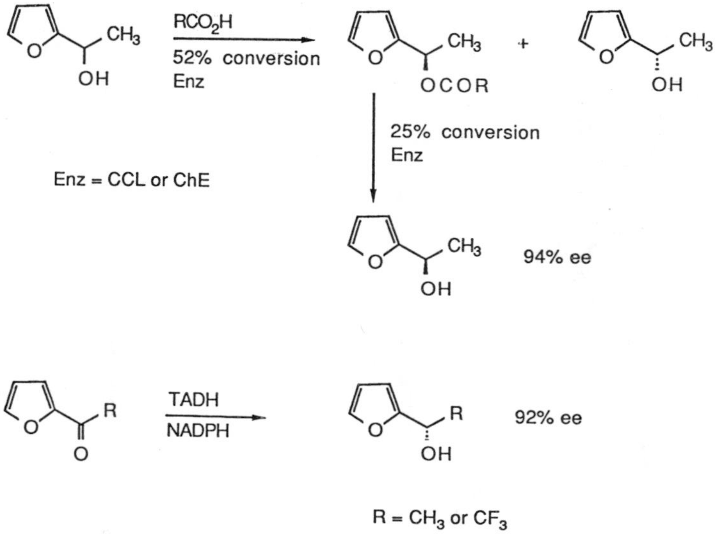

Figure 8.   Enzymatic preparation of (S)- and (R)-furyl methyl carbinol.   TADH, Thermoanaerobium brokii alcohol dehydrogenase (NADPH was regenerated by glucose/glucose dehydrogenase from Bacillus cereus obtained from Amano.); CCL, lipase from Candida cylindraceae; ChE, cholesterol esterase from Pseudomonas.

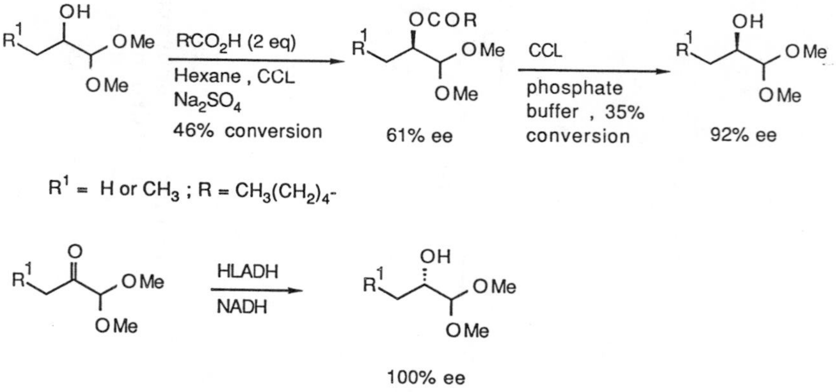

Figure 9.   Preparation of L- and D-lactaldehyde and L- and D-2-hydroxybutyraldehyde. HLADH, horse liver alcohol dehydrogenase (NADH was regenerated by glucose/glucose dehydrogenase).

byproduct. In a representative synthesis, ($\underline{S}$)-2-$\underline{O}$-benzylglycerol monoacetate was prepared from 2-$\underline{O}$-benzylglycerol catalyzed by a lipoprotein lipase from $\underline{Pseudomonas}$ in 90% yield and almost 100% ee (Figure 10).

The problem with enzyme catalyzed reversible transesterifications as an approach to biochemical resolution is that due to the reversible nature of the reaction, the enantiomeric excess of the desired product in the forward reaction decreases as the reverse reaction proceeds. As in hydrolytic reactions, the irreversible transesterification offers a better process for optimization of the transformation and for recovery of the product. Furthermore, there is no product inhibition observed in the irreversible process. Both enantiomers can be converted to glycerol acetonide with known procedures.

Double Kinetic Resolution. In an effort to resolve racemic glycerol acetonide via esterase-catalyzed hydrolysis of the corresponding esters, we have tried many different ester derivatives with chiral or achiral carboxyl components and many enzymes including lipases, esterases, and proteases from different sources. All the results showed low enantioselectivity. A double resolution strategy was then employed. The racemic acetate was hydrolyzed by pig liver esterase (PLE) to 60% completion. The unreacted ester recovered was then subjected to porcine pancreatic lipase (PPL)-catalyzed hydrolysis up to 44% conversion to obtain the hydrolysis product in 70% ee (Figure 11). Since PPL and PLE have an opposite enantioselectivity in this case, the unreacted substrate recovered with a low ee in the first reaction is converted to the alcohol product with a much higher ee in the seond reaction. The enantiomeric product (the $\underline{R}$-enantiomer) was also prepared with the two enzymes used in a reverse order. The enantiomeric excess of course can be enhanced at the cost of chemical yield by increasing the extent of the first reaction and decreasing that of the second reaction. A formula useful for prediction of enantiomeric excess at different extents of conversion is available ($\underline{48}$) and can be used to optimize the process. Similarly, a single enzyme can be used for the double resolution ($\underline{49}$). For example, the alcohol product isolated from the PLE-catalyzed hydrolysis of the acetate ester at 40% conversion was recovered, acetylated with acetic anhydride, and the acetate ester was treated with the same enzyme (40% conversion) to give the alcohol product in 85% ee.

Determination of Enantiomeric Excess Without Separation of Products. To determine the enantioselectivity in a given enzyme-catalyzed kinetic resolution, the reaction product very often must be isolated before the determination. This is very time-consuming when one

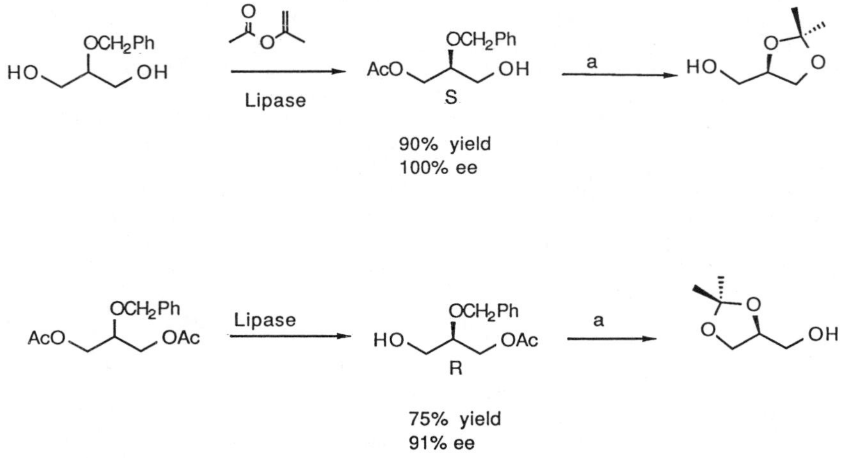

Figure 10.   Preparation of chiral glycerol derivatives and D- and L-glycerol acetonide. a: 1. H_2/Pd-C; 2. methyl isopropenyl ether/pyridinium p-toluensulfonate; 3. K_2CO_3/MeOH

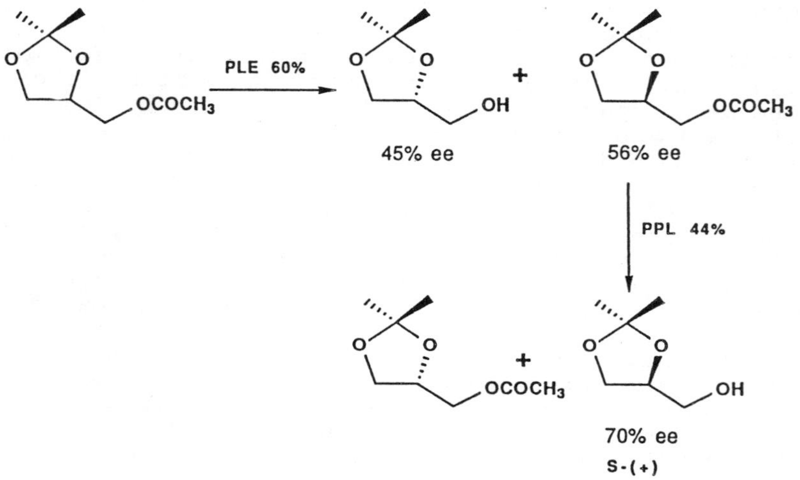

Figure 11. Double kinetic resolution using enzymes with opposite enantioselectivity. PPL, porcine pancreatic lipase; PLE, pig liver esterase.

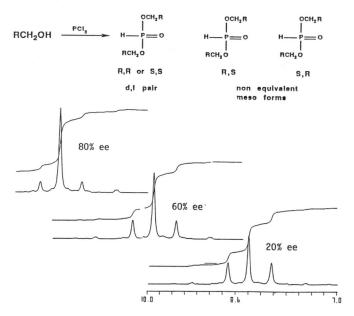

Figure 12. ^{31}P NMR determination of the enantiomeric excess of glycerol acetonide without separation of the unreacted ester. Approximately 1.35 mmol of a mixture of glycerol acetonide and the acetate ester was dissolved in a 10 mm NMR tube. To the solution was added 0.45 mmol of PCl$_3$. The tube was shaken and quickly uncapped to allow the escape of the HCl gas formed during the reaction. After the recapped tube stood for 30 min at room temp, 0.5 mL of CDCl$_3$ was added, and the NMR spectrum was recorded on a Varian 200 MHz instrument. The ee was calculated from Horean's formula (ee^2 = (K - 1)/(K + 1), Vigneron, J.P.; Dhaenens, M.; Horeau, A. Tetrahedron, 1973, 29, 1055-60) where K represents the ratio of the integrated peak areas of the d,l pair (the centered peak) to the meso isomers (the two outside peaks). The spectra were recorded from the phosphite derivatives of resolved glycerol acetonide at different extents of conversions with H$_3$PO$_4$ (0 ppm) as external standard.

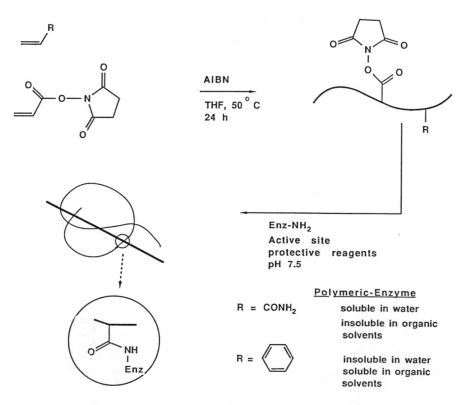

Figure 13. Preparation of immobilized enzymes with different solubilities in aqueous solutions and organic solvents. Procedure: A mixture of an enzyme (3 mg) and the polymer (10 mg) was incubated at pH 7.5 for 20 min. Ammonium phosphate (0.1 M, pH 7, 1 mL) was then added to react with the remaining active ester. After 20 min, the solution was ready for use, or lyophilization to give the immobilized enzyme as a powder to be used for reaction in organic solvents. Each gram of the polymer contains approximately 0.7 mmol of the active ester.

intends to optimize the resolution condition. To overcome this problem, we have used the ^{31}P NMR technique developed by Kellogg (50) to determine the ee of the alcohol product in the presence of the unreacted ester (Figure 12). Despite the fact that the OH group is one carbon away from the chiral center, the technique is very sensitive and can be used to determine the ee up to 98%.

Enzyme Immobilization

All the enzymes used in the work described above are quite stable at room temperature and can be used in a free form. They can also be used in an immobilized form to improve the stability and to facilitate the recovery. Many immobilization techniques are available today (25). The recent procedure developed by Whitesides et al using water-insoluble, cross-linked poly(acrylamide-acryloxysuccinimide) appears to be very useful and applicable to many enzymes (37). We have found that the non-crosslinked polymer can be used directly for immobilization in the absence of the diamine cross-linking reagent. Reaction of an enzyme with the reactive polymer produces an immobilized enzyme which is soluble in aqueous solutions but insoluble in organic solvents. Many enzymes have been immobilized by this way and the stability of each enzyme is enhanced by a factor of greater than 100. Horse liver alcohol dehydrogenase and FDP aldolase, for example, have been successfully immobilized and showed a marked increase in stability.

In a like manner, a co-polymer of styrene and acryloxysuccinimide with a 10 to 1 ratio was prepared. Enzymes immobilized on this type of polymer had different physical properties. They are soluble in organic solvents such as dioxane and DMF, but insoluble in aqueous solutions. Lipases and cholesterol esterase immobilized on this type of polymer are very stable and active in several organic solvents, and have been used in several enantioselective transformations. The protocols for the immobilization are depicted in Figure 13.

Conclusion

The enzymatic approach to the synthesis of carbohydrates and their precursors is practical for certain types of sugars. The next stage of our investigation will be to focus on the modification of these readily available sugars to agents of interest. Improvement of enzyme properties for synthetic application is also of interest.

Acknowledgment

Support of the work by the NSF, the Robert A. Welch Foundation and the Searle Scholars Program are gratefully acknowledged.

Literature Cited

1. Wong, C.-H.; Whitesides,G.M. J. Org. Chem., 1983,
    48, 3199.
2. Wong, C.-H.; Mazenod, F.P.; Whitesides, G.M. ibid.,
    1983, 48, 3493.
3. Durrwachter, J.R.; Drueckhammer, D.G.; Nozaki, K.;
    Sweers, H.M.; Wong, C.-H. J. Am. Chem. Soc., 1986,
    108, 7812.
4. Bednarski, M.D.; Waldlmann, H.S.; Whitesides, G.M.
    Tetrahedron Lett., 1986, 27, 5807.
5. Auge, C.; Gautheron, C. J.C.S. Chem. Soc., 1987, 859
    and ref. cited. Bednarski, M.D.; Chenault, H.K.;
    Simon, E.S.; Whitesides, G.M. J. Am. Chem. Soc.
    1987, 109, 1283-4.
6. Reimer, L.M.; Conley, D.L.; Pompliano, D.L.; Frost,
    J.W., J Am. Chem. Soc. 1986, 108, 1080.
7. Mocali, A.; Aldinucci, D.; Paoletti, F. Carbohyrd.
    Res. 1985, 141, 288-93.
8. Kapusinski, M.; Franke. F.P.; Flanigan, I.; MacLeod,
    Jr.; Williams, J.F. ibid 1985, 140, 69-79.
9. Brossmer, R.; Rose, U.; Kasper, D.; Smith, T.S.;
    Grasmuk, H.; Unger, F.M., Biochem. Biophys. Res.
    Comm. 1980, 96, 1281-9.
10. Wong, C.-H.; Drueckhammer, D.G.; Sweers, H.M., J.
    Am. Chem.Soc. 1985, 107, 4028.
11. Drueckhammer, D.G.; Barbas, C.F.; Nozaki, K.; Wong,
    C.-H.; Wood, C.Y.; Ciufolini, M.A., J. Org. Chem.
    1988, 53, 1607-11.
12. Boos, W. Method Enzymol, 1982, 89, 57-64.
13. Ooi, Y.; Hashimoto, T.; Mitsuo, N.; Satoh, T.
    Tetrahedron Lett. 1984, 25, 2241-4.
14. Ajisaka, K.; Nishida, H.; Fujimoto, H. Biotech.
    Lett. 1987, 9, 43-8.
15. Straathot, A.J.J.; Kieboom, A.P.G.; Van Bekkum, H.
    Carbohydr. Res. 1986, 146, 154-9.
16. Hedbys, L.; Larsson, P.O.; Mosbach, K.; Svensson, S.
    Biochem. Biophys. Res. 1984, 123, 8-15.
17. Wong, C.-H., Ph.D. thesis, Massachusetts Institute
    of Technology, 1982, p. 158-189.
18. Wong, C.-H.; Haynie, S.L.; Whitesides, G.M., J. Org.
    Chem. 1982, 47, 5416-8.
19. Rosevear, P.R.; Nunez, H.A.; Barker, R.,
    Biochemistry, 1982, 21, 421-31.
20. Auge, C.; Mathieu, C.; Merienne, C. Carbohydr. Res.
    1986, 151, 147-56.
21. Sabesan, S.; Paulson, J.C., J. Am. Chem. Soc. 1986,
    108, 2068-80.
22. Theim, J.; Treder, W. Angew. Chem. Int. Ed. Engl.
    1986, 25, 1096-7.
23. Sweers, H.M.; Wong, C.-H., J. Am. Chem. Soc. 1986,
    108, 6421-2.
24. Theridos, M.; Klibanov, A.M. ibid, 1987, 109, 3977-
    80.
25. Whitesides, G.M.; Wong, C.-H., Angew Chem. Int. Ed.
    Engl. 1985, 24, 617-38.

26. Wong, C.-H. in "Enzymes as Catalysts in Organic Synthesis" (Schneider, M.P. ed.), D. Reidel Publishing Co., 1986, p. 199-216.

27. Effenberger, F.; Straub, A. Tetrahedron Lett. 1987, 28, 1631-4.

28. Unger, F.M. Adv. Carbohydr. Chem. Biochem. 1981, 38, 323-88.

29. Danishefsky, S.J.; DeNinno, M.P.; Angew Chem. Int. Ed. Engl. 1987, 26, 15-23.

30. Paquet, F.; Sinaÿ, P. J. Am. Chem. Soc. 1984, 23, 430-1.

31. Schmidt, R.R.; Betz, R. Angew Chem. Int. Ed. Engl. 1984, 23, 430-1.

32. Claesson, A.; Luthman, K.; Gustafsson, K.; Bondesson, G. Biochem. Biophys. Res. Comm. 1987, 143, 1063-8.

33. Hammond, S.M.; Claesson, A.; Jansson, A.M.; Larsson, L.G.; Pring, B.G.; Town, C.M.; Ekstron, B. Nature, 1987, 327; 730-2; Golgman, R.C.; Kohlbrenner, W.E.; Larty, P.; Pernet, A. Nature, 1987, 329, 162-4.

34. Kohlbrenner, W.E.; Fesik, S.W., J. Biol. Chem. 1985, 260. 14695-00.

35. Luthman, K.; Orbe, M.; Wagluund, T.; Claesson, A. J. Org. Chem. 1987, 52, 3777-84.

36. Ray, P.H. Method Enzymol, 1982, 83, 525-8.

37. Pollak, A.; Blumenfeld, H.; Wax, M; Baughn, R.L.; Whitesides, M. J. Am. Chem. Soc. 1980, 102, 6324.

38. Hirschbeing, G.L.; Mazenod, F.P.; Whitesides, G.M., J. Org.Chem. 1982, 47, 3765.

39. Rosenbrook, W. Jr.; Riley, D.A.; Larty, P.A. Tetrahedron Lett. 1985, 26, 3-6.

40. Bednarski, M.D.; Crans, D.C.; Dicosimo, R.; Simon, E.S.; Stein, P.D.; Whitesides, G.M. Tetrahedron Lett. 1988, 29, 427-30.

41. Herrler, C.; Rott, R.; Klenk, H.D.; Muller, H.P.; Shukla, A.K.; Schauer, R. Embo J. 1985, 4, 1503.

42. Augé, C.; David, S.; Gautheron, C.; Veyrières, A. Tetrahedron Lett. 1985, 26, 2439.

43. Jones, J.B.; Tetrahedron 1986, 42, 3351-3403.

44. McGarvey, G.J.; Kimura, M.; Oh, T.; Williams, J.M., J. Carbohydr. Chem. 1984, 3, 125-88.

45. Hanessian, S. Total Synthesis of Natural Products: The Chivon Approach, Pergamon Press, 1983.

46. Achnatowitcz, O.; Bukowski, P.; Szechner, B.; Zwierzchowska, A. Tetrahedron, 1971, 27, 1973-96.

47. Breitgoff, D.; Lamen, K.; Schneider, M.P. J.C.S. Chem. Comm. 1986, 1523-4.

48. Chen, C.S.; Fujimoto, Y.; Girdaukas, G.; Sih, C.J. J. Am.Chem. Soc. 1982, 104, 7294-8; Chen, C.-S.; Wu, S-H.; Girdaukas, G.; Sih, C.J. ibid 1987, 109, 2812-7.

49. Chen, C.S.; Sih, C.J. J. Am. Chem. Soc. 1982, 104, 7294-5.

50. Feringa, B.L.; Strijtveen, B.; Kellogg, R. J. Org. Chem. 1986, 51, 5486.

RECEIVED September 19, 1988

INDEXES

Author Index

Affiliation Index

Subject Index

Production by Meg Marshall and Rebecca Hunsicker
Indexing by Deborah H. Steiner

Elements typeset by Hot Type Ltd., Washington, DC
Printed and bound by Maple Press, York, PA